THE
SPACE
BOOK

Books by Jim Bell

Asteroid Rendezvous: NEAR Shoemaker's Adventures at Eros

Mars 3-D: A Rover's-Eye View of the Red Planet

The Martian Surface: Composition, Mineralogy, and Physical Properties

Moon 3-D: The Lunar Surface Comes to Life

Postcards from Mars: The First Photographer on the Red Planet

THE SPACE BOOK

FROM THE BEGINNING TO THE END OF TIME,
250 MILESTONES IN THE HISTORY OF SPACE & ASTRONOMY

Jim Bell

STERLING
New York

For my many teachers and mentors, for their patience, wisdom, and insistence that we must learn from the struggles of those who came before us; and for my children and many students, who graciously put up with me constantly trying to pass that lesson on to them.

STERLING
New York

An Imprint of Sterling Publishing
387 Park Avenue South
New York, NY 10016

ISBN 978-1-4027-8071-4

Distributed in Canada by Sterling Publishing
c/o Canadian Manda Group, 165 Dufferin Street
Toronto, Ontario, Canada M6K 3H6
Distributed in the United Kingdom by GMC Distribution Services
Castle Place, 166 High Street, Lewes, East Sussex, England BN7 1XU
Distributed in Australia by Capricorn Link (Australia) Pty. Ltd.
P.O. Box 704, Windsor, NSW 2756, Australia

For information about custom editions, special sales, and premium and corporate purchases, please contact Sterling Special Sales at 800-805-5489 or specialsales@sterlingpublishing.com.

Manufactured in China

2 4 6 8 10 9 7 5 3 1

www.sterlingpublishing.com

It suddenly struck me that that tiny pea, pretty and blue, was the Earth. I put up my thumb and shut one eye, and my thumb blotted out the planet Earth. I didn't feel like a giant. I felt very, very small.

—Neil Armstrong

It is difficult to say what is impossible, for the dream of yesterday is the hope of today and reality of tomorrow.

—Robert Goddard

Contents

Introduction

It is basically impossible to summarize the entire history of astronomy and space exploration in just 250 milestones, but I'm not going to let that stop me from trying! I work in a field that has a rich and exciting history. Chronicling that history is a daunting task, but, from the perspective of a space enthusiast who was lucky enough to pursue a career in space sciences, it is an embarrassment of riches. In the last 50 years alone, we have been witness to one of the most profound and important bursts of human exploration in history: the Space Age. People have left the planet (some are living off-planet right now!), and a dozen have walked on the Moon. Using robotic proxies and giant telescopes—some launched into space—we have been able to see, up close, the alien landscapes of all the classically known planets, visit asteroids and comets, and view the cosmos in all its glory.

All of this has been made possible because we have, as Sir Isaac Newton put it best, "stood on the shoulders of giants." No appreciation for the wondrous discoveries of modern astronomy and space exploration would be complete without a thoughtful consideration of the foundations of modern science and experimentation that were built by our ancestors. Many of their achievements were attained at great personal or professional cost, and many others were not recognized as important until decades—even centuries—later. Where it has been impossible or impractical to recognize the specific individuals responsible for these contributions, I have included entries that at least acknowledge the importance of key groups of people in setting the stage for future achievements. Examples include the sky maps still preserved in the caves of some of the earliest humans; the Sumerians' contributions to the birth of cosmology 5,000 to 7,000 years ago; the still-mysterious Stone Age civilizations responsible for constructing ancient sky observatories like Stonehenge; the careful chroniclers of celestial events from the Chinese Xià, Shāng, and Zhōu dynasties (2100 BCE–256 BCE); and the various schools of mathematics and astronomy from early Egyptian, Indian, Arab, Persian, and Mayan societies that have exerted such a strong influence on modern astronomy, astrophysics, and cosmology.

It is, of course, possible to identify and recognize specific individuals who have played critical roles in the development of scientific thought in general, or physics and astronomy in particular. No accounting of the history of the science upon which modern astronomy is based would be complete, for example, without mentioning the lasting contributions of ancient philosophers, mathematicians, and astronomers, such as Pythagoras, Plato, Aristotle, Aristarchus, Eratosthenes, Hipparchus, and Ptolemy. In more recent times, scientists like Nicolaus Copernicus, Galileo Galilei, Johannes Kepler, Isaac Newton, Albert Einstein, Edwin Hubble, Stephen Hawking, and Carl Sagan have all become household names, famous for incredible advances in modern physics, astronomy, and space science. These giants are featured prominently in many entries in this book.

But many others—perhaps famous only in textbooks—have been responsible for important advances or discoveries, and their works also represent critical milestones. These eminent scientists include Christiaan Huygens, discoverer of Saturn's "thin, flat ring" and large moon, Titan; Giovanni Cassini, who discovered Jupiter's Great Red Spot, Saturn's moon Iapetus, and the true nature of Saturn's rings; Edmond Halley, whose eponymous periodic comet returns to the inner solar system every 76 years; the last of the pretelescopic giants of astronomy, Tycho Brahe, whose data enabled Johannes Kepler's discovery of the laws of planetary motion; Charles Messier, a prolific comet hunter who first catalogued more than 100 of the most famous nebulae in the sky; the mathematician Joseph-Louis Lagrange, who predicted the existence of the special gravitational balance points in space that are now named after him; William Herschel, the discoverer of Uranus and several of its moons; the spectroscopy pioneers Joseph von Fraunhofer, Christian Doppler, and Armand Fizeau, who provided the foundations for astronomers to measure both the compositions and velocities of celestial objects; the discoverers of radioactivity, Pierre and Marie Curie, and their colleague Henri Becquerel; the inadvertent father of quantum mechanics, Max Planck; Harlow Shapley, one of the first astronomers to truly grasp the phenomenal size of the Milky Way; liquid-fueled rocketry pioneer Robert Goddard; the astrophysicist and "cosmic web" codiscoverer Margaret Geller; and Eugene Shoemaker, the planetary scientist who helped recognize the importance of impact cratering on our planet and others. Thus I have tried to chronicle these and other important contributors to the advancement of astronomy, astrophysics, planetary science, and space exploration who might not have quite reached the pinnacle of scientific fame among the general public.

And then there are the forgotten—or, at least, the undeservedly neglected—men and women who have made discoveries, developed new theories, performed paradigm-shifting experiments, or simply slogged away to find some critical scientific needle in a haystack, but who, for whatever reason, have not received the public attention or scientific accolades that befit their contributions. These more obscure geniuses include the sixth-century Indian mathematician and astronomer Aryabhata; the venerable eighth-century calendar expert Bede of Jarrow; the tenth-century Arabic star mapper

'Abd al-Rahmān al-Sūfī; the heretical Giordano Bruno, burned at the stake in 1600 for asserting the existence of other habitable worlds; the seventeenth-century Danish astronomer Ole Røemer, who made the first accurate measurement of the speed of light; English astronomer and predictor of the 1639 transit of Venus, Jeremiah Horrocks; German physicist Ernst Chladni, who correctly deduced the extraterrestrial origin of meteorites in 1794; Arthur Eddington, a British astrophysicist who was among the first people to understand the insides of stars, and American radio engineer Karl Jansky, who in 1931 had an idea for an experiment that led to the invention of radio astronomy.

The unsung also include a number of extremely influential female astronomers who often had to work harder than their male colleagues to overcome the biases and prejudices of a male-dominated field. These noteworthy women include Caroline Herschel, younger sister of William Herschel and an accomplished late-eighteenth-century British comet hunter and star mapper; the world's first female professor of astronomy, Maria Mitchell; and the early-twentieth-century female "human computer" colleagues at Harvard, including Annie Jump Cannon and Henrietta Swan Leavitt, who developed the classification scheme for stars still widely in use today and who discovered so-called standard candle stars used to estimate distances in the universe. I've tried to mention many other important but often overlooked astronomers, physicists, philosophers, and engineers throughout this book, though I'm afraid I'm still not giving them the credit they deserve. As a professional astronomer and planetary scientist, I'm embarrassed to admit that even I hadn't heard of some of these amazing scientists prior to doing the research for this book.

I noticed partway through the research that the number of individuals being singled out for mention was decreasing over time, especially in the entries after the 1950s—the start of the Space Age. This reflects, in my opinion, a recent trend in astronomy and space exploration—and perhaps all scientific fields. Science and exploration used to be fairly individualistic enterprises, usually practiced by wealthy men who worked alone, often under a monarch or patron of some kind and often in fierce competition with other wealthy gentleman scientists. There were exceptions, of course: notable collaborations (such as that between Tycho Brahe and Johannes Kepler, or among Pierre and Marie Curie and Henri Becquerel) and research groups (for example, al-Tūsī's thirteenth-century research team at the Marāgheh Observatory in Iran, or the sixteenth-century Kerala school of mathematics in India) certainly existed. But overall, before World War II, most of the scientific advances in my field were primarily made by individuals.

In contrast, as technology advanced in the latter half of the twentieth century, more and more advances in physics, astronomy, and space exploration began to fall under the realm of what many now call Big Science. Big Science is a group or team enterprise; individuals have expertise in specific parts of the project, but the project spans such a wide range of disciplines that no one team member is expert in all of it. An early relevant example in physics was the US Army's Manhattan Project of the 1940s, aimed at developing the first atomic weapons. Experts were needed with engineering, materials, and aeronautics expertise, and the army also needed to find scientists who were the world's leaders in understanding nuclear reactions at extremely high temperatures and pressures. Of course, many of those scientists were astronomers who had been, just a few years earlier, developing those skills by figuring out how stars shine. Other early Big Science projects that relied on teams of individuals with astrophysical or space science expertise included the development of military radar systems and of rockets, such as intercontinental ballistic missiles for suborbital flight and Earth-orbiting satellites for military and civilian use.

The civilian history of astronomy-related Big Science has been dominated by the achievements of the US National Aeronautics and Space Administration (NASA), formed in 1957. This book is chock-full of NASA milestone achievements in human and robotic space science and exploration, and very few of those achievements can be directly associated with an individual. Indeed, my own experience with NASA robotic astronomy and planetary science missions—using the Hubble Space Telescope or working with instruments on orbiters around the Moon, Mars, and asteroids and on the Mars rovers *Spirit*, *Opportunity*, and *Curiosity*—has reinforced my realization that most cutting-edge modern astronomy and space exploration work requires large teams of people to succeed. The ranges of expertise required are impressive. A Mars rover mission, for example, requires planetary scientists (including physicists, chemists, mathematicians, geologists, astronomers, meteorologists, and even biologists), computer scientists and programmers, an enormous variety of engineers (including those specializing in software, materials, propulsion, power, thermal, communications, electronics, systems, and others), and management, financial, and administrative support staff. Similar ranges of expertise are needed to build, launch, and operate space telescopes, space shuttles, big particle detectors and colliders, and the International Space Station (by some estimates, the most expensive and complex project ever attempted by humans). Further, these kinds of Big Science projects can each cost hundreds of millions to tens of billions of dollars or more over their lifetimes. Individuals are usually not singled out when these kinds of projects succeed or fail, because the collective efforts of the team were required to get the job done. The Soviet Union's success in space exploration projects in the 1960s and 1970s was the result of a similar team-oriented (though more military-run) formula. Recently, the nineteen-nation European Space Agency and the nations of Canada, Japan, Brazil, South Korea, India, and China have become bigger players in international astronomy-oriented Big Science projects in addition to smaller astronomy and space exploration projects of their own.

Just as challenging as identifying key individuals has been identifying key events in the history of astronomy and space exploration. Some, like the formation of the Earth and planets, or the first humans in space, or the first humans to land on the Moon, are no-brainers. But most events fall into a continuum of importance that varies from one person to the next (more on that in a bit). Pinning down the exact dates for some of these events and putting them into a simple chronologic listing is also difficult, either because they are only best guesses of prehistoric occurrences (for example, when life emerged on Earth), or because they occurred over a wide span of time (for example, the formation of the first stars and galaxies), or because they are predicted to occur at some uncertain future time—such as the end of the universe! In cases where the chronological timing of key events is uncertain or broad, or both, I have indicated uncertainty by placing a *c.* (the Latin abbreviation for *circa*, meaning "about") in front of the date listed.

The timing of historical and especially of modern events is usually much more accurately known, but there is still a significant challenge in identifying *which* events— out of a seemingly infinite number of scientific discoveries, theories, inventions, and missions in astronomy and space exploration over the last few centuries, and particularly over the last 50 years—to include in such a compendium. It is perhaps inevitable, then, that a bias would creep into any attempt to focus on just a subset of these incredible achievements, and I will be the first to confess that such a bias exists in my own compilation of milestones: I am a solar-system snob. My passion at work is to study planets and moons and asteroids and comets—what, to many other astronomers, are effectively just little bits of leftover debris that didn't happen to fall into the newly forming Sun 4.5 to 5 billion years ago. It's true that the Sun is 99.86 percent of the mass of the solar system (and that Jupiter is most of the rest), but it's also true that the remaining 0.14 percent is incredibly interesting—partly because life developed and thrives on at least one speck of that debris, and may have existed (or perhaps still exists) on others. When my astrophysics or cosmology friends lament the fact that I have to focus my research on such insignificant, nearby objects, it's easy to counter with the fact that the latest discoveries in extrasolar planet research are showing that solar systems are probably common around other stars, too. Our solar system may be one of millions—or, more likely, billions   in our galaxy. And yet we do not know if any of them harbors life, as ours does. That makes us very special, even if we are very small.

As you voyage through this history of astronomy and space exploration, you may detect this bias, among my collection of milestones, toward discoveries and theories and adventures related to our nearest neighbors in space: our solar system. To me it's a good bias to have, partly because solar system objects are what we know the most about scientifically, and partly because it is important to get to know one's neighborhood in order to understand and appreciate the bigger community. That is, the physics and chemistry and celestial mechanics and geology and spectroscopy and engineering and other skills needed to explore *our* solar system—with telescopes, robotic spacecraft, high-speed computer simulations, cutting-edge laboratory experiments, or human exploration crews—provide the foundation for exploring our neighboring stars, our Milky Way galaxy, our nearby galaxies, and the cosmos, now or in the distant future. To me, these pivotal moments are most worthy of being called milestones in space exploration: when a point of light is resolved into a truly unique world (and there are more than 50 relatively large and millions of smaller unique worlds in our neighborhood), or when we visit these worlds for the first time, either virtually through the eyes of our robotic emissaries or in person. Our solar system is sort of like our playground. By getting to know the worlds around us, we are dipping our toes into what Carl Sagan famously referred to as the "shores of the cosmic ocean," preparing for that time when we will someday wade farther out into the water.

Finally, it's important to point out that this collection of milestones in the history of astronomy and space exploration is certainly not exhaustive or complete. Pragmatic limitations on the length and size of this book restricted the collection to just 250 entries, representing only a fraction of the people, historic discoveries, and paradigm-shifting events that have characterized this exciting field over the course of time—indeed, over the entire history of space and time. Different authors might certainly have assembled a different set of milestones, but all would have faced the same dilemma: how to decide which ones *not* to include? When I set about outlining this project, I decided to try to cover not only the many phenomenal achievements of the Space Age but also to include and acknowledge a sampling of the many fundamental achievements from scientists of antiquity, spanning the ancient empires of Mesopotamia, China, India, Egypt, Europe, and the Americas. As well, I wanted to make sure that some of the major achievements from the Middle Ages, Renaissance, and more recent history, from preindustrial times to the Industrial Revolution, were also captured. In attempting to balance the timeline, I may have shortchanged many deserving people, discoveries, or events from more modern times, for which I ask your forgiveness and indulgence. As I wrote in the beginning, it is basically impossible to summarize the entire history of astronomy and space exploration by choosing just 250 milestones. But let's not let that stop us!

Acknowledgments

I owe a debt of gratitude to the many colleagues and mentors who, knowingly and unknowingly, fueled my interest in the history of astronomy, planetary science, and space exploration. Among the most influential of these are the late Carl Sagan, Jim Pollack, and Leonard Martin, all of whom were also outstanding scientists. I am very grateful to the many friends, colleagues, new acquaintances, and anonymous Samaritans who graciously agreed to provide their beautiful photographs or artwork for this project. I extend a huge thank you to the creators of and worldwide editors/contributors to Wikipedia (to which I contribute financially) for creating an incredible research tool and jumping-off point for additional exploration of both historical and current topics. I thank Michael Bourret at Dystel & Goderich and Melanie Madden at Sterling for their never-wavering support of this seemingly never-ending project. I also thank my astronomy colleagues Rachel Bean and Margaret Geller for their reviews of a number of entries in areas relatively far from my own astronomical expertise. Finally, my biggest thanks and love go to my wife, Maureen, who helped enormously in the photo research for this book and who has been phenomenally patient during its long gestation period. In the words of Voltaire, "How infinitesimal is the importance of anything I can do, but how infinitely important it is that I should do it."

This image, showing the inside surface of the heat shield of NASA's Curiosity rover, was obtained during its descent to the surface of Mars on August 6, 2012. The Mars Descent Imager instrument, known as MARDI, shows the 15-foot (4.5-meter) heat shield at about 50 feet (16 meters) from the spacecraft.

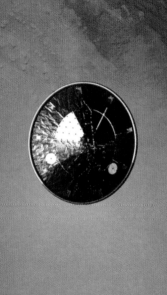

Big Bang

Edwin Hubble (1889–1953)

There's no better place to start considering the broad sweep of astronomical history than the beginning—that is, the *actual beginning* of both space and time. Twentieth-century astronomers such as **Edwin Hubble** discovered that the universe is expanding by observing that large-scale structures like galaxies are all moving away from each other, in any direction that we look. This means that, in the past, the universe was smaller and that, at some point in the far distant past, everything started out as a single point of space and time: a singularity. Years of careful observations by the **Hubble Space Telescope** and other facilities have revealed that the universe was born in a violent explosion of this singularity about 13.7 billion years ago.

The details of big bang theory—as it was initially dubbed by astronomers in the 1930s—have been rigorously tested with decades of astronomical observations, laboratory experiments, and mathematical modeling by cosmologists and astronomers who specifically focus their research on the origin and evolution of the universe. What we have learned about the early history of our universe from these studies is impressive: within the first second of the universe's existence, the temperature dropped from a million billion degrees to "only" 10 billion degrees, and *all* of the universe's present supply of protons (hydrogen atoms) and neutrons formed out of this primordial plasma. By the time the universe was only three minutes old, helium and other light elements had been formed from hydrogen in the same kind of **Nuclear Fusion** process that still occurs today deep inside of stars.

It's mind-blowing to think about both space and time being *created* at a single instant, 13.7 billion years ago. What caused the explosion? What was there before the big bang? Cosmologists tell us that we can't really ask that question because time itself was created in the big bang. It's also humbling to realize that the most abundant element within each of our bodies—hydrogen—was created in the very first second that ever was. We are ancient!

SEE ALSO Hubble's Law (1929), Nuclear Fusion (1939), Hubble Space Telescope (1990).

Graphically depicting the beginning of the universe is just as challenging as trying to understand it! Here, an artist has fancifully captured the idea that the big bang was triggered by a collision with another three-dimensional universe that had been hidden in higher dimensions.

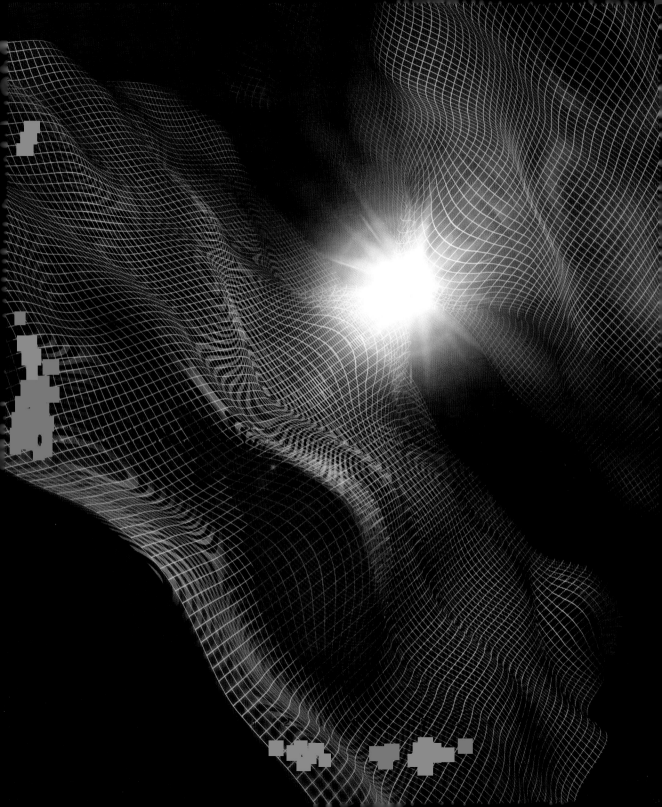

Recombination Era

The universe's early years were a time of intense heat, pressure, and radiation. All of space was bathed in the primordial light of highly ionized atoms and subatomic particles, interacting, colliding, decaying, and recombining at temperatures of millions of degrees. This period in cosmic history is often referred to as the radiation era. By the time the universe was about 10,000 years old, the expansion of space and the decay of many energetic particles had cooled the cosmos to "only" about 12,000 kelvins (kelvins, or K, are a measure of the temperature above absolute zero). This was an important threshold, because as the universe continued to cool, the total energy from heat and ionizing radiation became less than the total so-called rest mass energy of matter itself, embodied in physicist **Albert Einstein**'s famous equation $E = mc^2$. Still, for hundreds of thousands of years longer, the universe was essentially just an opaque, dense, high-energy soup of constantly colliding ionized protons and electrons. But as the expansion and cooling continued, radiation energy continued to decrease as compared to rest mass energy.

By about 400,000 years after the **Big Bang,** the temperature had dropped to only a few thousand kelvins—low enough to allow electrons to be captured (deionized) into stable hydrogen atoms and for multiple hydrogen nuclei to form the universe's first molecules: hydrogen gas, or H^2. This period in the universe's early history is known as the recombination era.

The cool thing about recombination is that it allowed the universe's remaining radiation—mostly high-energy photons and other subatomic particles—to decouple from matter and thus to finally travel, relatively unimpeded, through space. The universe grew colder and darker over the next few hundred million years, a time which cosmologists have dubbed "the dark ages." The residual 3-kelvin glow of the early universe's joyously freed radiation energy, known as the **Cosmic Microwave Background,** can still be detected today.

SEE ALSO Big Bang (c. 13.7 Billion BCE), Einstein's "Miracle Year" (1905), Cosmic Microwave Background (1964), Mapping the Cosmic Microwave Background (1992), Age of the Universe (2001).

NASA's Wilkinson Microwave Anisotropy Probe (WMAP) satellite generated this sky map of the residual heat left over after the early universe's initial expansion. The small fluctuations in temperature seen here—only a few hundred-millionths of a degree—acted as the seeds for the first stars and galaxies in the universe.

First Stars

Every dark age ends with a Renaissance, and the early history of the universe is no exception. Cosmologists believe that the so-called dark ages lasted about 100 to 200 million years, after which time molecular hydrogen and other molecules formed during the recombination era began to gravitationally clump together—perhaps from the effects of turbulence, but no one really knows why. The clumps of gas acted as seeds, gravitationally attracting more gas, making the clumps grow bigger and bigger until they eventually became enormous clouds of hydrogen that started to grow warm inside from the increasing pressure of the surrounding gas. Give a cloud a nudge—say, from the gravitational pull of another nearby cloud—and it will move and, eventually, start to spin. At some point, maybe 300 to 400 million years after the **Big Bang**, the temperatures in the centers of some of these huge, slowly spinning clouds of gas grew to be millions of degrees, as in the first three minutes after the big bang. The temperatures and pressures inside these spherical clouds became high enough to fuse hydrogen into helium, and the first stars were born. The dark ages were over!

The first stars, sometimes called Population III stars by astronomers, were more than just quirky local phenomena, though. They were enormous—perhaps one hundred to one thousand times more massive than our Sun—and they had a huge influence on their stellar neighborhood, radiating prodigious amounts of energy out into the surrounding clumps and clouds of hydrogen, heating them externally, and freeing the electrons that had been captured at the beginning of the dark ages. This time is known as the era of reionization, because the universe once again began to glow—not from the light and heat of creation but, like today, from the light and heat of the stars.

SEE ALSO Big Bang (c. 13.7 Billion BCE), Recombination Era (c. 13.7 Billion BCE), Eddington's Mass-Luminosity Relation (1924), Nuclear Fusion (1939).

In this supercomputer simulation, ionized hydrogen bubbles (blue) and molecular hydrogen clouds (green) form the first organized, large-scale structures in the early universe, eventually collapsing to form the first stars.

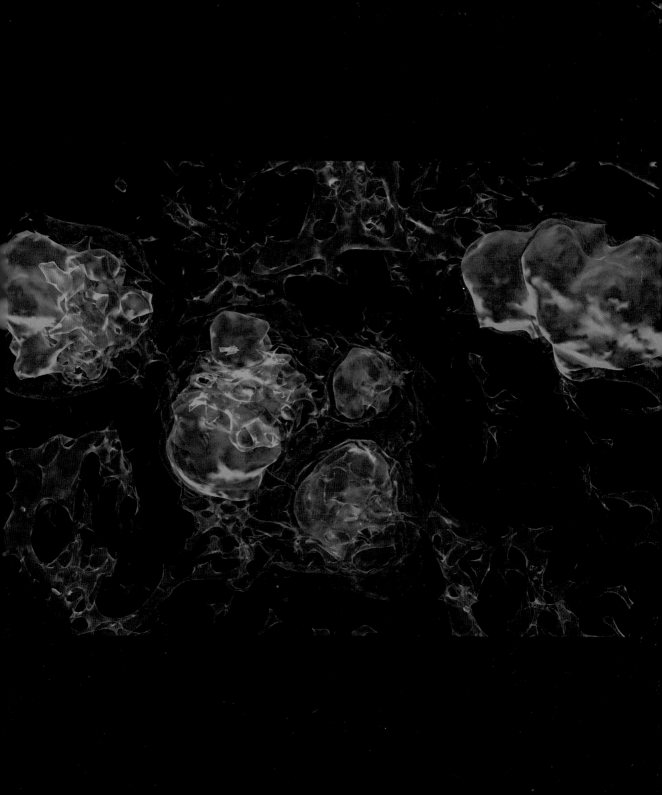

Milky Way

Astronomers define a galaxy as a gravitationally bound system of stars, gas, dust, and other more mysterious components (see **Dark Matter**), all moving collectively through the cosmos as if they were a single object. Once the first stars had formed, it was only a matter of time—not much time, in fact—that many of them would inevitably become attracted by each other's gravity and form clusters, then clusters of clusters, and eventually huge congregations of stars orbiting their common center of gravity.

Our own Milky Way galaxy consists of an estimated 400 billion stars and has a structure that is typical of the class of so-called barred spiral galaxies seen throughout the universe (see **Spiral Galaxies**). The Milky Way has a crowded central semispherical bulge of stars surrounded by a flatter, spiral-shaped disk of stars (including the Sun), gas, and dust, all of which is surrounded by a diffuse spherical halo of older stars, star clusters, and two smaller companion galaxies. It's an enormous structure, nearly 100,000 light-years (the distance light travels in a year, or about a billion billion miles) wide and 1,000 light-years thick in the disk. Our Sun is about halfway out from the galactic center, and one galactic-year orbit takes about 250 million Earth years.

Astronomers don't know exactly when the Milky Way was formed. The oldest known stars in the galaxy are in the halo and are about 13.2 billion years old. The oldest stars in the disk are younger—about 8–9 billion years old. It is likely that the different parts of the Milky Way formed at different times, although the basic structure appears to have been set in motion very early.

Our ancient ancestors were awed by the bright whitish band that dominated their night sky, often envisioning it in creation myths as a river of light and life. Though we now know that we are inside a massive, choreographed gathering of stars looking out, it's still easy to find awe in the scale and majesty of our home galaxy.

SEE ALSO Dark Matter (1933), Spiral Galaxies (1959).

Wide-angle photo of the Milky Way galaxy's Sagittarius arm. Light from billions of stars causes the bright, diffuse glow of the galaxy; dark dust in the disk blocks some of that starlight from our view. A meteor can be seen streaking through the scene near the bottom.

Solar Nebula

Star formation is a messy process. As giant molecular clouds collapse, almost all of the cloud's gas and dust eventually falls into the central protostar—almost. A tiny fraction of the gas and dust remains in orbit around the forming star, and as the whole system spins and cools, that residual cloud of debris slowly flattens into a disk of gas, dust, and (farther from the star) ice. During this phase of star formation, it appears that all young stars start out with an accompanying disk, often called a solar nebular disk.

The nebula from which our own Sun eventually formed probably started collapsing about 5 billion years ago, though the exact timing is uncertain. Observations indicate that sun-like stars typically take about 100 million years to form, and that nebular disks form in only about 1 million years around young stars. Once the disk is formed, it changes rapidly, with tiny dust and/or ice grains colliding, sticking to each other, and growing into marble-size particles, in a process (called accretion) that computer models indicate takes only a few thousand years. These small particles collide with others, sometimes sticking together, and the process appears to continue on in a poorly understood, runaway fashion until, within perhaps only a few more million years, planetesimals (kilometer-sized clumps of dusty, icy, rocky, and/or metallic grains) and then asteroids 100–1,000 kilometers in size have formed.

Solar nebular disks don't seem to last long; most of the dust accretes or is dispersed within about 10 million years. Close to the star, it's too warm for ice to condense, so the planetesimals are mostly rocky and too small to gravitationally hold on to much gas. Farther out, ice and dust can be accreted into larger planetesimals, with enough mass to accrete huge amounts of gas as well, eventually growing into "gas giant" planets. Exactly how such messy beginnings lead to such elegant planetary systems, and in such little time, is currently a topic of much debate and speculation among astronomers.

SEE ALSO First Stars (c. 13.5 Billion BCE), Violent Proto-Sun (c. 4.6 Billion BCE), Circumstellar Disks (1984), First Extrasolar Planets (1992).

Space artist Don Dixon's conception of the proto-Sun and its solar nebular disk, the spinning cloud of gas and dust and ice from which all our solar system's planets, moons, asteroids, and comets formed.

Violent Proto-Sun

Star birth, like childbirth, can be a rather intense and messy event that involves a lot of energy. Even before they get hot and dense enough to start **Nuclear Fusion** of hydrogen into helium, newly forming protostars can emit huge amounts of energy as they gravitationally contract during their 100-million-year gestation period. Some of these baby stars funnel their energy into solar system–size jets of gas, dust, and charged particles, possibly collimated and heated by strong magnetic fields from the star, or from material falling in from the associated nebular disk, or both.

Astronomers have identified many examples of violent jets of material being emitted from very young protostellar objects, often called T Tauri stars after the prototype example. In fact, the star T Taurus is very much like what astronomers believe the young Sun was like, suggesting that our own star went through a similar short and violent period of intense jetting and other high-energy activity before it started to stably fuse hydrogen and settle down into its long, relatively quiescent life on the so-called **Main Sequence.**

Evidence for whether the Sun went through such a violent early T Tauri phase may be preserved in some of the solar system's oldest materials: the ordinary chondrite meteorites. These rocks, which occasionally fall to Earth, are the oldest-known solids in the solar system, and they help to determine the age of the Sun and the timescale for the formation of the planets. These meteorites often contain large percentages of chondrules—small spheres of mineral grains that were once molten droplets of rock before cooling and accreting into larger grains, planetesimals, and asteroids. The source of energy for chondrule melting in the early solar system is unknown, but one possibility is high-energy outbursts and jets from the young Sun.

The deeper and more accurately astronomers peer into space, the more evidence they find for jets and disks around newly forming stars, suggesting that these features are a crucial part of star formation. A violent youth may be a normal, essential part of the life cycle of a typical star.

SEE ALSO Meteorites Come from Space (1794), Main Sequence (1910), Nuclear Fusion (1939).

A young T Tauri–like protostar is embedded in a cloud of dust in the lower left of this Hubble Space Telescope photograph. Called HH-47, it is emitting a spiraling jet of ionized gas and dust 1.2 trillion miles (2 trillion kilometers) long into space (from lower left to upper right).

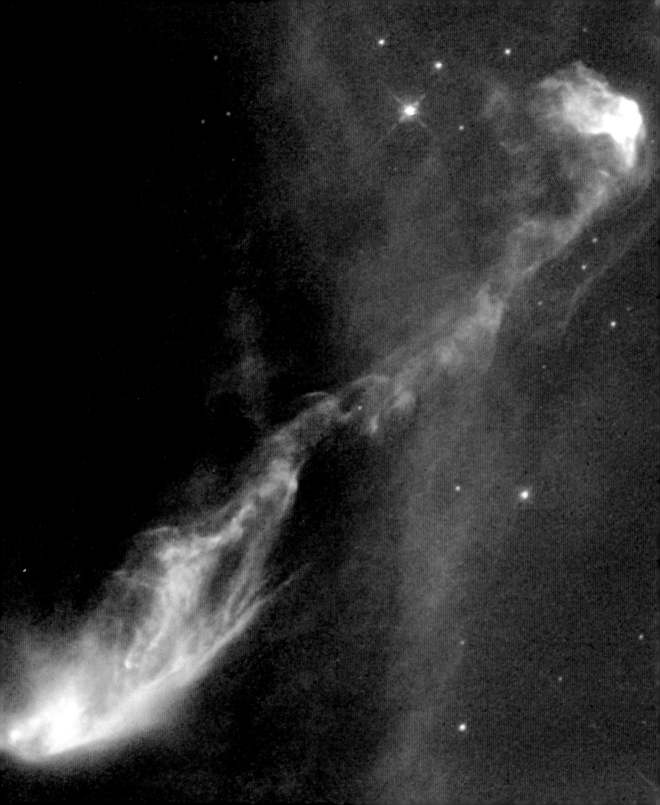

Birth of the Sun

The temperature and pressure in the central region of the **Solar Nebula** grew dramatically for about 100 million years, until they passed a threshold where hydrogen atoms were packed so tightly that they underwent **Nuclear Fusion,** becoming helium and releasing some energy as light and heat. Thus, our Sun was born!

We tend to think of the Sun as special, and rightly so—the Sun is critical to the creation and continuing survival of all life on our planet. It's harder to think of the Sun as typical, average, even mundane, but in many ways it is. Our star is one of more than *10 billion trillion* (1 with 22 zeros after it) stars in the known universe, all of which appear to be the natural result of matter—mostly hydrogen—interacting with gravity at high pressures and temperatures and releasing enormous amounts of energy into their surrounding space. Stars are truly the engines of our universe.

Once stars are born, they live relatively stable lives and then die, often in relatively predictable and sometimes spectacular ways. The Sun is no different. It will keep fusing hydrogen atoms into helium atoms for another 5 billion years or so. When the hydrogen runs out, the Sun will shed its outer layers (engulfing the Earth and the other inner planets) and start fusing helium in its core. When the helium runs out, the Sun slowly fades to a **White Dwarf** and then dims to a cinder.

Astronomers have been able to deduce that about one to three new stars are born each year, and about one to three old stars die each year in our Milky Way galaxy. If we extrapolate to all known galaxies and do a little math, that means that something like 500 million stars are born and 500 million stars die each day in the universe. It's a staggering and humbling thought that should make us appreciate even more every one of these precious days in the life of our own star, the Sun.

SEE ALSO Chinese Observe "Guest Star" (185), "Daytime Star" Observed (1054), Planetary Nebulae (1764), White Dwarfs (1862), Nuclear Fusion (1939).

An ultraviolet image of our local star, the Sun, taken by NASA's Solar Dynamics Observatory UV space telescope. Streamers, loops, hotter spots (brighter), and cooler spots (darker) are all evidence of an extremely active, though quite typical, middle-aged star.

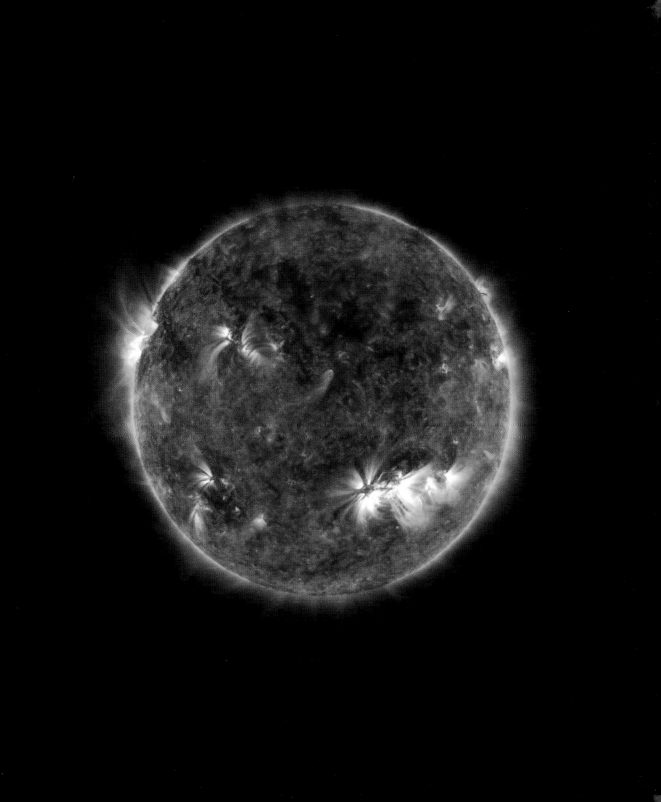

Mercury

All the planets in our solar system formed around the same time, about 4.5 billion years ago, as the **Solar Nebula** cooled and tiny grains condensed, collided, stuck together, and eventually grew into a small number of big objects. In the warm zone close to the Sun, the planets were rocky. Farther out, beyond the "snow line," they were mixtures of rock, ice, and gas.

Mercury is the closest in of the so-called terrestrial planets, with a diameter of 3,032 miles (4,880 kilometers); **Earth**'s diameter, by comparison, is 7,926 miles (12,756 kilometers). Mercury orbits the Sun at an average distance of only 0.38 astronomical units (or AU; 1 AU = 93 million miles = 150 million kilometers = Earth's average orbital distance from the Sun). Mercury is the Roman name of the Greek god Hermes, the fleet-footed messenger. The planet was aptly named: even the ancients knew that Mercury takes only 88 days to complete a circuit in the sky, which we now know represents its orbital period around the Sun.

Mercury is a small world of harsh extremes and curious enigmas. There is no atmosphere, and temperatures range from only 90 kelvins in permanently shadowed craters near the poles to more than 700 kelvins (above the melting point of lead) in the harsh midday sunlight. Earth-based radar observations indicate that there may be ice in those polar craters. Mercury has a very high density and a large iron core that spans 75 percent of the planet's radius. The core might be partially molten, perhaps explaining Mercury's weak magnetic field (1 percent as strong as Earth's). Images from the two space missions that have encountered Mercury (*Mariner 10* in 1974–1975 and *MESSENGER* since 2008) reveal a heavily cratered surface and some evidence of ancient volcanic activity similar to the Moon's. Perhaps most surprising, the planet preserves a network of large tectonic thrust faults (scarps) that seem to indicate that Mercury may have been completely molten early in its history and then shrank by a few percent when it cooled.

SEE ALSO Solar Nebula (c. 5 Billion BCE), Earth (c. 4.5 Billion BCE), Kirkwood Gaps (1857), Habitable Super Earths? (2007), *MESSENGER* at Mercury (2011).

NASA's MESSENGER spacecraft flew past Mercury three times in order to set itself up to go into orbit around the planet in 2011. This third flyby image was taken in January 2008 and revealed many never-before-seen craters and other features.

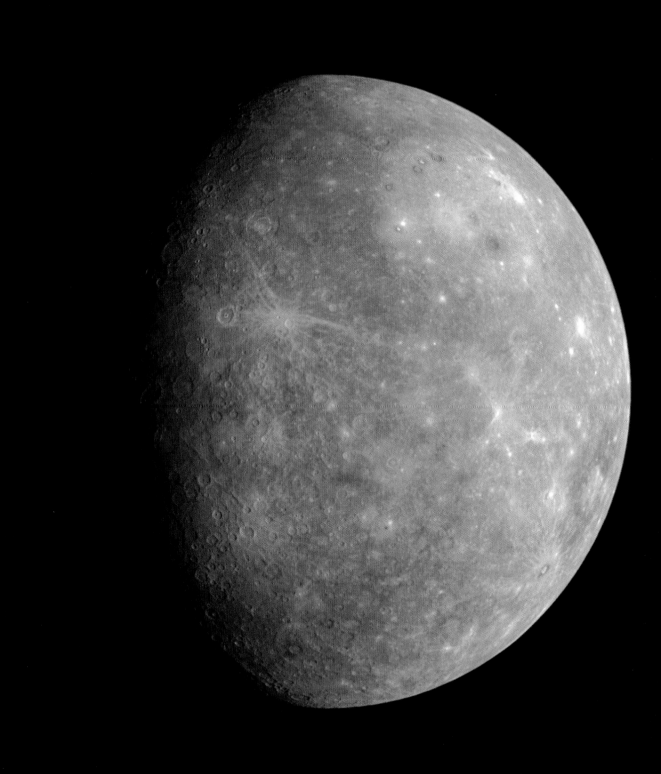

Venus

It's fun to ponder the relative importance of "nature versus nurture" in determining the origin of people's personalities and characteristics. Twins, for example, make great case studies. Well, the same is true for planets, and one of the best examples to consider is Venus, a near-twin of Earth in some ways but profoundly different in others.

Venus is only about 5 percent smaller than Earth and has about the same density—meaning that it is essentially a rocky, terrestrial planet very much like our own. Both planets have atmospheres, and Venus even orbits in the same general neighborhood of the inner solar system as we do, at an average distance of 0.72 astronomical units compared to Earth's 1.0. But that's where the similarities end. Venus is barely spinning, taking about 243 Earth days to spin once on its axis—backward! The Venusian atmosphere is much thicker than ours, with 90 times the pressure at its surface. That thick atmosphere sports violent upper-level wind speeds of more than 218 miles (350 kilometers) per hour and is almost entirely carbon dioxide, with only scant traces of the nitrogen dioxide, oxygen, and water found in Earth's atmosphere. The carbon dioxide molecule is transparent to visible light but is exceedingly good at trapping heat radiation (like a greenhouse), causing the surface of Venus to be very hot—more than 750 kelvins, or about 300 degrees hotter than an oven!

Astronomers are trying to understand how Earth and Venus ended up with such radically different surface conditions. Understanding carbon dioxide may be the key. Earth has as much carbon dioxide as Venus, but it dissolves in our oceans and is trapped in rocky carbonate minerals. Any ocean on early Venus, slightly closer to the Sun, would have since evaporated away, however, leaving no way to remove the carbon dioxide.

Venus is a case study of carbon dioxide gone wild and is a prime example of how studying other planets can help us understand what may be in store for our own world.

SEE ALSO Earth (c. 4.5 Billion BCE), Venus Transits the Sun (1639), *Venera 7* Lands on Venus (1970), Venus Mapped by *Magellan* (1990), Earth's Oceans Evaporate (~1 Billion).

In 2009 the European Space Agency's Venus Express orbiter acquired this false-color composite of infrared heat emitted from the planet's night side (lower left, red) and sunlight reflected from the planet's swirling day-side clouds (upper right).

Earth

Our home world is the largest of the terrestrial planets and the only one with a large natural satellite. To a geologist, it's a rocky volcanic world that has separated its interior into a thin, low-density crust, a thicker silicate mantle, and a high-density, partially molten iron core. To an atmospheric scientist, it's a planet with a thin nitrogen-oxygen-water vapor atmosphere buffered by an extensive liquid water ocean and polar ice cap system, all of which participate in large climate changes on seasonal to geologic timescales. To a biologist, it's heaven.

Earth is the only place in the universe where we know that life exists. Indeed, evidence from the fossil and geochemical record is that **Life on Earth** began almost as soon as it could, when the **Late Heavy Bombardment** of asteroids and meteorites quieted down. Earth's surface conditions appear to have remained relatively stable over the past four billion years, which, combined with our planet's favorable location in the so-called habitable zone, where temperatures remain moderate and water remains liquid, has enabled that life to thrive and evolve into countless unique forms.

Earth's crust is divided into a few dozen moving tectonic plates that essentially float on the upper mantle. Exciting geology—earthquakes and volcanoes and mountains and trenches—occurs at the plate boundaries. Most of the oceanic crust (70 percent of Earth's surface area) is very young, having erupted from mid-ocean-ridge volcanoes spanning a few hundred million years ago to today. Because of its youth, there are only a few hundred impact craters preserved on our planet's surface, in stark contrast to the battered face of our neighbor, the **Moon**.

The high amounts of oxygen, ozone, and methane in Earth's atmosphere are a sign of life that could be detected by alien astronomers studying our planet from afar. Indeed, these gases are exactly what astronomers are looking for today among the panoply of newly discovered **Extrasolar Planets.** Are there more Earths out there, waiting to be found and explored?

SEE ALSO Birth of the Moon (c. 4.5 Billion BCE), Late Heavy Bombardment (c. 4.1 Billion BCE), Life on Earth (c. 3.8 Billion BCE), First Extrasolar Planets (1992).

Digital portrait of Earth's Western Hemisphere on September 9, 1997, using data from a variety of NASA and National Oceanic and Atmospheric Administration orbital weather and geologic/ocean-monitoring satellites.

Mars

We may have to go no farther than the next planet out to find out if life exists—or ever existed—beyond Earth. Mars has seemingly always been the subject of fascination, from ancient times, when it was seen as a cosmic incarnation of the Roman god of war, to the twentieth century, when many imagined the planet to be the abode of Percival Lowell's desperate canal builders.

Mars is a small planet, about half the diameter of **Earth** and only about 15 percent of its volume. For further reference, the surface area of Mars is about the same as the surface area of all of the continents on Earth. On average, the planet orbits about 50 percent farther from the Sun than we do. The thin Martian carbon dioxide atmosphere (only 1 percent as thick as Earth's) can't trap much heat, so the surface is very cold. Daytime temperatures near the equator rarely rise above the freezing point of water, and nights near the poles routinely drop down to the freezing point of carbon dioxide (which is 150 kelvins, or about -190°F). Today Mars is a dusty world in a deep freeze.

And yet, spacecraft images, meteorites from Mars, and other data over nearly 50 years have shown that Mars is the most Earthlike place in the solar system (besides Earth itself), and that during its first few billion years, the Red Planet may have been a much warmer and wetter world. What happened? Possibilities include gradual cooling of the planet's core and solar wind or catastrophic impact destruction of the atmosphere. Determining how and why the planet's climate changed so dramatically is a hot topic of research.

We've learned enough about the Mars of 3 or 4 billion years ago to know that parts of the surface and subsurface were habitable to life. The next 50 years of Mars exploration will be all about expanding the search for habitable environments there and finding out if any were—or still are—inhabited.

SEE ALSO Earth (c. 4.5 Billion BCE), Deimos (1877), Phobos (1877), *Mars and Its Canals* (1906), First Mars Orbiters (1971), *Vikings* on Mars (1976), Life on Mars? (1996), First Rover on Mars (1997), Mars Global Surveyor (1997), *Spirit* and *Opportunity* on Mars (2004).

This Hubble Space Telescope photo of Mars was taken during the Red Planet's close approach to Earth in 1999. Dusty, more oxidized areas are orange; volcanic rocks and sand are brownish-black. The north polar water ice cap is at top; wispy bluish water-ice clouds and a polar storm system provide evidence of the planet's thin atmosphere.

Main Asteroid Belt

The terrestrial planets were rather quickly assembled about 4.5 billion years ago from small rocky and metallic building blocks called planetesimals, which condensed from the slowly cooling, warm inner regions of the **Solar Nebula.** Each of the growing planets "swept up" planetesimals along and near its orbital path, generally clearing out their orbital zones until the lack of new material to sweep up set the limit on the growth of these rocky worlds.

Beyond the orbit of Mars, however, the accretion of planetesimals and growth into larger planets was continually thwarted and disrupted by the strong gravitational influence of nearby Jupiter. Jupiter's influence made collisions between planetesimals more energetic, minimizing the gentle collisions that would allow them to stick together and grow, and close encounters with Jupiter itself ejected many of the planetesimals in the Mars–Jupiter zone. Thus, instead of a large planet in the region between Mars and Jupiter, we have a rather diffuse disk or belt of small rocky and metallic asteroids: the main asteroid belt.

Astronomers estimate that there might be more than a million asteroids larger than a half mile (about a kilometer) in size in the main belt. To date, the orbits, positions, and general characteristics of more than half a million of these are known, including the most massive two, **Ceres** and **Vesta.** These two, plus asteroids Pallas and Juno, account for more than half of the total mass of the entire main belt.

Asteroids aren't just randomly located, though. Jupiter's gravitational pull has cleared out many gaps in the main belt (see **Kirkwood Gaps**), and some groups of asteroids travel together in "families" that may represent the disrupted and slowly scattering remains of once-larger objects. **Jupiter's Trojan Asteroids** are two large groups of small bodies trapped in special orbits where Jupiter's gravity and the Sun's gravity balance each other out.

Small pieces of impact-shattered asteroid interiors fall to Earth all the time—we call them meteorites—and their ages and compositions provide an enormous amount of detailed information about the timing, formation, and evolution of our solar system.

SEE ALSO Solar Nebula (c. 5 Billion BCE), Meteorites Come from Space (1794), Discovery of Ceres (1801), Discovery of Vesta (1807), Kirkwood Gaps (1857), Jupiter's Trojan Asteroids (1906).

Overhead computer-generated plot of the inner solar system on August 14, 2006, out to the orbit of Jupiter (outer blue circle) with the Sun at center. Asteroids in the main belt are colored white. Orange dots are the Hilda asteroid family, and green dots are Jupiter's Trojan asteroids.

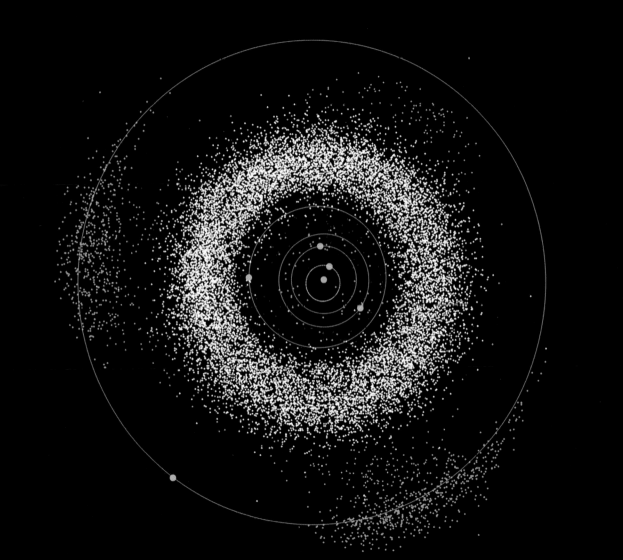

Jupiter

Our solar system is basically comprised of the Sun (about 99.8 percent), Jupiter (about 0.1 percent), and everything else. Jupiter is truly the king of the planetary realm, with more than twice the mass of all the other planets combined. Sixty-three known moons and a series of faint rings orbit this colossal world. Jupiter's diameter is 23 Earths across; if it were hollow, more than a thousand Earths would fit inside.

Partly because of its enormous size and its orbital position at the inner edge of the outer solar system (around 5.2 astronomical units), Jupiter is the fourth-brightest object in our night sky after the Sun, Moon, and Venus. Jupiter is also luminous because its visible surface is made up of bright clouds. Indeed, there is no "surface" visible on Jupiter or any of the other giant outer planets—everything we see is cloud or haze, made of exotic and sometimes colorful chemical compounds like methane, ethane, ammonium hydrosulfide, and phosphine. Winds traveling several hundred miles per hour twist the clouds into horizontal belts, and giant Earth-size storm systems, such as the **Great Red Spot,** have churned for many hundreds of years.

Below the clouds, Jupiter's pressure and temperature increase dramatically, but the chemistry is, on average, much simpler: Jupiter is about 75 percent hydrogen and 25 percent helium, just like the Sun. In fact, if the **Solar Nebula** had been bigger and Jupiter had formed with about 50 to 80 times more mass, it would have become a star.

Jupiter's formation has had a major influence on the architecture of the solar system, perturbing the orbits of the other giant planets, preventing a planet from forming in the region of the **Main Asteroid Belt,** and gravitationally scattering asteroids and comets on orbits that caused impacts with other planets in the **Late Heavy Bombardment.** Some objects were even flung into the **Kuiper Belt** or out of the solar system entirely! Today Jupiter is a gravitational magnet, still occasionally drawing in small bodies such as **Comet SL-9,** which split up and smashed into its cloud tops in 1994.

SEE ALSO Solar Nebula (c. 5 Billion BCE), Main Asteroid Belt (c. 4.5 Billion BCE), Late Heavy Bombardment (c. 4.1 Billion BCE), Great Red Spot (1665), Kuiper Belt Objects (1992), Comet SL-9 Slams into Jupiter (1994), *Galileo* Orbits Jupiter (1995).

True-color mosaic of Jupiter and the Great Red Spot obtained in 2000 by the NASA Cassini spacecraft when it flew past the giant planet for a gravitational assist on its way to Saturn.

Saturn

There might be no more enchanting experience to a fan of astronomy than viewing Saturn and its glorious rings through a small telescope. The scene is almost surreal: a shimmering, egg-shaped orb hanging against the blackness of space and girded by what seems like an incredibly delicate, thin disk of material almost twice as wide as the planet itself. It is truly one of the gems of the sky.

Saturn is the second largest of the gas giants, more than nine times wider and nearly one hundred times as massive as Earth. The flat disk that circles the planet's equator is, of course, the famous **Rings of Saturn**. Composed mostly of ice, the ring system is probably no more than 22 to 33 yards (20 to 30 meters) thick. No one knows whether the rings of Saturn are an ancient, primordial feature, or whether they are a relatively new feature, perhaps formed from the catastrophic breakup of a former icy moon. Accompanying Saturn are 62 known moons, hundreds of smaller "moonlets" embedded in the rings, and billions of ring particles ranging from the size of houses and cars to specks of dust. Saturn's largest moon, **Titan,** is larger than Mercury and is the only moon in the solar system with a thick atmosphere.

Saturn's clouds and haze bands are fainter and less colorful than Jupiter's, although the composition of the atmosphere is fairly similar. Perhaps the biggest chemical difference between Saturn and Jupiter is that, for reasons not fully understood, Saturn has a little less helium relative to hydrogen, making it less "solar" than Jupiter. Another mystery is why the wind speeds on Saturn are much higher than on Jupiter, or anywhere else in the solar system—more than 1,120 miles (1,800 kilometers) per hour in places! Detailed studies of Saturn and Jupiter by the *Pioneer*, *Voyager*, *Galileo*, and *Cassini* spacecraft show us that not all gas giants are the same. As we discover more gas giants among the **Extrasolar Planets**, those worlds, too, are likely to be both lovely and enigmatic.

SEE ALSO Titan (1655), Saturn Has Rings (1659), *Pioneer 10* at Jupiter (1973), *Pioneer 11* at Saturn (1979), *Voyager* Saturn Encounters (1980, 1981), First Extrasolar Planets (1992), *Galileo* Orbits Jupiter (1995), *Cassini* Explores Saturn (2004).

This spectacular photo of Saturn and its rings was taken by NASA's Cassini spacecraft as it approached the planet in 2007.

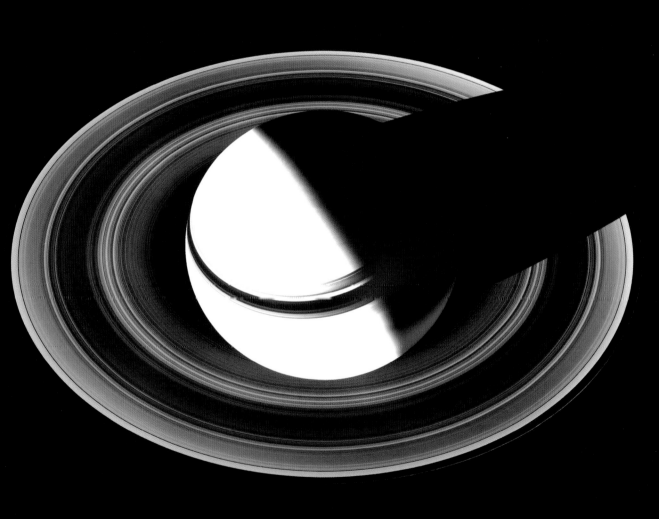

Uranus

William Herschel (1738–1822)

Our solar system's seventh planet, unlike the first six, was not known to the ancients. Uranus (pronounced YUR-uh-nus by astronomers) was discovered in 1781 by telescopic observations of the English astronomer Sir William Herschel. Indeed, it had been observed by many other astronomers as early as 1690, but because of its extremely slow motion across the sky (an 84-year orbit period), it was mistaken for a star. Because Uranus has an average orbital distance of about 19 astronomical units (**Saturn**, the next closest planet to the Sun, has an average orbital distance of about 9.5 astronomical units), its discovery instantly doubled the size of the solar system.

At 4 times the diameter and 15 times the mass of **Earth,** Uranus is classified as a giant planet, but it is much smaller than planetary cousins **Jupiter** and Saturn. Still, the atmosphere of Uranus contains mostly hydrogen and helium, and the planet's distinctive blue-green color is caused by methane clouds and hazes in the upper atmosphere. Storms on Uranus are rare, and the cloud and haze bands are usually quite faint. Uranus has a different overall planetary composition from Jupiter and Saturn, however, with significant amounts of ice and rock in the deep interior. In fact, the ratio of ice and rock to gas is so much higher in Uranus (and **Neptune**), as compared to Jupiter and Saturn, that the planet is more appropriately called an ice giant instead of a gas giant.

As discovered by telescopic observations and the *Voyager 2* **flyby** in 1986, Uranus has 5 large moons and 22 smaller moons, all of them dark and icy. The planet also sports a series of about a dozen thin, dark, icy rings, possibly formed from a relatively recent breakup of one or more small moons.

Perhaps the strangest thing about Uranus is that its spin axis is tilted on its side by about 98 degrees relative to the ecliptic (the plane of Earth's orbit around the Sun). The unusual axial tilt of Uranus may be a result of a grazing giant impact or a close encounter with Jupiter long ago. Whatever the case may be, it is yet another of our solar system's many unsolved mysteries.

SEE ALSO Earth (c. 4.5 Billion BCE), Saturn (c. 4.5 Billion BCE), Jupiter (c. 4.5 Billion BCE), Neptune (c. 4.5 Billion BCE), Discovery of Uranus (1781), Titania (1787), Oberon (1787), Discovery of Neptune (1846), Ariel (1851), Umbriel (1851), Miranda (1948), *Voyager 2* at Uranus (1986).

The Keck telescope in Hawaii obtained this spectacular false-color infrared composite image of Uranus and its rings in 2004. Rare white storm clouds can be seen in the atmosphere.

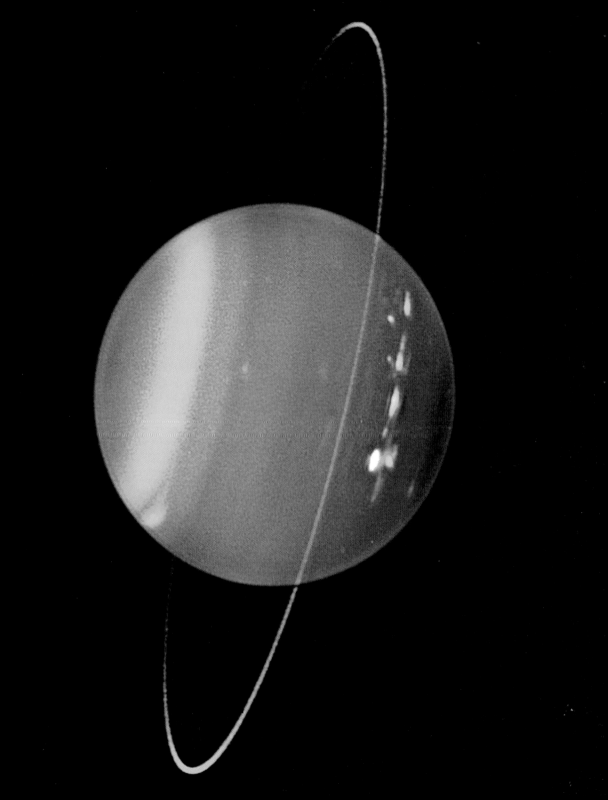

Neptune

If **Earth** and **Venus** can be considered fraternal twin planets, **Uranus** and Neptune
are more like identical twins. Both are denizens of the deep outer solar system, with
Neptune, out at 30 astronomical units, taking approximately 165 Earth years to orbit the
Sun once. Both ice giants are about the same size and mass (Neptune is slightly heavier,
at 17 Earth masses), and both have a similar composition: about 80 percent hydrogen,
19 percent helium, and trace amounts of methane and other hydrocarbons. As it does
on Uranus, methane gives Neptune its beautiful azure color.

Neptune, like Uranus, is another ice giant world with a modest number of icy
satellites (13) and a system of dark icy rings. From telescopic measurements, data
from the 1989 *Voyager 2* **flyby**, and laboratory studies, astronomers have deduced that
Neptune's gaseous atmosphere extends about 10 to 20 percent of the way to the center
of the planet. Then, as pressure and temperature increase, higher concentrations of
water, ammonia, and methane form a hot liquid mantle. Astronomers refer to this
zone as "icy" because the molecules there are thought to have originally come from
the mostly icy outer **Solar Nebula** planetesimals that were part of Neptune's original
building blocks. Some astronomers even think of this zone as a water-ammonia ocean,
and computer simulations suggest that a rain of diamonds fall through this ocean to the
planet's Earthlike core of rock, iron, and nickel.

It's puzzling to astronomers that the ice giants reside in the far outer solar system,
because there may not have been enough solar nebula material at those distances to
form them. One explanation may be that they formed closer to the Sun and slowly
migrated out, perhaps gravitationally nudged by Jupiter and/or Saturn. We think of the
solar system today as stable, running like clockwork, but when the planets were forming,
it was much more violent and chaotic.

SEE ALSO Solar Nebula (c. 5 Billion BCE), Venus (c. 4.5 Billion BCE), Earth (c. 4.5 Billion BCE), Uranus
(c. 4.5 Billion BCE), Discovery of Uranus (1781), Discovery of Neptune (1846), Triton (1846), *Voyager 2* at
Neptune (1989).

*Voyager 2 photo of Neptune's Great Dark Spot, a smaller dark spot to the south, and fast-moving, wispy, cirrus-
like white clouds. Winds in excess of 1,300 miles (2,100 kilometers) per hour were measured in these atmospheric
storm systems.*

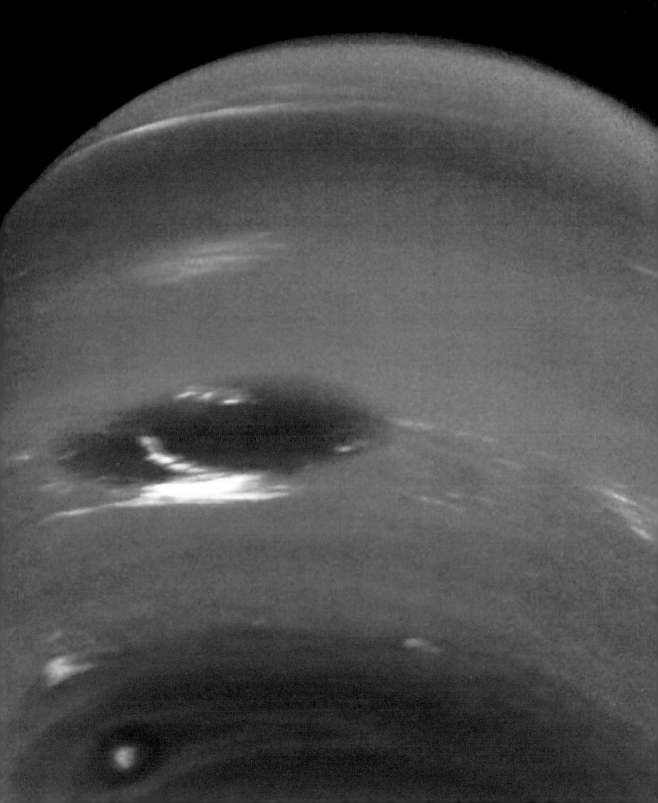

Pluto and the Kuiper Belt

Gerard P. Kuiper (1905–1973)

Most of the rock and ice from the **Solar Nebula** that didn't fall into the young Sun eventually went into building **Jupiter** and, to a lesser extent, the other big planets of our solar system. But there were still some leftover building blocks that never got incorporated into planets, like the small (0.6–6 miles [1–10 kilometers]) rocky **planetesimals** in the **Main Asteroid Belt** that Jupiter's gravity prevented from growing into full-fledged planets, and similar icy ones beyond Neptune that were simply too far apart and that collided too rarely to grow into large planets. This latter class of so-called trans-Neptunian objects is of special interest, partly because the first one ever discovered is also the most famous: Pluto.

Pluto is a small, icy, rocky world in an elliptical orbit between about 30 to 50 astronomical units (AU). It is only about 20 percent the mass and 35 percent the volume of our **Moon**, and yet it has a large icy moon of its own, **Charon**, along with at least four other smaller icy moons, and a thin comet-like atmosphere of nitrogen, methane, and carbon monoxide.

Since the early 1990s, astronomers have discovered many more "Plutos" out beyond Neptune, orbiting within a doughnut-shaped disk called the Kuiper belt, named after the Dutch-American astronomer Gerard P. Kuiper. Beyond the Kuiper belt, which contains small icy bodies that formed in the zone between about 30 and 55 AU, another "scattered disk" consists of icy bodies that formed closer to the Sun but were flung out there by gravitational encounters with Jupiter to between 30 and 100 AU. More than 1,100 trans-Neptunian objects are now known. As it became clear that there are huge numbers of Pluto-like objects in the Kuiper belt and scattered disk, the International Astronomical Union demoted Pluto and other similar objects in 2006 to dwarf-planet status.

Pluto is the last well-known body in our solar system not yet visited by a space mission. That will change in 2015, when the *New Horizons* mission flies past Pluto and its moons and reveals them as new worlds rather than just fuzzy points of light.

SEE ALSO Solar Nebula (c. 5 Billion BCE), Main Asteroid Belt (c. 4.5 Billion BCE), Jupiter (c. 4.5 Billion BCE), Discovery of Pluto (1930), Charon (1978), Kuiper Belt Objects (1992), Demotion of Pluto (2006), Pluto Revealed! (2015).

Artist's rendering of the orbits of the four giant planets (circles) and the tilted, elliptical orbit of Pluto. The dots represent the doughnut-shaped cloud of small trans-Neptunian objects known as the Kuiper belt, of which Pluto was the first member to be discovered.

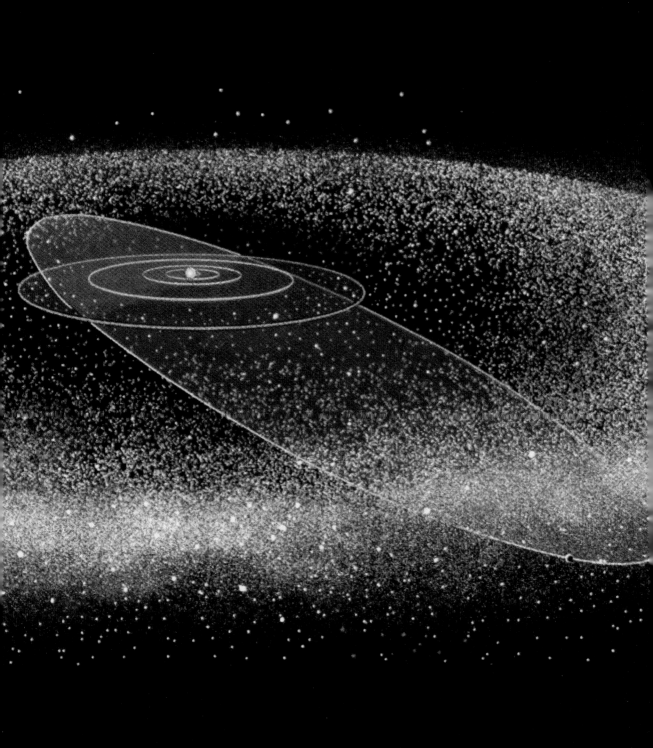

Birth of the Moon

Earth is unique among the terrestrial planets in having a very large natural satellite. But where did our Moon come from? Many possible ideas have been considered by astronomers. One idea is that the Moon formed in orbit around our planet at the same time and in the same way that **Earth** formed: by the accretion of rocky and metallic planetesimals condensed out of the warm inner **Solar Nebula**. Another idea is that the early (molten) Earth was spinning so fast that a blob of it shed off (fissioned) and went into orbit, forming the Moon. Yet another hypothesis proposes that the Moon was formed somewhere else in the inner solar system and was captured by Earth's gravity.

These ideas competed for supremacy until the *Apollo* missions brought Moon rocks and other information back to Earth and revealed that none of them fit the actual data about the Moon. The co-accretion model predicted that the Moon would have the same basic age and composition as Earth, but it does not: the Moon has a much lower density, much less iron, and appears to have formed 30–50 million years after Earth and other planets formed. Fission required the early Earth to be spinning too fast, and the capture model suggested that there was no way to dissipate all the energy a free-flying Moon would have had to lose in order to get captured into orbit.

In the 1990s, planetary scientists proposed the giant impact model: if the early Earth had been struck obliquely in a giant impact by a Mars-size protoplanet, computer simulations show that enough of Earth's low density, iron-poor mantle could have been melted and launched into orbit to eventually cool, accrete, and form the Moon. The composition, density, and even age of the Moon matched the model's predictions. Indeed, the giant impact model is still the best explanation for the origin of our Moon.

SEE ALSO Solar Nebula (c. 5 Billion BCE), Earth (c. 4.5 Billion BCE), Late Heavy Bombardment (c. 4.1 Billion BCE), First on the Moon (1969), Roving on the Moon (1971), Lunar Highlands (1972), Last on the Moon (1972).

Artist's conception of the grazing impact of a Mars-size body with the proto-Earth more than four billion years ago. A giant impact like this is believed to have led to the formation of our Moon.

Late Heavy Bombardment

All the planets, including Earth, have been hammered by a veritable rain of asteroids and comets throughout geologic history, and the rate of impact back in the early days of the solar system was many orders of magnitude higher than it is now. The record of that early cosmic impact history is not preserved on Earth because most of our planet's surface is covered by younger volcanic deposits or has been eroded away by the action of wind, water, and ice. The surface of the Moon, on the other hand, is much more revealing, and the huge number of lunar impact craters and basins provides a stark reminder of just how battered Earth's surface must have once been.

One of the major legacies of the *Apollo* missions is the ability to determine the absolute ages of specific impact cratering events using **Radioactive Dating** of lunar samples. The results indicate dates for large lunar impact events of around 3.8 to 4.1 billion years BCE—a surprising discovery, considering that all the major planets formed around 4.5 billion years ago. Many planetary scientists believe that the simplest explanation is that the Moon—and, by inference, Earth—went through a period of intense impact cratering about 400 to 700 million years after their initial formation.

The cause of this relatively brief spike in what was an otherwise decreasing rate of impacts is unknown, but some have speculated that **Jupiter**'s gravitational nudging—the same nudging that influenced the putative migration of **Uranus** and **Neptune** and that likely propelled the trans-Neptunian objects out into the **Kuiper Belt**—may also have diverted some asteroids and comets into the inner solar system. If true, the resulting cataclysm certainly wreaked havoc on the terrestrial planets and no doubt had a profound influence on the development and stability of life on our home world.

SEE ALSO Jupiter (c. 4.5 Billion BCE), Uranus (c. 4.5 Billion BCE), Neptune (c. 4.5 Billion BCE), Pluto and the Kuiper Belt (c. 4.5 Billion BCE), Life on Earth (3.8 Billion BCE), Radioactivity (1896), First on the Moon (1969), Roving on the Moon (1971), Lunar Highlands (1972), Last on the Moon (1972).

Artist's conception of the so-called late heavy bombardment of asteroids and comets hypothesized to have impacted the Earth and Moon around 3.8 to 4.1 billion years ago. The rate of impacts and number of objects impacting at once is exaggerated here for dramatic effect.

Life on Earth

No one knows exactly how, when, or why life first appeared on planet **Earth,** but we know that almost as soon as it could, it did. The oldest signs of life on Earth are chemical, not fossil, and are inferred as evidence because all known life on this planet is based on a common chemical architecture. Specifically, certain biogeochemical processes and reactions that are common to all life on Earth—reactions involving certain amino acids commonly associated with DNA or RNA, for example—create recognizable patterns in the isotopes of carbon and some other elements. Life prefers to use (and create) certain materials, in essence, and anomalous chemistries, like the occurrence of extra carbon-12 (^{12}C) compared to carbon-13 (^{13}C) in some 3.8-billion-year-old rocks from Greenland, provide circumstantial but controversial chemofossil evidence for life very early in our planet's history.

The oldest known fossil evidence of microbial life on our planet is dated at around 3.5 billion years old and is preserved in the layers of ancient stromatolites, which are rock and mineral structures built up by colonies of simple organisms such as blue-green algae. Stromatolites still form in places such as Shark Bay in Western Australia, making them among the oldest life forms on our planet.

Recent studies of the very earliest period of Earth's history, the Hadean (4.5–3.8 billion years ago), provide evidence that oceans and continents may have formed much earlier than previously thought, and that conditions may have been suitable for life just a few hundred million years after our planet formed. The **Late Heavy Bombardment** of 3.8 to 4.1 billion years ago may have killed off earlier life forms, or perhaps just frustrated their attempts to flourish. Whatever the case may be, soon after Earth's crust cooled, the oceans formed, the late heavy bombardment ended, and Earth became stable enough to support life. The fact that it thrived and began to evolve into so many niches is remarkable. Now astronomers, planetary scientists, and astrobiologists are searching for evidence of life on other Earthlike worlds.

SEE ALSO Earth (c. 4.5 Billion BCE), Late Heavy Bombardment (c. 4.1 Billion BCE).

Cross-sectional view of a stromatolite fossil; the reddish layers are thought to be fossilized remains of the ancient blue-green algae that are some of the oldest preserved evidence for life on Earth. This particular piece, from the Ord Range of Western Australia, is about 2.4 inches (6 centimeters) tall.

Cambrian Explosion

The first life forms to emerge on Earth were simple, single-celled microbes that were able to tap into chemical and thermal energy sources in the environment of early Earth. For the first 3 billion years of Earth's history, in fact, life was apparently dominated by single-celled organisms, occasionally organized into colonies like those found in stromatolites. Around 550 million years ago, however, in what is often called the Cambrian explosion, the diversity of life on Earth began to dramatically increase. As such, many of the ancestors of today's modern plants and animals appear rather early in the fossil record, geologically speaking. Biologists are actively debating possible reasons for the rather sudden and dramatic diversification of life on Earth at the Precambrian/Cambrian boundary.

Biologists are also trying to understand the reasons for many rather sudden, drastic decreases in the number of species and species diversity in the fossil record. The most dramatic of these occurred at the boundary between the Permian and Triassic geologic periods, around 250 million years ago. Across a span of perhaps only a million years, about 70 percent of all land species and 96 percent of all oceanic species died off, a period informally called the "great dying" and the "mother of all mass extinctions." It took more than 100 million years for the diversity of life on Earth to once again reach pre-Permian levels.

Such a massive loss of life on Earth could have been triggered by climate change, although many believe that the rapidity is difficult to explain that way. Alternatively, a catastrophe such as a large impact event or an enormous volcanic outpouring, which may also have triggered other climatic and geologic catastrophes, could be responsible. Biologists, geologists, and astronomers are still searching for clues.

SEE ALSO Life on Earth (c. 3.8 Billion BCE), Dinosaur-Killing Impact (65 Million BCE), Arizona Impact (50,000 BCE).

The Permian extinction 250 million years ago saw Earth's largest decrease in the number of species and diversity of life. The view from Yavapai Point across the Grand Canyon exposes nearly 2 billion years of Earth's geologic record for detailed study.

Dinosaur-Killing Impact

Luis W. Alvarez (1911–1988), **Walter Alvarez** (b. 1940)

The role of large impacts in catastrophically changing Earth's climate and biosphere was not fully appreciated until a father-and-son geologic team—Luis and Walter Alvarez—and their colleagues found what they believe to be smoking-gun evidence that the impact of a large asteroid with Earth was probably responsible for causing the extinction of the dinosaurs and many other species about 65 million years ago. The key was the discovery of a thin layer of sediment all around the world enriched in the rare element iridium and dating to what is now defined as the boundary between the Cretaceous and Tertiary geologic periods (often abbreviated as the K–T boundary). Iridium is a metal in the platinum family that often bonds with iron in rocks and minerals. Most of Earth's iron (and, presumably, iridium) sank into the deep mantle and core when Earth was forming, so a globally distributed, iridium-rich crustal deposit all of the same age is quite an anomaly. The Alvarez hypothesis was that the iridium came from a large metal-bearing asteroid that impacted Earth and vaporized, dramatically changing the climate and wreaking havoc on many plant and animal species.

The impact lifted vaporized rock and dust into the atmosphere and set off large-scale fires that added soot and smoke, blotting out the Sun and lowering global surface temperatures for years. While the effect was not as large as that of the Permian-Triassic extinction, species that depend on sunlight and photosynthesis for their survival were decimated at the K–T boundary, destroying the base of many other predators' food chains. Some species, such as mammals and birds capable of burrowing or subsisting on insects, carrion, or other nonplant food-chain staples, weren't severely disrupted by the event. When the dust cleared (literally), the survivors were apparently able to expand into niches that they couldn't occupy before.

The idea that the dinosaurs were killed off by an asteroid impact is a hypothesis that is constantly being tested. Other geologic and climatic effects, like a dramatic sea level drop or massive eruptions of volcanic rock, both of which occurred around the same time but before the hypothesized impact, may also have contributed to the environmental conditions that wiped out so many species.

SEE ALSO Cambrian Explosion (550 Million BCE), Arizona Impact (50,000 BCE), Tunguska Explosion (1908), Comet SL-9 Slams into Jupiter (1994).

An artist's conception of a large asteroid crashing into the Earth marks the precise moment of the end of the Cretaceous and the beginning of the Tertiary period of geologic history, about 65 million years ago.

Homo Sapiens

Humans of the species *Homo sapiens* are relatively new on Earth. Our earliest appearance in the fossil record, judging by archaeological sites in Africa, dates back only about 200,000 years. Fossil evidence shows that *Homo sapiens* coexisted for a time with our closely related subspecies, the Neanderthals, and that evidence for Neanderthal traits disappeared from the record around 30,000 years ago.

We are a persistent lot, excelling at survival with tools, language, long memories, and hard-won experience. Our history and evolution reflect an inquisitiveness and a desire for more intangible nourishments of the soul, too, which may explain why music, dance, and art have been such an important part of the human experience.

I remember seeing the 17,000-year-old Paleolithic cave paintings of the Dordogne region of France and being amazed that our ancestors could make the time for art amid what must have been a constant struggle just to stay alive. But they didn't just paint animals, plants, and other mundane things from their daily lives. Many archaeologists now believe that some of the dots, lines, and perhaps even animal figures were representations of constellations or other features of the night sky. If so, then not only are these the oldest paintings on Earth but they are also the oldest sky maps, painted by the world's first astronomers.

The emergence of the human race may not seem worthy of note as a significant milestone in the history of astronomy. We are, after all, just one species, on one planet, orbiting a typical star that is in a fairly ordinary part of an average spiral galaxy. Our planet may be one of billions capable of supporting life, and we could be one of countless numbers of sentient species among the stars. But it might also be the case that we are the only intelligent, self-aware, technological species—and civilization—in all the cosmos. That latter possibility is humbling, daunting, and perhaps even a little scary, but it reminds us that we should indeed celebrate as truly extraordinary the appearance of a species that, through its achievements, provides a way for the universe to know itself.

SEE ALSO Birth of Cosmology (5000 BCE), SETI (1960), First Extrasolar Planets (1992), Planets around Other Suns (1995), Habitable Super Earths? (2007).

Part of a reconstructed painting from the famous caves of Lascaux in southwestern France depicts a prehistoric horse and other symbols that some think might be representations of stars and constellations in the night sky.

Arizona Impact

Grove Karl Gilbert (1843–1918), **Daniel Barringer** (1860–1929), **Eugene M. Shoemaker** (1928–1997)

We know that the Earth has been slammed by asteroid and comet impacts throughout its history—we need only look at the record of the **Late Heavy Bombardment** preserved in the ancient, airless surface of our celestial neighbor the Moon. Given the erasure of impact craters on our planet via the constant erosion of wind and water and the continuous renewal of the seafloor through plate tectonics and volcanism, it's no wonder that it took terrestrial geologists a long time to realize the importance of impact cratering as a geologic process on Earth and other terrestrial planets, moons, and asteroids.

One of the best natural laboratories for that realization is the Meteor Crater (also known as Barringer Crater or Canyon Diablo Crater), just east of Flagstaff, Arizona. The crater is about 1,200 meters (3,900 feet) wide and 170 meters (557 feet) deep. Up until the 1960s, geologists heatedly debated the origin of this feature. In the 1890s, G. K. Gilbert, one of the first advocates for the impact origin of the circular craters on the Moon, argued that the lack of significant debris from the impactor itself meant that the crater was formed by an explosive volcano. In the early twentieth century, mining engineer Daniel Barringer purchased the crater and spent years drilling in vain for what he believed was a giant buried iron meteorite from an impact event. Finally, geologist Eugene M. Shoemaker, who had studied craters created by the US government's nuclear test program in Nevada, confirmed the crater's origin as an impact feature, based primarily on the discovery of certain forms of quartz minerals that could only form in the high-pressure, high-temperature environment of an impact rather than a volcano.

Since then, more than two hundred other impact craters have been recognized on Earth—most larger but much less preserved than Meteor Crater. Lab and computer studies of impact physics now reveal that impactors—in this case, an approximately 164-foot (50-meter) iron-rich asteroid traveling at more than 6 miles (10 kilometers) per second—are almost always completely vaporized on impact, resolving the question about the lack of debris in the area around the crater.

SEE ALSO Late Heavy Bombardment (c. 4.1 Billion BCE), Tunguska Explosion (1908), Comet SL-9 Slams into Jupiter (1994).

View from the rim of Meteor Crater, a hole in the Arizona desert about ¾ mile (1.2 kilometers) wide, created about 50,000 years ago by the impact of a small iron-rich asteroid traveling more than 6 miles (10 kilometers) per second.

Birth of Cosmology

In Greek, *kosmos* means "the universe," and thus our modern word *cosmology* refers to the study of the nature, origin, and evolution of the universe. In the classical context, a society's cosmology refers to its worldview or its way of thinking about where its people came from, why they are there, and where they are going. Civilizations throughout human history have created and nourished their cosmologies through creation stories, mythology, religion, philosophy, and, most recently, science.

We often hear (or read) such platitudes about how humanity has always been looking to the stars, or how our distant ancestors must have pondered the heavens in this way or that. While it's fun to speculate, it's impossible to know what prehistoric people were really thinking because (by definition) there's no record of prehistory. That's one reason why the oldest archaeological artifacts that depict or represent astronomical themes are so important: they provide some real data with which to try to understand how ancient people viewed the universe.

The oldest preserved evidence of a civilization pondering the heavens comes from the Sumerians, in their partial star maps or pieces of crude astronomical instruments that some scholars believe date to between 5,000 and 7,000 years ago. Even the scant fragments of information available from that time reveal a significant degree of sophistication in the Sumerians' understanding of the motions of the Sun, Moon, major planets, and stars. Perhaps this is not surprising: the Sumerians built the first city-states supported by the cultivation of crops by a year-round, nonmigratory population. Knowing how to read the sky translated directly into knowing when to plant, irrigate, and harvest, and a stable food supply gave them time to invent writing, arithmetic, geometry, and algebra.

Sumerian cosmology appears to have been the first to make gods of the heavenly bodies, a practice inherited by later Babylonian, Greek, Roman, and other cosmologists. Sumerian cosmology also espoused the idea of many heavens and many Earths in what was a decidedly nongeocentric universe. It's a worldview that resonates—surprisingly—with modern cosmological thinking, as the reality seems to be a universe without any center at all and apparently brimming with many Earths.

SEE ALSO Big Bang (c. 13.7 Billion BCE), Greek Geocentrism (c. 400 BCE).

Reconstruction of an ancient Sumerian star chart from 3300 BCE known as the planisphere of Nineveh, which is believed to be one of the oldest astronomical instruments and data sets ever discovered.

Ancient Observatories

While ancient peoples were clearly aware of the sky, it wasn't until the Bronze Age (about 3000–600 BCE) that large-scale, often astronomically themed monuments began to appear. The most famous of these is the prehistoric monument of Stonehenge in southern England, which is just one of many ancient stone circles, burial mounds, and other earthwork structures found around the world with cultural, religious, and/or astronomical importance.

Stonehenge's construction is impressive—especially the 25-ton lintel stones somehow perched atop the 13-foot (4-meter), 50-ton standing post stones. Modern experiments and simulations have shown that building such structures was possible using Neolithic and Bronze Age tools and methods—neither magic nor architecturally gifted aliens were required. Still, it must have been near the limit of the available technology to build such unprecedented structures.

It also appears to have been an impressive feat of prehistoric design. Detailed examination of the orientations of some of the site's various stones, postholes, pits, paths, and ridges has been interpreted by some archaeologists as evidence that Stonehenge was an astronomical observatory, designed in some ways as a giant sundial to mark the passing of the seasons and to reckon the specific dates of the winter and summer solstices. While the details of the monument's use as an observatory are the subject of active research and debate, there is broad consensus among both archaeologists and astronomers that the basic alignment of the structure was designed with the paths of the Sun and Moon in mind.

Other examples of prehistoric astronomical observatories include the Newgrange and Maeshowe burial mounds in Ireland and Scotland, respectively, aligned so that only the rising winter solstice Sun fills their inner tombs with light; solar-aligned trilithons and passage mounds in Portugal; and the stacked *taula* stones on the Spanish island of Minorca. The civilizations that built these remarkable monuments perhaps as far back as 5,000 years ago left no written records about themselves or their traditions and beliefs. They left lasting records of stone and earth, however, which reveal how much they must have valued their knowledge of the heavens.

SEE ALSO Egyptian Astronomy (c. 2500 BCE).

Some of the sarsen (sandstone) trilithons (stone posts about 25 feet [8 meters] tall, topped by horizontal lintels) and smaller bluestone markers within the inner circle of the prehistoric Stonehenge megalithic structure in southern England.

Egyptian Astronomy

The great pyramids of Giza are a monument to the technological (and labor management) prowess of ancient Egyptian civilization. They are also testaments to the designers' astronomical skill, which figured prominently in Egyptian society and religion 4,500 years ago.

Because the earth's spin axis slowly precesses, or wobbles like a spinning top, back in 2500 BCE Polaris was not the North Star. Indeed, much like the skies near our south celestial pole today, there was no bright star near the north celestial pole in those days. To the pharaohs, astrologers, and commoners, the sky at night appeared to spin around a vortex-like dark hole, thought to be a gateway to the heavens. In ancient Egypt, this gateway was located about 30 degrees above the northern horizon, and so the pyramids were carefully aligned to the north, with a small shaft leading from the pharaoh's main burial chamber to the outside, pointing directly into the center of the gateway. If the plan was to join the gods in the afterlife, why not go in through the main door?

Egyptian astrologers also played an important role in developing a rather sophisticated calendar system that was already well established by the time the pyramids were being built. A new year was defined by the first sighting of the brightest star in the sky, Sirius (Sopdet to the Egyptians), just before sunrise in midsummer. The year was divided into 12 months of 30 days each, with 5 extra days of worship or parties tacked onto the end for a 365-day year. They also knew from carefully observing and recording star positions on different dates that they needed to add an extra day every fourth year— what we call a leap day—to keep their calendar synced to the motions of the sky. The predawn rising times of a number of bright stars were tracked in order to determine times for major religious festivals, as well as to plan for the annual floods of the Nile.

The pyramid shape itself may even represent a facet of ancient Egyptian cosmology, as some myths claim that the god of creation, Atum, lived within a pyramid that, along with the land, had emerged from the primordial ocean.

SEE ALSO Ancient Observatories (c. 3000 BCE).

The great pyramids of Giza, burial places of the pharaohs and astronomical pointers to the presumed gateway to the heavens at the north celestial pole. These were the largest human-made structures in the world for nearly 4,000 years.

Astronomy in China

Interest in and use of astronomy developed independently in cultures around the world. The roots of Chinese astronomy date back to prehistoric and Bronze Age times, based on archaeological evidence of star and constellation names preserved in ancient burial sites. Like other civilizations before and since, the early Chinese were intimately tuned in to the cycles and motions of the Sun, Moon, and stars, and each of the major dynasties of early China enlisted astronomers to develop their own elaborate solar- and lunar-based calendar systems. They were diligent and thoughtful in their astronomical observations, carefully recording evidence of changes in the sky, including solar and lunar eclipses, sunspots, the motions of the planets, and the appearance of new comets or supernovae. Astronomers today still use unique records kept by Chinese astronomers from the Xià, Shāng, and Zhōu dynasties (about 2100–250 BCE) for historical astronomical research.

Chinese astronomers developed new and accurate instruments for observing the heavens, including large celestial globes and armillary spheres that were used to map the stars and constellations and track the motions and brightnesses of planets and "**Guest Stars**" such as comets and novae. More sophisticated versions of these instruments were used right up until the introduction of the telescope in the seventeenth century to develop Chinese theories of planetary motion that rivaled those being developed using similar methods by Western astronomers such as **Tycho Brahe** and **Johannes Kepler**.

The early Chinese appear to have had a number of complex cosmological models of the universe. Some models envisioned the heavens as a dome or celestial sphere (as in early Western cosmology), while others viewed the heavens as infinite and somewhat chaotic. Long before it was fashionable in the West, some early Chinese astronomers had deduced that the Moon and other celestial bodies were spherical. The potential clash with the prevailing worldview of the Earth as flat didn't seem to be a problem because the emphasis of early Chinese astronomical studies—consistent with the prevalent concepts and practices of Confucianism—was on careful observations of the way the universe simply was, or at least appeared to be.

SEE ALSO Chinese Observe "Guest Star" (185), "Daytime Star" Observed (1054), Brahe's "Nova Stella" (1572), Three Laws of Planetary Motion (1619).

Illustration from 1675 by Ferdinand Verbiest depicting an early Chinese astronomer-priest, along with several astronomical instruments, including a celestial globe and an astrolabe quadrant (foreground).

智多星呉用

東溪村の人
字ハ學究道
號ヲ加亮先生トいふ
陣法ハ孔明太公望よ
不芳陰謀ハ死蟲
ふも勝まり
梁山泊乃
軍師ふり

Earth Is Round!

Pythagoras (c. 570–495 BCE)

We take it for granted: the Earth is a beautiful, blue, spherical marble set against the blackness of space. But without the relatively recent benefit of being able to go out into space and look back, someone had to advance the idea that the Earth might be round rather than flat, as it appears to anyone on the ground. By many accounts, that someone was Pythagoras of Samos, a sixth-century-BCE philosopher, mathematician, and part-time astronomer from Greece, also famous for his Pythagorean theorem in geometry.

The argument made by Pythagoras and his followers for a spherical Earth was an indirect one based on a variety of observations. For example, sailors traveling south from Greece reported seeing southern constellations higher in the sky the farther south they went. Expeditions that departed for destinations along the African coast south of the equator, for example, reported that the Sun shone from the north rather than from the south (as it does in Greece). Another important piece of evidence came from observing lunar eclipses: when the full Moon passes directly behind the Earth relative to the Sun, the Earth's curved shadow is clearly visible as it eclipses the Moon.

It is a matter of some debate whether Pythagoras himself actually "discovered" that the Earth is spherical or whether he was simply the most outspoken (and famous) advocate of what was becoming relatively common wisdom among educated people of early Greek civilization. Regardless, the issue would be proven in another 250 years or so in the experiments of **Eratosthenes**, and nearly 2,500 years later the first astronauts to leave Earth orbit, aboard the Apollo 8 mission, would share with the world the first glorious photos of our beautiful, spherical, blue marble in space.

SEE ALSO Eratosthenes Measures the Earth (c. 250 BCE).

ABOVE: *One piece of evidence that the Earth is round is the curved shadow of the Earth cast onto the Moon during a lunar eclipse, such as this one, observed in 2008 from Greece.* RIGHT: *Our precious blue planet is a sphere of rock and metal, covered by a thin layer of air and (in many places) liquid water. To our distant ancestors, it wasn't obvious that the world was round.*

Greek Geocentrism

Plato (427–347 BCE), Aristotle (384–322 BCE)

Ancient Greece established a number of important legacies that strongly influenced Western civilization for thousands of years. These included a cosmological worldview that was derived primarily from the teachings and writings of two of the most prominent thinkers of the classical era: the mathematician and philosopher Plato, and his student Aristotle, who mastered almost all of the arts and sciences of the day. Plato and Aristotle created the basic foundations upon which modern Western philosophy and science—including physics and astronomy—were built.

The basic focus of ancient Greek astronomical thinking (and science in general) was to try to seek mathematical, physical explanations and models for observed phenomena. It was natural to turn to the geometry and trigonometry pioneered by Pythagoras for solutions. Plato used geometry to divide the universe into two realms: the fixed sphere of the Earth, and the nested, constantly moving spheres of the Sun, Moon, five known planets, and known stars—all of which revolved around a motionless Earth. To the region of the cosmos beyond the sublunary sphere—the region between Earth and the Moon composed of the basic elements earth, water, air, and fire—Aristotle added the element aether, which he believed constituted the rotating heavenly spheres that contained the stars and planets.

This geocentric view of the cosmos was a major feature of Greek cosmology. Furthermore, the quest for symmetry and simplicity implied that the heavenly spheres should move in uniform circular motions or combinations thereof, an interpretation consistent with much of the available astronomical data of the time. But Plato's model could not explain all of the observed motions of the sky. The concept of perfect circular motions was expanded upon by the Egyptian astronomer **Ptolemy** during the Roman Empire, but it would be challenged by later astronomers until the seventeenth century, when the observational and theoretical work of **Copernicus** and **Kepler** officially put the geocentric model to rest.

SEE ALSO Earth Is Round! (c. 500 BCE), Ptolemy's *Almagest* (c. 150 BCE), Copernicus's *De Revolutionibus* (1543), Three Laws of Planetary Motion (1619).

A Renaissance-era depiction of the geocentric celestial spheres model of the cosmos. The outer shell is labeled "The Firey Heavens, Home of God and All of the Chosen Ones."

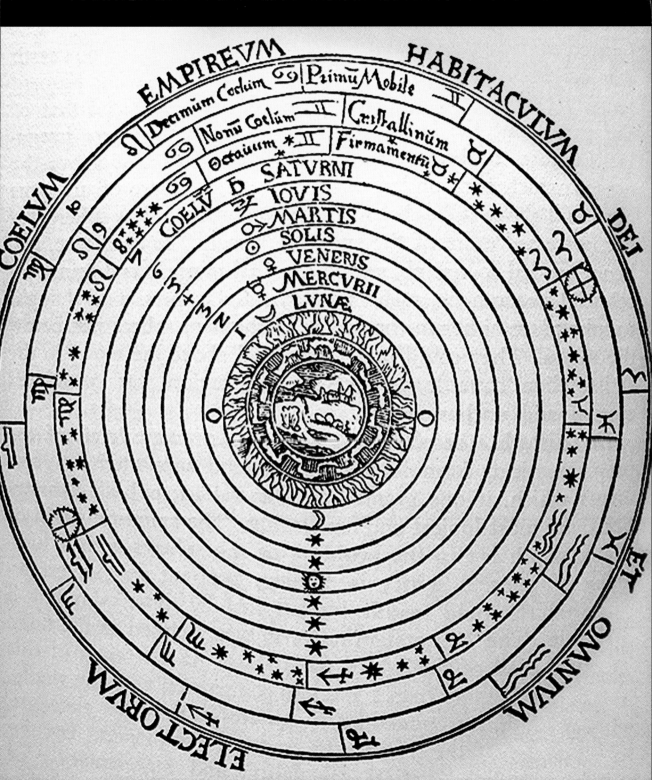

Western Astrology

Alexander the Great (356–323 BCE), **Ptolemy** (c. 90–168)

Astrology is the belief that human and earthly matters are influenced or predetermined by the positions and motions of the Sun, Moon, planets, and stars at the time of a person's birth or at other key points in a person's life. Most civilizations in human history have adopted some form of this belief. Some scholars believe that the roots of Western astrology extend back to the earliest records of Sumerian cosmology in the sixth millennium bce, when priests turned to the celestial sphere for signs that might help them to ascertain the will of the gods (see **Birth of Cosmology**). It was during the development of ancient Greek cosmology, however, that astrology transformed into what we would recognize today.

The Greeks transformed astrology from what had been an attempt by priests and kings during Babylonian times to understand the heavens' role in the success of the upcoming crop season, or a potential war, or some other stately affair to a very personal enterprise: deriving the horoscopes of individuals—from kings to commoners—for events ranging from the historic to the mundane. A key force in the development of astronomy was Alexander the Great, the Greek king who studied under Aristotle and created and ruled over an expansive empire from northern Africa through the Mediterranean and Middle East. Alexander established the world's foremost center of learning at the great Library of Alexandria in Egypt, where the practice of horoscopic astrology first began.

Over the next few centuries, and particularly under the influence of the Egyptian astronomer **Ptolemy**, the twelve classic constellations of the zodiac were established, and the basic functions and roles of the planets in all of science—from astronomy to medicine to zoology—were defined. Ptolemy's geocentric, astrology-driven cosmology would influence and dominate Western astronomical thinking for more than 1,300 years.

Belief in astrology is an easy notion to dismiss in the modern world. And yet newspapers still carry daily horoscope columns, and many people still flock to modern-day astrologers (or at least to astrological websites) for advice and so-called predictions about their futures. Maybe it's all just a bit of fun, but there must be something in human nature that thirsts for this kind of perceived order in the universe.

SEE ALSO Birth of Cosmology (c. 5000 BCE), Greek Geocentrism (c. 400 BCE), Ptolemy's *Almagest* (c. 150).

The astronomer Ptolemy is being instructed on how to observe the heavens by the Greek goddess Astrologia, using an astrolabe and a giant armillary sphere, in this 1515 illustration by Erhardt Schön.

Sun-Centered Cosmos

Aristarchus (c. 310–230 BCE)

The geocentric model of Plato and Aristotle permeated ancient Greek thinking about the cosmos. And why not? Everyone could see that the Sun, Moon, and stars rotated around the Earth. Scholars added other supposedly irrefutable evidence: the Moon went through phases consistent with it orbiting our planet. If the Earth spun on its own axis, why was nothing flung off the surface? None of the stars showed any observed parallax, or shifting of position relative to other stars, that they would show if the Earth were moving in its own orbit. Case closed!

There were doubters and skeptics, however. The earliest one on record was the astronomer and mathematician Aristarchus, from the Greek island of Samos, who challenged the nearly 200-year-old common wisdom of his esteemed Greek colleagues by making detailed naked-eye observations of the Sun and Moon and trying to interpret them in a geocentric context. His methods were limited by the acuity of the human eye, but nonetheless he was able to deduce from geometrical calculations that the Sun was at least 20 times farther away than the Moon (the actual value is 400). He then deduced that, because the Sun and Moon have about the same apparent angular diameter in the sky, the Sun's diameter must be at least 20 times larger than the Moon's diameter and 7 times Earth's diameter. Thus, according to his reasoning, the Sun's volume was more than 300 times the volume of the Earth (the actual value is about a million). It must have seemed foolish to him, then, that such a giant Sun would be indentured to such a relatively tiny planet like the Earth, instead of the other way around. Naturally, he advanced the idea that the Earth and other planets orbit the Sun and that the stars are so far away that no parallax could be observed. Aristarchus's universe was a much larger universe than anyone had described before.

Like most revolutionary ideas, Aristarchus's idea of a Sun-centered cosmos was met with ridicule and disdain by most of his colleagues; 250 years later, the idea was effectively crushed by the geocentric teachings and writings of Ptolemy. Aristarchus had planted a critical seed of doubt, but it would not germinate until the sixteenth century.

SEE ALSO Earth Is Round! (c. 500 BCE), Greek Geocentrism (c. 400 BCE), Eratosthenes Measures the Earth (c. 250 BCE), Ptolemy's *Almagest* (c. 150), Copernicus's *De Revolutionibus* (1543).

Copy of a section of Aristarchus's original third-century BCE calculations of the relative sizes of the Sun, Earth, and Moon, used to help support his then-radical notion of a heliocentric cosmos.

[illegible]

ἡ ΝΛ πρὸς ΛΞ ἡ ΛΝ ἄρα ἐστὶν ἡ διάμετρος τοῦ διορίζοντος ἐν τῇ σελήνῃ τό τε σκιερὸν καὶ τὸ λαμπρὸν καὶ ἡ ΒΝ ἐφαπτέσθω τοῦ ΛΚΘ κύκλου ἀπὸ ἄρα τοῦ Β [illegible] καὶ τὴν ΛΝ διάμετρον τοῦ διορίζοντος ἐν τῇ σελήνῃ τό τε σκιερὸν καὶ τὸ λαμπρὸν καὶ [illegible] ἡ ΝΛ ἴση ἐστίν. [illegible] ἡ ΞΗ τῆς ΛΝ ἐλάσσων ἐστὶν [illegible] ΑΓ ΓΗ καὶ [illegible] ἡ ΑΓ [illegible] ἐλάσσων ἐστὶν ἡ Β καὶ [illegible] ἡ ΓΛ [illegible] ἐστὶν [illegible] ΒΛ παράλληλος ἐστὶν τῇ ΞΗ [illegible] ἐστὶν ἡ [illegible] ΛΞΗ [illegible] τῇ [illegible] ΓΛΝ γωνίᾳ καὶ ἐστὶν ὀρθὴ ἡ ΝΛ τῇ ΛΞ [illegible] ΑΓ τῇ ΓΗ ὅμοιον ἄρα ἐστὶ τὸ ΞΝΛ τρίγωνον τῷ ΑΛΓ τριγώνῳ ἐστὶν ὡς ἡ ΞΗ πρὸς τὴν ΝΛ, οὕτως ἡ ΝΛ πρὸς τὴν ΛΓ [illegible] ΝΛ πρὸς τὴν ΛΓ μείζονα λόγον ἔχει ἤπερ τὰ ΠΘ πρὸς ΜΕ [illegible] τὸ [illegible] ΝΛ πρὸς [illegible] ΛΓ μείζονα λόγον ἔχει ἤπερ τὰ ΖΛΚΑ πρὸς τὰ ΒΙΚΕ καὶ τὸ [illegible] ΞΝ ἄρα πρὸς [illegible] ΝΛ μείζονα λόγον ἔχει ἤπερ τὰ ΖΛΚΑ πρὸς τὰ ΒΙΚΕ καὶ ἡ ΞΝ πρὸς [illegible] μείζονα λόγον ἔχει ἤπερ ΖΛΚΑ πρὸς ΔΠΗ [illegible] καὶ τὰ ΖΛΚΑ πρὸς ΔΠ μείζονα λόγον ἤπερ τὰ ΠΗ πρὸς ΜΕ ἄρα ἡ ΝΞ πρὸς [illegible] μείζονα λόγον ἔχει ἤπερ τὰ ΠΗ πρὸς τὰ ΜΕ [illegible] [illegible] ὑπὸ τῆς [illegible] ἐν τῇ [illegible] σελήνῃ [illegible] τοῦ κύκλου [illegible] ἄρα [illegible] διαμέτρου τοῦ διορίζοντος ἐν τῇ σελήνῃ τό τε σκιερὸν καὶ τὸ λαμπρὸν [illegible] διαμέτρου τῆς σελήνης ἐλάσσων μέν ἐστιν ἢ Β μείζονα δὲ λόγον ἔχει ἢ ὃν τὰ ΠΗ πρὸς ΜΕ:

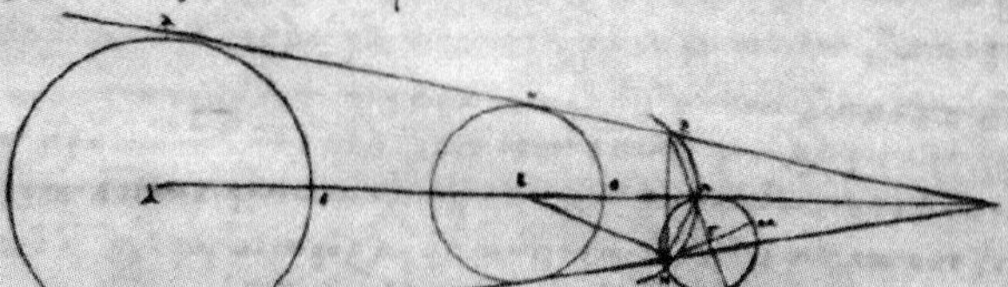

Τῶν αὐτῶν ὑποκειμένων [illegible] ἀπὸ τοῦ Α τῇ ΑΒ πρὸς ὀρθὰς ἡ ΠΡ [illegible] ὅτι ἡ ΞΗ τῆς διαμέτρου τοῦ ἡλίου ἐλάσσων μέν ἐστιν ἢ [illegible] μείζονα δὲ λόγον ἔχει πρὸς αὐτὴν ἢ ὃν ταῦτα ΚΕ πρὸς [illegible]

[illegible]

Eratosthenes Measures the Earth

Plato (427–347 BCE), **Aristotle** (384–322 BCE), **Eratosthenes** (c. 276–195 BCE)

The Greeks had generally accepted the fact that the **Earth Is Round** at least as far back as the time of Pythagoras, but estimates of the actual size of the Earth varied widely. Plato had guessed the Earth's circumference to be around 44,000 miles (70,000 kilometers), corresponding to a diameter of about 14,000 miles (22,000 kilometers), and Archimedes had estimated a circumference of about 34,000 miles (55,000 kilometers) and diameter of 109,000 miles (17,500 kilometers). To make a more accurate determination, Eratosthenes, a mathematician, astronomer, and the third chief librarian of Alexandria, devised a simple experiment that was akin to treating the Earth as a giant sundial.

Eratosthenes had learned that at noon on the summer solstice in the southern Egyptian city of Syene, the Sun was almost exactly overhead (at the zenith), so posts in the ground did not cast any shadows. He also knew that in his own city of Alexandria in the north of Egypt, posts in the ground did cast (small) shadows at noon on the summer solstice. He made some measurements and determined that the Sun was a little over 7 degrees south of the zenith in Alexandria. This corresponds to about ¹⁄₅₀th of the circumference of a circle, so he surmised that the circumference of the Earth was about 50 times the distance between Alexandria and Syene. With a distance of about 5,000 stadia (the stadium was an ancient Egyptian and Greek unit of measure) between Alexandria and Syene, he estimated the circumference of the Earth at about 250,000 stadia, or 25,000 miles (40,000 kilometers). Assuming that 1 stadium was about 175 yards (160 meters) to Eratosthenes, this yields a circumference of about 25,000 miles (40,000 kilometers), which, given the various uncertainties and assumptions involved in the measurements, is essentially the correct answer.

Eratosthenes is widely regarded as the father of geography—indeed, he coined the word. It seems appropriate, then, that he was the first to accurately determine the size of the Earth. His method is also a fabulous example of the power of a simple, well-timed experiment. Archimedes had once quipped about levers, "Give me a place to stand, and I will move the Earth." Eratosthenes could easily have retorted, "Give me a few sticks and some shadows, and I will measure the Earth."

SEE ALSO Earth Is Round! (c. 500 BCE), Greek Geocentrism (c. 400 BCE), Sun-Centered Cosmos (c. 280 BCE).

Cartoon illustrating Eratosthenes's simple method for measuring the circumference of the Earth. A vertical post in Syene (lower inset) casts no shadow at noon on the summer solstice. The same post in Alexandria (upper inset), however, casts a shadow that indicates it is ¹⁄₅₀ of a circle away from Syene.

Stellar Magnitude

Hipparchus (c. 190–120 BCE)

Some stars are bright, others are dim. Anyone who's looked up at a clear night sky can determine that, but until the middle of the second century BCE, no one had tried to quantify just how bright is bright or how dim is dim. The first to do so was the Greek astronomer and mathematician Hipparchus, who is also credited with creating the first comprehensive star catalog.

Hipparchus decided to assign the stars a brightness using a scale of magnitudes that ranged from 1 (for the 20 brightest stars) to 6 (the faintest stars visible to the naked eye), with each step in the scale corresponding to stars that he judged to be half as bright as the previous step. Modern astronomers still use a brightness scale based on the original from Hipparchus, with some tweaks. In the mid-nineteenth century, the magnitude system was redefined such that 5 steps in magnitude correspond to a factor of 100 in brightness (thus each step is formally brighter or dimmer than the next step by a factor of 2.5). The scale was expanded beyond 6 steps, and linked to the brightness of the bright star Vega, which was defined as having a magnitude of 0. Extremely bright stars, planets,

the Moon, and the Sun were all assigned negative magnitudes; extremely faint stars, only visible with the world's most advanced telescopes, pushed the maximum magnitude limit toward 30. It seems odd that dimmer stars have higher magnitudes, but astronomers have gotten used to a system that's been around for over 2,100 years.

In addition to defining the way star brightness is measured, Hipparchus also invented trigonometry, discovered the precession of the earth's spin axis, and obtained the most accurate measurements to date of the relative positions (astrometry) of the bright stars. In recognition of his stellar (literally and figuratively) observing skills, a 1989 orbiting space astrometry satellite was named in his honor.

SEE ALSO Ptolemy's *Almagest* (c. 150), Andromeda Sighted (964), Globular Clusters (1665), Proper Motion of Stars (1718).

The Globular Cluster NGC 6397, right, photographed by the Hubble Space Telescope. The brightest stars here have a stellar magnitude of about +10. The inset shows a small part of this rich field of stars; the faint red dot in the center is the faintest red dwarf star ever photographed, at magnitude +26.

First Computer

Sometime near the end of the first century BCE, a Roman ship carrying a large supply of ancient Greek artifacts sank in the Mediterranean, off the coast of the Greek island of Antikythera. Two thousand years later, in 1901, divers came across the wreck and discovered, among the artifacts, the corroded remains of what may be the world's oldest computer, now known as the Antikythera mechanism.

At first archaeologists thought that the mechanism might have been a mechanical clock because of its dozens of small gears. This would have been an amazing find in its own right, because the workmanship was said to be comparable to mechanical clockwork in seventeenth-century Europe. But after decades of cleaning and additional study, the device was discovered to be much more than a clock; it appears to have been a sophisticated mechanical astronomical computer and calendar that could be used to determine the past and future positions of the Sun, Moon, and planets in the sky, to predict eclipses, and to display the phases of the Moon. It is the oldest (by 1,500 years!) example of an orrery, a clockwork-like planetarium model of the solar system, superbly designed and built.

The discoveries about this device are remarkable on so many levels. They reveal an extremely precise and detailed knowledge of planetary motions by the ancient Greeks, including subtle variations in the Moon's velocity over the course of each month. The flawlessly worked mechanisms suggest that many of these devices had been built, and the compact size suggested that they were intended to be portable.

We are accustomed to technological advances and scientific discoveries being closely linked in the modern world, so perhaps it should not be surprising to learn that the same appears to have been true in ancient Greco-Roman times. Still, it is both humbling and jarring to discover evidence, from a single artifact, that a civilization was far more technologically advanced than we had thought before.

SEE ALSO Ancient Observatories (c. 3000 BCE), Finding Easter (c. 700).

RIGHT: *A modern reconstruction of the Antikythera mechanism, based on detailed study and analysis of the badly damaged remaining fragments.* LEFT: *One of the surviving fragments, recovered in 1901 from an ancient Greek shipwreck.*

Julian Calendar

Julius Caesar (100–44 BCE)

Like other past civilizations that were tuned in to the skies, the Romans had developed a calendar system that had strong astronomical connections. The calendar system that they originally designed in the eighth century BCE was the source of constant confusion, however, partly because it was cobbled together from pieces borrowed from the Greeks and others before them. For example, the year had 10 months of 30 or 31 days for a total of 304 days—the remaining 61-plus days needed to make up an actual trip around the Sun were brushed under the rug as "winter." A later change added two new winter months (January and February) but still came to only 355 days total per year. To keep the calendar lined up with the seasons, a leap month was occasionally added in by the high priests, but the decision to add extra days to a given year was often arbitrary and politically motivated. The situation got so muddled that many ordinary Romans had no idea what day, year, or month it was.

In fact, the Roman calendar system was such a confused mess when Julius Caesar came to power in 49 BCE that he ordered a reform that would align the calendar more with the motions of the Sun rather than with the affairs of men. Days were added to some of the 12 months to bring the total number of days in a year to 365, and he decreed that every fourth year an extra leap day would be added to the end of February, making the average length of the year 365.25 days, which is close to the actual length of a solar year—365.242 days. Caesar's reform of the calendar took effect on January 1 in the year 45 BCE (or, to the Romans, 709 years after the founding of Rome), after the priests had to make the year 46 BCE 445 days long to try to fix all the problems that had accumulated before the reform.

The Julian calendar worked well for a long time because it was only 0.008 days (about 11 minutes) per year different from a true solar year. By the sixteenth century, however, those 11 minutes per year had added up to a significant shift between the calendar year and the solar year, and so a further tweak, called the **Gregorian Calendar** reform, was needed to resync the calendar with the seasons.

SEE ALSO Egyptian Astronomy (c. 2500 BCE), Gregorian Calendar (1582).

Ancient Roman calendars were sometimes carved into stone blocks, as in this marble version that shows day names and astrological symbols for the months of April through September.

...LLA...N...

<table>
<tr><td>Ʃ VF</td><td>M TITIVS</td><td>M ANNIV MELIO</td></tr>
<tr><td></td><td>C N POMPEIVS</td><td>C SENTIVS SATVRNI</td></tr>
<tr><td></td><td>IMP CAESAR AVGINI</td><td>Ʃ VF M VINICIVS</td></tr>
<tr><td>Ʃ VF</td><td>C ANTISTIVS</td><td>L CORNE C MENTE</td></tr>
<tr><td></td><td>M TVLLIVS</td><td>C FVRNI SICILI</td></tr>
<tr><td></td><td>L SAENIVS</td><td>L DOMITI VS L DOTIC</td></tr>
<tr><td>Ʃ VF</td><td>IMP CAESAR C CAESARE AVG F</td><td>Ʃ VF SIANVS</td></tr>
<tr><td></td><td>L OPPI VALERI</td><td>A DIDIVS L LISO</td></tr>
<tr><td></td><td>IMP CAESAR DIVI F GE III</td><td>M LOLLIVS SATV IGV</td></tr>
<tr><td></td><td>IMP C CAESAR AVG GE III</td><td>TI CLAVDI QVI IGIL</td></tr>
<tr><td></td><td>IMP CAESAR F AVG F VI</td><td>M VALERIVS DVLVIC</td></tr>
<tr><td></td><td>IMP CAESAR VI AVG F VI</td><td>CVRTISIVS</td></tr>
<tr><td></td><td>IMP CAESAR AVG COND EA</td><td>Ʃ VF C C ANNIVS</td></tr>
<tr><td></td><td>IMP CAESAR AV DISO</td><td>L VOLVSIVS</td></tr>
</table>

Ptolemy's *Almagest*

Ptolemy (c. 90–168)

For nearly 700 years (c. 300 BCE–400 CE), the great Library of Alexandria in Egypt was the academic center of the world. Most of the famous Greek and Roman scholars that we celebrate today as pioneers of mathematics or astronomy or other fields either visited or worked there, overseeing an amazing collection of hundreds of thousands of scrolls and books. Among them was the most famous book in all of classical astronomy, the *Almagest*, published around the year 150 by the Egyptian mathematician and astronomer Claudius Ptolemaeus (Ptolemy).

Ptolemy was what we might today call a big-picture kind of scientist. He mastered a variety of astronomical instruments and made important observations, but his greatest skill appears to have been culling and synthesizing the work of his predecessors over the previous 800 years or so to assemble a new, comprehensive view of the cosmos in a single book. The *Almagest* had 13 separate sections, which included a detailed description of Ptolemy's geocentric cosmology, expanding on the ideas originally proposed by Plato, Aristotle, and others and describing the orbits of the planets as moving in small circular paths (epicycles) that in turn are moving on larger circular paths (deferents) with the Earth at the center. The system was complicated, but fundamentally the stars and planets traveled only on perfect spherical or circular paths, a beauty and symmetry that fit the available data and that likely appealed to Ptolemy as befitting the work of a perfect Creator.

The book also included what he called handy tables to calculate the risings and settings of planets and stars, and chapters that covered the general motions of the Sun, Moon, and planets, eclipses, precession, and observing tools and methods. A massive (for the day) catalog of more than a thousand stars was also included, based on Hipparchus's star catalog and **Stellar Magnitude** system.

Sadly, Ptolemy's big-picture view of the solar system was wrong. But because it was still so remarkably good at describing and predicting the motions observed in the sky, the *Almagest* and its geocentric depiction of the cosmos would remain the definitive source of astronomical information for more than a thousand years.

SEE ALSO Greek Geocentrism (c. 400 BCE), Western Astrology (c. 400 BCE), Sun-Centered Cosmos (c. 280 BCE), Stellar Magnitude (c. 150 BCE), Copernicus's *De Revolutionibus* (1543), Brahe's "Nova Stella" (1572), Three Laws of Planetary Motion (1619).

A 1559 woodcut depicting Ptolemy observing the altitudes of the Sun, Moon, and stars using a parallactic ruler that he describes in the Almagest.

It is made of 3. pea-
ces, beyng 4. ſquare:
As in the Picture
where A. F. is the
firſt peace or rule.
A.D. The ſeconde.
G.D. the third rule.
E. The Foote of the
ſtaffe.
C.F. The Plumrule.
C.B. The ioyntes, in
which the ſecond &
thirdRulers are mo-
ued.
K.L. The ſighte ho-
les.
I. The Sonne.
H. The Zenit, or ver-
ticall pointe.
M. N. The Noone-
ſtead Lyne.

PTOLOMEVS.

Chinese Observe "Guest Star"

Ancient Chinese astronomers were meticulous observers of the skies. As historians have pointed out, because official Chinese astronomy research was often conducted by cadres of full-time, court-appointed civil servants as opposed to individual scholars, they were much more systematic and thorough about surveying the skies for changes than their Roman, Greek, or Babylonian counterparts and predecessors were. Thus, when something new happened in the sky, the Chinese noticed and recorded the observation, and the recording became part of the imperial dynastic records, many of which are still preserved.

A prime example was the sudden appearance of what they called a guest star in the southern skies in the year 185. This appearance was recorded by Chinese astronomers as a notable event in the surviving annals of the Eastern Han dynasty (25–220 CE).

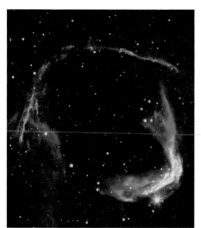

Although no drawings were included, the description of the location of the guest star and the fact that it faded from view over a period of about six months convinced modern astronomers that the Chinese had made the first recorded observations of a supernova. Modern optical, radio, and X-ray telescopes trained on that location reveal a semispherical gaseous nebula called RCW 86, which appears to be the expanded remains of that stellar explosion from more than 1,800 years ago.

Many other guest stars were recorded in ancient Chinese astronomical drawings that are still preserved today. Among the most interesting are drawings that show objects with a bright, round "head" and one or more feathery or spiky "tails." These objects were referred to as "broom stars" by the Chinese, and they are now widely interpreted as bright comets with long tails of gas and dust. In fact, prominent comets observed by Chinese astronomers in 240 BCE and 12 BCE, and in 141, 684, and 837 CE, are all likely to have been observations of the same comet, eventually recognized in 1682 as the 76-year periodic **Halley's Comet**. The careful, methodical sky watching and record keeping of early Chinese astronomers has proven to be a rich treasure trove of data for studies by both historians and astronomers.

SEE ALSO Astronomy in China (c. 2100 BCE), "Daytime Star" Observed (1054), Halley's Comet (1682), Miss Mitchell's Comet (1847), Tunguska Explosion (1908).

LEFT: *A modern view of the remains of the supernova explosion of 185.* RIGHT: *Ancient Chinese drawings of the changing appearances of several comets, as recorded in the* Bamboo Annals, *a historical accounting of China covering the period from about 2400 BCE to 300 BCE.*

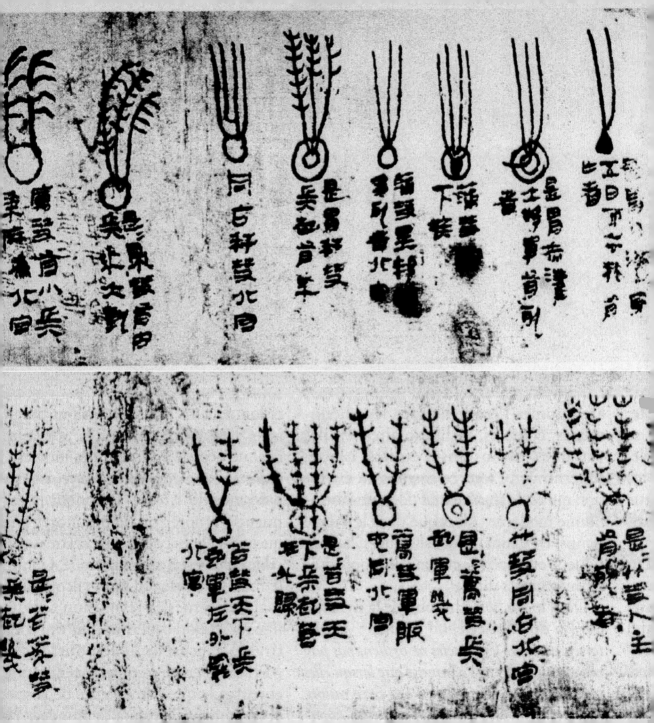

Aryabhatiya

Aryabhata (476–550)

The roots of astronomy in India were closely connected to the development of religion, as they were in most other early civilizations. Astronomical knowledge formed the foundation of early calendar systems that established dates for Hindu religious observances, or seasonal planting and harvesting. Just as in the West, advances over time in instrumentation used by clerics and early astronomers for charting the positions of the Sun, Moon, and stars led to the development of more sophisticated cosmological schemes for explaining and predicting the motions of the sky. The earliest of the great thinkers to make lasting contributions to the development of astronomy in India was the mathematician and astronomer Aryabhata, author of the oldest surviving Indian treatise on math and astronomy, the *Aryabhatiya*, published around the year 500.

Aryabhata essentially summarized all of mathematics in verse format in the *Aryabhatiya*, and included convenient new sine tables to use in trigonometric calculations. In the astronomy sections of the *Aryabhatiya*, Aryabhata explained eclipses as the logical result of shadows from the Earth or the Moon, rather than the work of sky demons. Using some of his newly developed spherical trigonometry calculations and eclipse measurements, he calculated the circumference of the Earth to an accuracy of less than 0.2 percent of its actual value, significantly improving upon the previous best estimate of the Earth's circumference by **Eratosthenes** 750 years earlier.

Perhaps the most revolutionary aspect of Aryabhata's thinking, though, was his claim that the Earth is not fixed in space but spins on its own axis, and that it is the stars in the celestial sphere that are fixed in space. Indian cosmology of Aryabhata's time was definitively geocentric; while his own (accurate) calculations of planetary positions relied on orbits and epicycles similar to those in **Ptolemy's Almagest**, Aryabhata was the first to advocate elliptical rather than circular paths for the planets. There are also suggestions in the *Aryabhatiya*, debated by scholars interpreting the original language and meaning, that Aryabhata believed in a **Sun-Centered Cosmos**. Like those of Aristarchus in Greece, however, such radical heliocentric ideas would take more than a thousand years to catch on in India.

SEE ALSO Sun-Centered Cosmos (c. 280 BCE), Eratosthenes Measures the Earth (c. 250 BCE), Ptolemy's *Almagest* (c. 150 BCE).

Statue of the mathematician and astronomer Aryabhata at the Inter-University Centre for Astronomy and Astrophysics in Pune, India.

Finding Easter

Bede of Jarrow (c. 672–735)

Many of the world's religions celebrate special feast days and other holidays on dates that are not fixed on the calendar but instead vary depending on the season, phase of the Moon, rising of a particular star, or some other astronomical circumstance. This meant that the priests, monks, or other religious leaders who were responsible for determining the dates of these events in advance often had to be trained in, or at least be up to date on, the latest astronomy and/or mathematics.

The calendar position of Easter, the Christian celebration of Christ's resurrection, is perhaps one of the most confusing of the world's major movable religious holidays to determine. In theory, Easter is held on the first Sunday after the full Moon following the spring equinox. In practice, however, predicting the date of the full Moon or of the equinox was so confusing that astronomers of the Middle Ages invented a special word just to describe the act of calculating the calendar date of Easter each year: *computus*.

Many styles of computus were in use in the early Middle Ages, each giving a different date for Easter. In the early eighth century, a monk named Bede of Jarrow— from the kingdom of Northumbria, in what is today part of northern England and Scotland—proposed a standardized computus that was described and widely circulated in his books *On Time* and *On the Reckoning of Time*. Bede's computus and his education in astronomy led him to discover that the date of Easter repeats in a 532-year cycle made up of 19-year lunar and 28-year solar cycles. Finally, the date of the most important Christian holiday of the year could be predicted.

Bede's computus, and his prowess as a historian and theologian, earned him the nickname Venerable Bede. His writings were also influential throughout the Middle Ages because they clarified a wide range of other practical solar, lunar, and tidal calendar calculations. He even dabbled in calculating the age of the Earth based on the book of Genesis and came up with 3952 BCE—not far from the 4004 BCE estimate popularized in the seventeenth century by the Anglican archbishop James Ussher.

SEE ALSO Greek Geocentrism (c. 400 BCE), Julian Calendar (45 BCE).

Part of a public painting near the Tyne Tunnel crossing in Jarrow, England, depicting the medieval Christian monk and astronomical scholar Bede of Jarrow.

Early Arabic Astronomy

Habash al-Hāsīb (c. 770–870), **Muhammad ibn Mūsā al-Khwārizmī** (c. 780–c. 850), **Muhammad ibn Jābir al-Harrānī al-Battānī** (c. 858–929), **Abū ar-Rayhān al-Bīrūnī** (973–1048)

Much of the modern language and methodology of astronomy and mathematics can be directly traced back to a several-centuries-long burst of genius and creativity in the arts and sciences in medieval Islam. Scientific development stagnated in Europe during this period, and so it was primarily the Arab world that became the heir to the Greco-Roman legacy of astronomy and mathematics.

Among the many early Arabic astronomers and mathematicians who made important new contributions were al-Khwārizmī, who founded modern algebra (*al-jabr*, or "completion" in Arabic) and developed new methods for calculating the positions of the Sun, Moon, and planets; al-Hāsīb, who calculated the best estimates yet made for the diameter and distance of the Moon and the diameter of the sun, and compiled his observations in *The Book of Bodies and Distances*; al-Battānī, who refined results from **Ptolemy's *Almagest*** and developed new methods of timing the first appearance of the Moon's crescent; and al-Bīrūnī, who invented new astronomical instruments and observing methods, and who (along with a number of other Arabic astronomers) hypothesized that a Sun-centered model of the solar system could fit the available observational data as well as the widely accepted Earth-centered model. Indeed, the work of these and other medieval Islamic astronomers went on to influence Renaissance Western astronomers such as **Brahe**, Kepler, **Copernicus**, and **Galileo**, and the eventual overthrow of Ptolemaic geocentrism in favor of a heliocentric cosmology.

In addition, almost all the noted astronomers and mathematicians of early Islam were working as part of teams in what were essentially the world's first research groups — part of the world's first system of state-run observatories and research institutes. This kind of collaborative environment enabled Arabic scientists to achieve significant advances in astronomy and other fields, and it is the basis for the way most science is done today.

SEE ALSO Greek Geocentrism (c. 400 BCE), Sun-Centered Cosmos (c. 280 BCE), Ptolemy's *Almagest* (c. 150), Andromeda Sighted (964), Experimental Astrophysics (c. 1000), Copernicus's *De Revolutionibus* (1543), Brahe's "Nova Stella" (1572), Galileo's *Starry Messenger* (1610), Three Laws of Planetary Motion (1619).

Illustration by al-Bīrūnī of different phases of the Moon, from his astrological treatise Kitāb al-tafhīm.

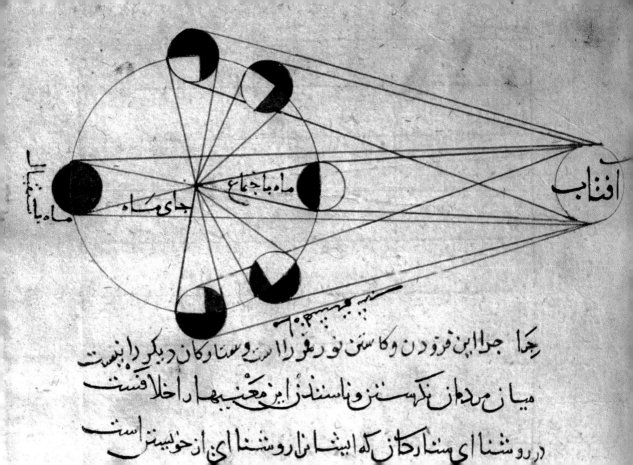

ماه یا با
ماه یا با

آفتاب

جای ماه

ماه با جماع

رجا جرم ابن فروذن وكاسن نور را دیگر را دن وسناوكان دیگری دا نیست
میان مردمان نگرسن وناسندن این معنبها را خلا فست
در وشنای سناركان كه ایشان را روشنای ئ از خویستن است

Andromeda Sighted

‘Abd al-Rahmān al-Sūfī (903–986)

Another important early astronomer from the Arab world was ‘Abd al-Rahmān al-Sūfī of Persia (modern-day Iran). Like most other astronomers of the Middle Ages, al-Sūfī was aware of the major aspects of classical Greek astronomy and cosmology, including **Ptolemy's *Almagest***, which he translated into Arabic. He and others sought to expand on Ptolemy's ideas, and to synthesize them with new observations and theories from **Early Arabic Astronomy**. His results were published around the year 964 in a landmark work called *The Book of Fixed Stars*.

Al-Sūfī's book was essentially a detailed map of the stars in the 48 then-known classical Greek constellations, using star data based on Ptolemy's older catalog in the *Almagest* and Hipparchus's **Stellar Magnitude** system, but refined or corrected using his and other newer observations of stellar brightnesses and colors. *The Book of Fixed Stars* uses the Arabic names for the bright stars in each constellation; we still use many of these star names—including Altair, Betelegeuse, Deneb, Rigel, and Vega—today.

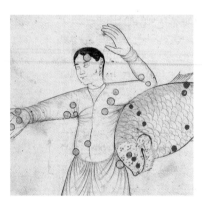

In the section of al-Sūfī's book devoted to the constellations Andromeda and Pisces, he notes a "little cloud" detected among the major stars. Although he could never have known it, al-Sūfī is widely believed to have made the first recorded observation of the Andromeda galaxy, the nearest (about 2 million light-years distant) spiral galaxy to the **Milky Way**. Also known as Messier object 31, Andromeda is nearly eight times larger than the full Moon, but it is extremely faint and thus takes excellent eyesight and patience to detect. Al-Sūfī was also the first to detect other faint star clusters, nebulae, and "clouds," including one of the Milky Way's faint elliptical companion galaxies in the southern skies, an object that, 550 years later, was named the Large Magellanic Cloud because it was prominently noted and popularized in Europe after Ferdinand Magellan's voyage around the world in 1519.

SEE ALSO Ptolemy's *Almagest* (c. 150), Early Arabic Astronomy (c. 825), Messier Catalog (1771).

LEFT: *The inset above shows a drawing of part of the constellations Andromeda and Pisces from al-Sufi's 964 Book of Fixed Stars.* RIGHT: *A modern digital astronomical photograph of the Andromeda galaxy, viewed in ultraviolet light from the NASA Galaxy Evolution Explorer satellite.*

Experimental Astrophysics

Abū ‘Alī al-Hasan Ibn al-Haytham (965–1040), **Abū ar-Rayhān al-Bīrūnī** (973–1048)

Ancient Greek scholars had predominantly philosophical or theoretical interests in science, especially astronomy. In contrast, the Arabic style of astronomy and mathematics that came to dominate the Middle Ages was more focused on inventing new instruments or methods, acquiring new observations, and using the data to develop new ways to address practical religious or other societal needs. This way of thinking was new: observe, record, analyze, interpret, hypothesize, repeat. It proved to be very effective, and it is basically the root of the modern scientific method.

The origins of this new, observationally focused approach to understanding the cosmos can be traced back to a small number of prominent Arab and Persian mathematicians from around the turn of the first millennium CE. One was Abū ‘Alī al-Hasan Ibn al-Haytham, a Muslim physicist and mathematician who specialized in many fields and who advocated experimentation and critical tests of prevailing theories, rather than reliance on speculation or natural philosophy (he was a critic of Ptolemy). He wrote in his *Book of Optics*, "In all we do, our purpose should be balanced not arbitrary, the search for truth, not support of opinions."

Around the same time, the Persian scholar Abū ar-Rayhān al-Bīrūnī, another expert in a wide range of physical and social sciences, was advocating a similar approach to experimentation in astronomy and other fields, introducing new approaches such as repeat experiments and analysis of random and systematic errors on derived results. "I do not shun the truth from whatever source it comes," he wrote in his encyclopedic canon of science, *Kitāb al-qānūn al-masādī*.

In many ways, al-Haytham, al-Bīrūnī, and many other such polymaths of the Middle Ages were the world's first scientists—passionate about observation and discovery across a wide range of fields, skeptical of so-called truths that could not be verified, and intrinsically self-critical. Such traits have gone on to serve scientists well for more than a thousand years.

SEE ALSO Earth Is Round! (c. 500 BCE), Ptolemy's *Almagest* (c. 150), Early Arabic Astronomy (c. 825).

Arab astronomers study the heavens in this medieval European print from a commentary on Cicero's first-century-BCE Somnium Scipionis (Dream of Scipio).

c. 1000

Mayan Astronomy

Prehistoric and Middle Age astronomy was not only studied and practiced by Europeans or Asians. Indeed, a rich astronomical tradition had emerged in Mesoamerica going back to at least 2000 BCE in complex and advanced indigenous civilizations like the Maya, Olmec, Toltec, Mississippian, and other related cultures. Few written records remain from these civilizations, however, partly because many were lost or destroyed during later European conquest.

For the Mayan civilization (peaking from c. 2000 BCE to 900 CE), only four surviving books are available to assess the level of scientific knowledge of this once-dominant Mesoamerican culture. One of those books, from the late period of Mayan history shortly before European contact, is called the *Dresden Codex* (after the location where it is currently archived); it provides fascinating and revealing evidence that Mayan astronomy had reached a level of advancement and sophistication comparable to that of the Greeks, Arabs, and other early societies.

The *Dresden Codex* is part history and part mythology, but it is mostly a series of detailed astronomical tables for charting and predicting the motions of the Sun, Moon, Venus, and the other known planets. After deciphering the glyphs and numeric symbols, archaeoastronomers determined that the 74 pages of illustrated tables track the cycles of Venus (which repeats its pattern of rising and setting every 584 days) and the Moon (857 full Moons repeat every 25,377 days). The tables could also be used to predict eclipses, as the Mayans recognized the various repeating eclipse cycles to a much higher level of precision than their Babylonian and Greek colleagues. They could also apparently predict lunar and planetary conjunctions with great accuracy. Knowledge of these periodicities in the heavens to such a high level of precision must have required centuries of careful, detailed observations and sophisticated instruments. Once the Mayans discovered the cycles, the tables could be used essentially forever to predict the heavens.

What did the Mayans use this information for? Much remains a mystery, but historians have identified many potential religious, agricultural, social, and even military events and traditions tied to their astronomically derived calendar system.

SEE ALSO Egyptian Astronomy (c. 2500 BCE), Astronomy in China (c. 2100 BCE), Greek Geocentrism (c. 400 BCE), Early Arabic Astronomy (c. 825).

Part of page 49 of the Dresden Codex, *one of three known surviving books from the Mayans, depicting part of the cycle of appearances and disappearances of Venus and the Moon goddess Ixchel.*

"Daytime Star" Observed

By the end of the early Middle Ages, a number of societies on the planet had flourishing or nascent communities of astronomers and mathematicians who were tuned in to the heavens. It is perhaps of little surprise, then, that when a new star suddenly and dramatically appeared in the sky in the constellation Taurus in the year 1054, many of those societies took notice.

Chinese astronomers first recorded the appearance of a **"Guest Star"** on July 4. Their observations were corroborated by Persian, Arab, Japanese, and Korean observers; the event was even recorded in rock paintings by Native American Anasazi artists. Europeans, still mired in the Dark Ages, didn't seem to record the event. Chinese observers saw the new star in the daytime for 23 days, and at night for 653 days, until it faded away. At maximum brightness, it was estimated to be around magnitude −6 or −7, brighter than everything in the sky except the Sun and Moon.

We now know that what these medieval astronomers had observed was a supernova—the violent, catastrophic explosion of a massive progenitor star about 6,300 light-years from Earth that ran out of fuel for **Nuclear Fusion** and collapsed in on itself,

releasing an enormous amount of gravitational energy that expelled the star's outer layers into space at enormous speeds, perhaps as high as 10 percent of the speed of light. More than 650 years after the supernova faded from view, eighteenth-century astronomers first detected the crab-shaped emission nebula of ionized gases being heated by the explosion's shock wave. In the late 1960s, radio astronomers discovered that the compressed core of the original star had become a rapidly spinning (30 times per second) neutron star, or **Pulsar**, that is probably only about 12 miles across but has a mass of about 1.5 to 2 times that of the Sun. The careful records of this event by early astronomers helped to establish the previously unknown link between supernovae, emission nebulae, and neutron stars.

SEE ALSO Birth of the Sun (c. 4.6 Billion BCE), Astronomy in China (c. 2100 BCE), Chinese Observe "Guest Star" (185), Nuclear Fusion (1939), Pulsars (1967).

LEFT: *The inset shows an Anasazi petrograph depicting a hand, crescent Moon, and the new "guest star" of 1054.*
RIGHT: *Hubble Space Telescope mosaic of the Crab Nebula, a six-light-year-wide expanding remnant of ionized gas from a violent supernova explosion seen by medieval astronomers in the year 1054.*

De Sphaera

John of Sacrobosco (c. 1195–c. 1256)

The slow emergence of western Europe from the Dark Ages coincided with the
founding of the world's first universities (in Bologna and Oxford) near the end of the
eleventh century. As more institutions of higher learning were founded, a need arose for
scholarly works to serve as textbooks for the students. Printing books in medieval Europe
was expensive and cumbersome, and so only a small number of standard textbooks
proliferated in certain fields.

The first standard textbook for astronomy that was used throughout western
Europe was a tract called *De Sphaera*, published c. 1230 by the English monk and
astronomer John of Sacrobosco. Sacrobosco taught at the University of Paris and was a
firm believer in Ptolemaic cosmology. Much of *De Sphaera* is a summary and review of
the *Almagest,* but the text is also augmented with more "modern" ideas and discoveries
from **Arabic Astronomy** and their nascent field of **Experimental Astrophysics**, which
was far more advanced than European astronomy in the Middle Ages.

In addition to its review of Ptolemy, *De Sphaera* also included illustrated
definitions of the celestial spheres and circles (likely intended for use as a tutorial for
students learning how to use an armillary sphere), reviews of rising and setting times
and circumstances for bright stars and the Sun, and descriptions of the motion of the
Sun and planets using Ptolemy's epicycle and deferent model. Sacrobosco made it
unambiguous that the Earth is a spherical body and provided accurate explanations for
solar and lunar eclipses.

If there had been a medieval best-seller list for books, *De Sphaera* would have
been on it for hundreds of years. Many copies were transcribed by hand between the
thirteenth and fifteenth centuries (hundreds of manuscripts survive today), and after
the first printed edition appeared in 1472, more than 90 editions were printed over the
following 200 years. It is described by historians as having been required reading in
university astronomy classes well into the seventeenth century.

SEE ALSO Ptolemy's *Almagest* (c. 150), Early Arabic Astronomy (c. 825), Experimental Astrophysics (c. 1000).

Pages 90 and 91 from a sixteenth-century edition of Sacrobosco's astronomy textbook, De Sphaera *(1230 CE),
outlining Ptolemy's geocentric model for the cosmos.*

Theorica del Sole, & delli superiori. ♄. ♃. ♂. & inferiori. ♀. ☿. ☽. Imaginando il Sole essere nel luogo dell'epiciclo delli altri Pianeti

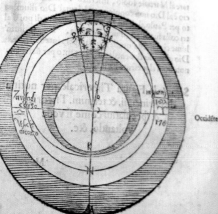

Oriète.

Occidète.

A. Centro del mondo.
B. Centro del deferente.
C. Centro del Equante.
D.e.f.g. Epiciclo, colli altri 6. Pianeti
e. Statione prima.
y. Statione seconda.
e.g. Arco della direttione.
g.d.f. Arco della Retrogradatione.
x. g.d.f. Corpo solare.
A. E. linea dell'auge, dell'orbe che

porta l'epiciclo.
A. N. linea dell'opposto dell'auge detta
L. M. Conuesso dell'orbe che porta
l'auge dell'eccentrico.
E. I. Concauo del dett'orbe.
E. I. & D. K. Conuesso & concauo
delorbe.
E. K. Circulo Equante.
E. Auge dell'Epiciclo.
D. Opposto della detta auge.

Theorica delle linee, & de i moti.

D. Centro del mondo.
C. Centro del deferente.
H. Centro del Equante.
H.g.i.f. Epiciclo.
Eclitica l'estremo circulo.
B. N. linea dell'auge.
G. Auge media, dell'epiciclo.
H. Auge vera dell'epiciclo.

A. Principio dell'ariete.
A. N. Auge nella seconda significatiõe, del'arco A. N.
D. L. Linea del mezo moto.
A. N. L. Arco del mezo moto.
D. K. Linea del vero moto dellepiciclo.
A. N. K. Vero moto dell'epiciclo.

Large Medieval Observatories

Nasīr al-Dīn al-Tūsī (1201–1274), **Hülegü Khan** (c. 1217–1265), **Ulūgh Beg** (1394–1449)

It is easy to think of astronomical observatories as modern-day inventions—enormous domes on high mountains, with giant telescopes and high tech computer equipment. But the concept of the observatory as a research institute and a shared access facility for teams of astronomers can trace its roots back to the first astronomical observatories that were established in the Islamic world and China in the Middle Ages.

Among the first major observatories in the world was the Marāgheh Observatory in northwestern Iran, established in 1259 by the Mongol ruler Hülegü Khan (grandson of Genghis) and directed by his court astronomer and mathematician Nasīr al-Dīn al-Tūsī. Marāgheh housed a huge library of more than 40,000 books, and al-Tūsī led a team of astronomers and students performing observations and calculations of planetary motions and Earth's precession that would later be used by Copernicus and others as key inputs for a new heliocentric cosmology. Around the same time, Hülegü's brother Kublai Khan established the Gaocheng Astronomical Observatory as the first such facility in China. Early Yuán dynasty (1279–1368) astronomers there made observations of the

Sun and planets and used an enormous stone sundial to reckon time and more accurately determine the length of the year. Inspired by Marāgheh (which had been destroyed by earthquakes), in 1420 the Timurid astronomer and mathematician Ulūgh Beg established a university and observatory in Samarkand, in what is now Uzbekistan. Samarkand Observatory, later named Ulūgh Beg Observatory, included early astronomical instruments such as astrolabes and armillary spheres and a giant stone sextant/meridian circle, with a radius of 131 feet (40 meters), carved into the mountainside—the largest of its kind in the world—for accurately measuring the positions of the Sun and stars. Astronomers at Ulūgh Beg updated Ptolemy's and al-Sūfī's star catalogs to account for precession, enabling consistently accurate predictions of eclipses and other celestial events.

SEE ALSO Ptolemy's *Almagest* (c. 150), Early Arabic Astronomy (c. 825), Andromeda Sighted (c. 964), Copernicus's *De Revolutionibus* (1543).

LEFT: *The Gaocheng Astronomical Observatory in China, established in 1276.* RIGHT: *The remains of the giant underground meridian circle, 6 feet (2 meters) wide, at Ulūgh Beg Observatory in Samarkand.*

Early Calculus

Mādhavan of Sangamagrāmam (c. 1350–c.1425), **Nīlakantha Somāyaji** (1444–1544)

Astronomical research in India through the Middle Ages was initially based on the early findings and writings of Aryabhata and other mathematicians and astronomers; it was ultimately expanded by the creation of dedicated research and teaching groups like the Kerala school of astronomy and mathematics, founded in the fourteenth century by the mathematician Mādhavan of Sangamagrāmam.

Mādhavan and subsequent Kerala mathematicians like Nīlakantha Somāyaji developed mathematical methods of estimating the motions of the planets based initially on geometry and trigonometry and later on newly developed techniques for modeling complex curves and mathematical shapes using combinations of functions. Among these shapes were parabolas, hyperbolas, and ellipses; their work on ellipses proved especially applicable to astronomy because they were able to show that Aryabhata's earlier conjecture was correct: the paths of the planets could be described by elliptical orbits. The new mathematical methods developed at Kerala that focused on series of functions were early versions of calculus, predating the European development of calculus some 200 years later by scientists like Isaac Newton.

Nīlakantha's work *Aryabhatiyabhasya* (a commentary on Aryabhata's *Aryabhatiya*), published around 1500, further demonstrated that a rotating Earth and a partially heliocentric solar system provided a more accurate way of fitting the planetary orbits. In his model, Mercury, Venus, Mars, Jupiter, and Saturn all orbited the Sun, but the Sun orbited Earth. A similar model was adopted by the sixteenth-century Danish astronomer **Tycho Brahe**, and some aspects of Nīlakantha's model are also consistent with the fully heliocentric cosmology proposed in 1543 by Polish astronomer **Nicolaus Copernicus**.

The contributions of the Kerala school, and perhaps of Indian mathematics and astronomy in general, may have previously been underappreciated in the West. It seems clear now that they should be counted among the "shoulders of giants" that supported the later discoveries of Copernicus, Newton, and others.

SEE ALSO Earth Is Round! (c. 500 BCE), *Aryabhatiya* (c. 500), Copernicus's *De Revolutionibus* (1543), Brahe's "Nova Stella" (1572).

Planetary orbital calculations by mathematicians from the Kerala school in southern India, active between the fourteenth and sixteenth centuries, fit a heliocentric model for the solar system. This figure shows some examples from modern Indian physicists reconstructing the geometry used by Kerala school astronomers.

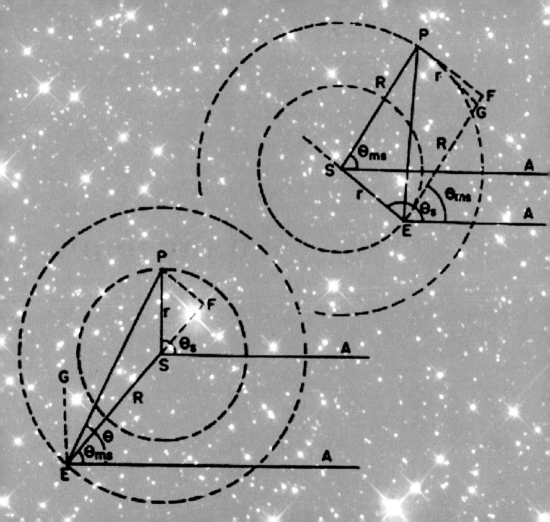

Copernicus's *De Revolutionibus*

Nicolaus Copernicus (1473–1543)

After nearly a thousand years of stagnation, the Renaissance in western Europe was truly an awakening—of art, music, culture, and science. With the founding of universities in Bologna, Oxford, Cambridge, Paris, Padua, and elsewhere, and learned clergy, such as John of Sacrobosco, introducing the advances in astronomy made by Arabic, Chinese, and Indian scholars during Europe's Dark Ages (and reintroducing the work of the Greeks and Romans), the stage was set for European science to flourish.

The first, and in some ways most important, Renaissance scientist to take that stage was Nicolaus Copernicus, a Polish canon (his uncle, who was also his patron, was a bishop), doctor, lawyer, economist, and part-time astronomer. As a community leader in Frombork, in northwestern Poland, Copernicus assumed many legal, administrative, and economic responsibilities as his nominal vocation. But he also made the time to conduct astronomical observations and to analyze his data, to read the classical and contemporary astronomical literature, and to ruminate on problems that he had

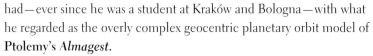

had—ever since he was a student at Kraków and Bologna—with what he regarded as the overly complex geocentric planetary orbit model of **Ptolemy's *Almagest.***

By 1514 he was circulating the basic outline of an alternative paradigm, a solar system with the Sun fixed at the center, the Earth and the other planets spinning on their axes and orbiting the Sun, and the Moon orbiting the Earth. But it wasn't until just before he died in 1543 that he finally published his theory as *De Revolutionibus Orbium Coelestium* (*On the Revolutions of the Heavenly Spheres*). Perhaps surprisingly, the book and its Sun-centered thesis did not generate much interest or controversy at the time. It would take more than 50 years and the supporting observations and interpretations of Brahe, Kepler, and Galileo (and Galileo's telescope) to make it apparent that *De Revolutionibus* had started what became widely known as the Copernican revolution in cosmology.

SEE ALSO Sun-Centered Cosmos (c. 280 BCE), Ptolemy's *Almagest* (c. 150), *De Sphaera* (1230), Early Calculus (c. 1500), Brahe's "Nova Stella" (1572), Galileo's *Starry Messenger* (1610), Three Laws of Planetary Motion (1619), Newton's Laws of Gravity and Motion (1687).

LEFT: *Painting of Copernicus by an unknown artist (1580).* RIGHT: *Illustration of the Copernican model of the solar system, from Andreas Cellarius's 1660 star atlas,* Harmonia Macrocosmica.

SCENO SYSTE COPER

GRAPHIA MATIS NICANI.

Brahe's "Nova Stella"

Tycho Brahe (1546–1601)

For astronomers before the seventeenth century, making the best possible naked-eye observations of the motions of celestial bodies meant having the best possible equipment and the keenest possible eyesight. Except for some Chinese astronomers, few had recognized the value of numerous, consistent, and systematic observations of the heavens to maximize the accuracy of their data. Perhaps the best Western practitioner of this "brute force" approach to beating down the errors in measuring planetary motions was the Renaissance astronomer and Danish nobleman Tycho Brahe.

Tycho's passion for astronomy was apparently ignited when he saw a solar eclipse as a teenager in 1560. His family's wealth and connections allowed him to pursue that passion and build world-class observing facilities. He immersed himself in cosmology, but he subscribed neither to Ptolemy's geocentric theory nor to Copernicus's heliocentric view. Rather, his careful observations of new comets and a new star that appeared in 1572 (Tycho coined the term *nova*) convinced him that the "fixed stars"

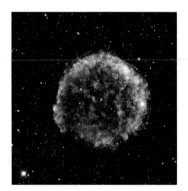

were not fixed, and that they and the planets did not move along the transparent crystalline spheres that the Ptolemaic and Copernican systems posited. His own cosmologic model was similar to that of Nīlakantha's from the Kerala school: a geo-heliocentric hybrid in which all the planets except Earth orbited the Sun, and the Sun and Moon orbited Earth.

Tycho set the standard for high-quality observations of planetary motions using the best instruments, a near-continuous data set over many decades, and a thorough description of the uncertainties in his data. His measurements were coveted by theorists, and he was particular about his choice of collaborators. In eventually choosing to work with Johannes Kepler, Tycho perhaps unknowingly guaranteed his immortality, as his amazing data set eventually led to Kepler's discovery of the fundamental **Laws of Planetary Motion**. Tycho was as eccentric as he was thorough, losing part of his nose in a duel and replacing it with prosthetics of copper or silver, and literally partying himself to death at the age of 54.

SEE ALSO Sun-Centered Cosmos (c. 280 BCE), Early Calculus (c. 1500), Copernicus's *De Revolutionibus* (1543), Galileo's *Starry Messenger* (1610), Three Laws of Planetary Motion (1619).

LEFT: *The inset shows a modern false-color X-ray and infrared photo of the expanding spherical shell of ionized gas that is the remnant of the supernova that Tycho and other astronomers observed in 1572.* RIGHT: *Portrait of Tycho Brahe from the National Museum of Denmark.*

Gregorian Calendar

Most of us think of a year as 12 months, or 365 days, and we don't give it much more thought. Once every four years, though, we're reminded that it's not quite true, that one trip around the Sun is more like 365.25 days, which is why the **Julian Calendar** reform of 45 BCE mandated adding February 29 as a leap day every fourth year to keep Caesar's calendar system synced with the Earth's motion around the Sun.

The Julian calendar's assumed vernal-equinox-to-vernal-equinox year of 365.25 days turned out to be too long by about 11 minutes per year (the actual value is about 365.24237 days). This didn't matter much during the Roman Empire, but 1,500 years later, the calendar date had drifted more than 10 days from the same solar, or seasonal, date that it had been given back in Caesar's time. The spring equinox as determined by astronomers was occurring around March 11 instead of March 21. This meant that the feast of Easter was drifting back into winter, which the Catholic Church didn't like. A fix was needed. Julius Caesar's 45 BCE calendar reform needed a reform.

The Roman Catholic Council of Trent decided to modify Caesar's calendar so that instead of one out of every four (0.25) years being a leap year, it would now be 97 out of every 400 (0.2425) years, established by designating only every century year divisible by 400 as a leap year. So the years 1700, 1800, and 1900 were not leap years, but the year 2000 was. They still had to fix the 10-day drift, so the Julian calendar was ended on October 4, 1582, and the new "Gregorian calendar" plan was put in place by Pope Gregory XIII on the next day, which was decreed to be October 15, 1582.

The modern Gregorian calendar, in wide global use today, stays true to the Earth's orbital position to better than one day out of every 7,600 years—good enough for the foreseeable future. However, the Earth's spin is slowing down slightly over time because of tidal friction, and so the length of the day is slowly getting longer. Thus, since 1972, the International Earth Rotation and Reference Systems Service (there really is such a service!) has occasionally added official leap seconds to the world's atomic clocks, to keep them in sync with the slowly changing spin, and thus the slowly changing apparent solar time, of our planet.

SEE ALSO Julian Calendar (45 BCE), Finding Easter (c. 700), *De Sphaera* (c. 1230), Earth's Rotation Speeds Up (1999).

A perpetual calendar—a type of mechanical date-calculating device—from c. 1600, used to predict the dates of upcoming feasts, holidays, and other events in the then-new Gregorian calendar system. In the background is the first page of Pope Gregory XIII's papal bull of 1582 announcing the new calendar.

CALENDARIVM
GREGORIANVM
PERPETVVM

Orbi Christian…   IIII. P. M. pro-
po   II.

GR   PVS

non est, vt quæ sa-
… finem optatum, Deo
… ad reliquam cogitatio-
… en exclusi rem totam ex
… pani Pontificis retulerunt.
Duo autem … num preces, laudesque diui-
nas festis, prof   … ter, alterum pertinet ad annnos

Paschâ, festorúmque ex eo pendentium … una metu metiendos: Atque illud quidem
felicis recordationis Pius … prædecessor noster … quotiendum curauit, atque edidit. Hoc vero, quod ni-
mirum exigit legitimam Calendarij restitutionem, is indi à Romanis Pontificibus prædecessoribus no-
stris, & sæpius tentatum est, verùm absolui, & ad exitum perduci ad hoc vsque tempus non potuit, quod

Mira Variables

David Fabricius (1564–1617), **Johannes Fabricius** (1587–1615), **Johannes Holwarda** (1618–1651)

Astronomers of the sixteenth century were aware that occasionally a seemingly ordinary star would flare up and dramatically increase in brightness—what Tycho Brahe had termed a *nova*. But no one had ever observed a star brighten, then dim, then brighten and dim again. That's what the Dutch-German pastor and part-time astronomer David Fabricius discovered from his observations of the star Omicron Ceti, also known as Mira, in 1596 and again in 1609. In 1638 the Dutch astronomer Johannes Holwarda discovered that Mira is a periodic or pulsating variable star, with a period of about 330 days.

When Fabricius discovered Mira, it was a magnitude 3 star that dimmed quickly (within a month or so) to beyond the human visual limit of magnitude 6. Later astronomers monitoring Mira telescopically over the years have seen it brighten to magnitude 2 and dim to magnitude 10, a factor of over 1,700 in brightness. We now know that Mira is a puffed-up red giant star about 350 times bigger than the Sun—if Mira were in our solar system it would extend out to the orbit of Mars! Its pulsations are part of normal stellar evolution for relatively low-mass stars near the end of their lives. Nearly 7,000 stars are now known that exhibit pulsations like this, with periods from around 100 to 1,000 days; they are collectively referred to as Mira variables.

Mira was discovered to be in a double star system in 1923, with a smaller white dwarf companion called Mira B. Recent **Hubble Space Telescope** and **Chandra X-Ray Observatory** images show that Mira B is gravitationally pulling gas off Mira A and into a solar nebula–like disk that may be accreting planets. Mira is a dying star, yet in its death throes it may be giving life to new planets.

Fabricius and his son Johannes are also credited with being the first astronomers to systematically observe sunspots and to use them to discover that the Sun rotates, as predicted by physicist Johannes Kepler, and has a rotation period of about 27 days. Coincidentally, there's a Mira connection there as well, as astronomers have discovered that Mira has starspots that, like sunspots, may be related to strong magnetic fields in the star's outer layers.

SEE ALSO Solar Nebula (c. 5 Billion BCE), Stellar Magnitude (c. 150 BCE), Three Laws of Planetary Motion (1619), Mizar-Alcor Sextuple System (1650), Solar Flares (1859), Main Sequence (1910), End of the Sun (5–7 Billion).

Artist's conception of the red giant star Mira A (right) and its companion star Mira B, a small white dwarf that is surrounded by a disk of gas and dust being drawn off the pulsating red giant.

Bruno's *On the Infinite Universe and Worlds*

Giordano Bruno (1548–1600)

The heliocentric view of the solar system promoted by Copernicus in 1543 was not widely accepted by his sixteenth-century peers. Although the idea that the Earth was not the center of the universe was inconsistent with the scriptures of the sixteenth-century Roman Catholic Church, Copernicus, a church canon, was, ironically, never the focus of much controversy regarding his views. Others would soon inherit that controversy, however.

One of Copernicanism's earliest and most vocal advocates was the late-sixteenth-century Italian philosopher, astronomer, and Dominican friar Giordano Bruno. Bruno appears to have been an outspoken advocate of a number of unorthodox and controversial views about science, religion, and natural philosophy. While not known for any particular observations, skills, or discoveries, Bruno eventually came to believe in a form of nongeocentrism far more extreme than Copernicus had espoused.

In his 1584 book, *De l'Infinito, Universo e Mondi* (*On the Infinite Universe and Worlds*), Bruno postulated that Earth was just one of an infinite number of inhabited planets orbiting an infinite number of stars, which are just suns like our own. To the Church, advocating such a plurality of worlds was mildly heretical; Bruno made it wholesale heresy with other brash demotions of the central tenets of Christian theology, such as the noncentrality of even God in his infinite universe. He fled persecution by the Inquisition for more than 15 years but was eventually arrested, tried, convicted, and burned at the stake in Rome in 1600.

It is tempting to romanticize Bruno simply as a scientific martyr, fighting for the truth against a dogmatic regime, especially because some of his ideas about cosmology and the plurality of worlds have turned out to be right. But others before him had held views at odds with the Church, as did others of his contemporaries (most famously, **Galileo**), without suffering as drastic a fate. Bruno's demise may not have been so much about his Copernicanism as it was about his confrontational style and his passion for outspoken criticism of authority and the so-called common wisdom.

SEE ALSO Copernicus's *De Revolutionibus* (1543), Galileo's *Starry Messenger* (1610), First Extrasolar Planets (1992).

Part of a bronze relief by Italian sculptor Ettore Ferrari (1845–1929) depicting the trial of Giordano Bruno by the Roman Inquisition in 1600.

First Astronomical Telescopes

Thomas Harriot (1560–1621), **Galileo Galilei** (1564–1642), **Hans Lippershey** (1570–1619), **Jacob Metius** (1571–1630), **Zacharias Jansen** (1580–1638)

The popular myth is that the Italian astronomer Galileo Galilei invented the telescope. While it is true that he invented an astronomical telescope of unprecedented resolving power for his time and used it in 1610 to make discoveries that would firmly establish the heliocentric nature of the solar system, the telescope itself had been invented several years earlier, and its basic principles and components actually date back to antiquity.

The earliest telescopes were so-called refracting instruments because they used combinations of concave and convex glass lenses to bend and magnify (or diminish, in the case of the earliest microscopes) the field of view. The knowledge that curved, transparent surfaces could magnify images was known to the Egyptians in the fifth century BCE, as well as to Greek and Roman scholars. Glass lenses (from the Latin word for "lentils," because of their shape) for use in spectacles to correct bad eyesight are thought to have been invented in the eleventh century or perhaps earlier in China or the Arab world; they were introduced in the West in the late 1200s. Spectacle making was a widely practiced craft in late medieval and early Renaissance Europe.

It was in that environment that at least three Dutch craftsmen and lens makers may have independently invented devices between about 1604 and 1608 that we would call telescopes. Hans Lippershey, Zacharias Jansen, and Jacob Metius all designed simple nested tubes—spyglasses—containing two lenses that would yield 2×–3× magnifications of a small field of view. Lippershey was granted a Dutch patent for his version, so he is often cited as the inventor of the telescope; the real history is still actively debated, however. Simple spyglasses began quickly selling in Europe in 1609.

Since the design was relatively simple and lenses were fairly easy to procure or grind, astronomers like Galileo, Thomas Harriot, and others began tinkering with and enhancing the invention and pointing it upward. Galileo made several versions of increasing power; his telescope of late 1609 achieved a magnification of about 20×, allowing him to see details of the heavens that no one had previously observed, and to change astronomy forever.

SEE ALSO Brahe's "Nova Stella" (1572), Copernicus's *De Revolutionibus* (1543), Galileo's *Starry Messenger* (1610).

This nineteenth-century painting by French artist Henry-Julien Detouche (1854–1913) depicts Galileo presenting an early version of his newly invented astronomical telescope to the doge of Venice, Leonardo Donato, in 1609.

Galileo's *Starry Messenger*

Galileo Galilei (1564–1642)

Revolutions often begin with a single pivotal, precipitating event from which springs an entire movement. In the case of the scientific revolution—a movement that continues to this day—the single event was arguably the publication of a small treatise called *Sidereus Nuncius* (*Starry Messenger*) in 1610 by the Renaissance Italian physicist, mathematician, and astronomer Galileo Galilei. *Starry Messenger* changed the history of astronomy.

Galileo constructed one of the first—and, at that time, the best—astronomical telescopes and used his instrument to observe the heavens in unprecedented detail. He was the first human being to see and track the satellites of Jupiter and to realize that not all planetary bodies orbit the Earth. He was the first person to see the phases of Venus and thus to know that it must be orbiting the Sun, and not the Earth. He was the first person to realize that the Moon has mountains, craters, and valleys and is not a perfect celestial sphere. Galileo's observations and his subsequent interpretations in *Starry Messenger* were a direct refutation of Aristotle's and Ptolemy's geocentric cosmology and provided compelling evidence, though not proof, of Copernicus's Sun-centered cosmology. To Galileo it was proof enough, however.

He continued to make observations with successively more powerful telescopes, recording the disks and features of the planets, and resolving the Milky Way into countless densely packed stars. His telescopes allowed him, and other contemporaries who confirmed and expanded on his work, the ability to see stars as faint as magnitudes 8 or 9, more than 15 times fainter than those visible to the naked eye.

He had powerful supporters in the Roman Catholic Church who initially supported his discoveries. But as **Giordano Bruno**'s execution had shown, the Church eventually came to regard Copernicanism as a threat. Galileo was spared, but he was forced into house arrest for the last nine years of his life.

SEE ALSO Stellar Magnitude (c. 150 BCE), Copernicus's *De Revolutionibus* (1543), Bruno's *On the Infinite Universe and Worlds* (1600), First Astronomical Telescopes (1608).

LEFT: *Galileo's phases of Venus.* RIGHT: *Examples of some of Galileo's 1610 sketches of craters, hills, and other features of the Moon.*

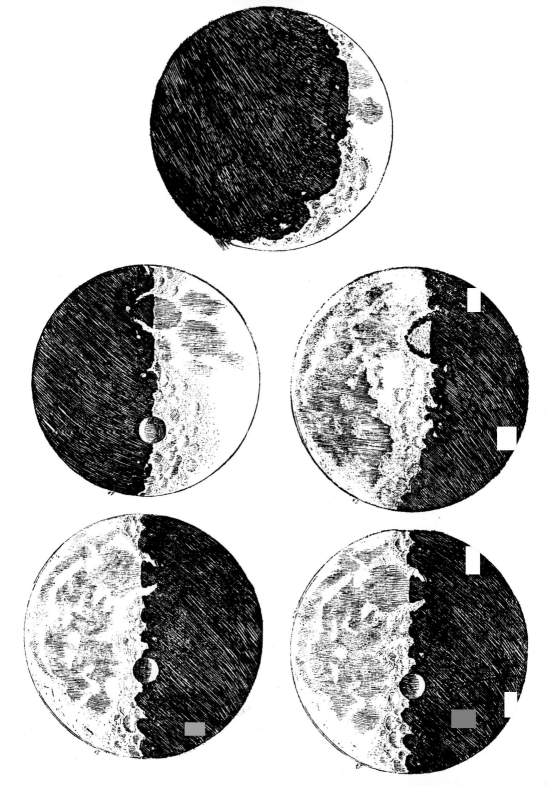

Io

Galileo Galilei (1564–1642), Simon Marius (1573–1624)

When Galileo first trained his telescope on Jupiter on January 7, 1610, he noticed what he described as "three fixed stars, totally invisible by their smallness." The stars were right next to Jupiter (two on one side and one on the other) and all in a straight line that passed through the middle of Jupiter's disk. It is fun to imagine his amazement when he saw four tiny stars on the next night instead of three—again all along the same line—and as he continued to watch over subsequent weeks, the little stars moved relative to the planet. It didn't take long for him to spot the pattern: they were orbiting Jupiter.

Galileo had discovered four new worlds, the first known moons of another planet besides our own. As their discoverer, he earned the right to name them, and in a politically astute move he decided to name them the Medician stars, after his patron and funding source Cosimo II de' Medici. Other contemporary astronomers hated the idea. The German astronomer Simon Marius, who claimed to have discovered the moons before Galileo, proposed names from Greek mythology: Io (a nymph who was seduced by Zeus), **Europa**, **Ganymede**, and **Callisto**. Galileo hated other naming options and eventually started calling the moons Jupiter I through IV. Astronomers used those designations well into the twentieth century, until Marius's more romantic naming scheme finally stuck. In honor of their discoverer, though, they are now collectively called the Galilean satellites.

Io (pronounced EYE-oh) is the closest in of the four large moons of Jupiter, with an orbital period of about 42 hours. Seven space missions (*Pioneers 10* and *11*, *Voyagers 1* and *2*, *Galileo*, *Cassini*, and *New Horizons*) have now studied Io up close, revealing it to have a diameter of 2,275 miles (3,660 kilometers)—slightly larger than Earth's Moon—and a surprisingly "rocky" density of 3.5 grams per cubic centimeter. The biggest surprise, however, was *Voyager 1*'s discovery of **Active Volcanoes on Io**, responsible for the moon's young surface being covered with red, orange, and black sulfur and silicate lava, and its thin atmosphere of sulfur dioxide. Strong tidal forces cause constant eruptions on Io, making it the most volcanically active world in the solar system.

SEE ALSO Europa (1610), Ganymede (1610), Callisto (1610), Speed of Light (1676), Active Volcanoes on Io (1979).

TOP: *Galileo's January 8, 1610, drawing of Jupiter's four bright satellites, with Io's position on that night highlighted by the red arrow.* RIGHT: *Jupiter's innermost moon, Io, photographed against the backdrop of Jupiter's clouds in 1996 by the NASA Galileo Jupiter orbiter spacecraft.*

Europa

* * * ○ *

Galileo Galilei (1564–1642), Giovanni Domenico Cassini (1625–1712)

1610

Another of the "small stars" that Galileo discovered orbiting Jupiter is Europa, named after a princess who was a lover of Zeus (the Greek version of the Roman god Jupiter). Galileo had seen Io and Europa together as a single "star" on January 7, 1610; by the next night they had moved enough for him to distinguish them as two objects.

Europa is the second-closest to Jupiter of the Galilean satellites and takes about three and a half days to orbit the giant planet. By monitoring the motions of Europa as well as Io, **Ganymede**, and **Callisto**, Galileo developed a way to accurately predict their movements, and proposed that the relative positions and timing of their occasional eclipses could be treated as a sort of natural celestial clock that could be used as a way to determine longitude. The Italian-French astronomer Giovanni Domenico Cassini verified Galileo's method in 1681, and it was used successfully by explorers such as Lewis and Clark.

Observations from the seven robotic space probes that have visited the Jovian system reveal that Europa is the smallest of the four large moons, with a diameter of 1,950 miles (3,140 kilometers), which is slightly smaller than Earth's Moon. A planetary body's mean density can be estimated by dividing its mass (determined from how much the body "bends" the trajectory of a passing space probe) by its volume (determined from photographs). Europa's density of about 3 grams per cubic centimeter is interpreted to mean that the satellite is mostly rocky, even though it has an icy surface.

That surface was found to be extraordinarily smooth, young (with very few impact craters), crisscrossed with intersecting reddish cracks and streaks, and broken up into many fractured and jumbled plates that appear to have moved tectonically relative to each other. Up close, the surface is reminiscent of sea ice. Indeed, several lines of evidence point to the existence of a deep liquid water **Ocean on Europa** under a relatively thin icy crust. The possibility of another ocean in our solar system, warmed by tidal energy and protected from Jupiter's harsh radiation, raises the prospect that Europa may be another abode for life.

SEE ALSO Galileo's *Starry Messenger* (1610), Io (1610), Ganymede (1610), Callisto (1610), An Ocean on Europa? (1979).

TOP: *Galileo's January 8, 1610, drawing of Jupiter's four bright satellites, with Europa's position on that night highlighted by the red arrow.* RIGHT: *The smallest of Jupiter's Galilean satellites, Europa, photographed in 1998 by NASA's Galileo spacecraft.*

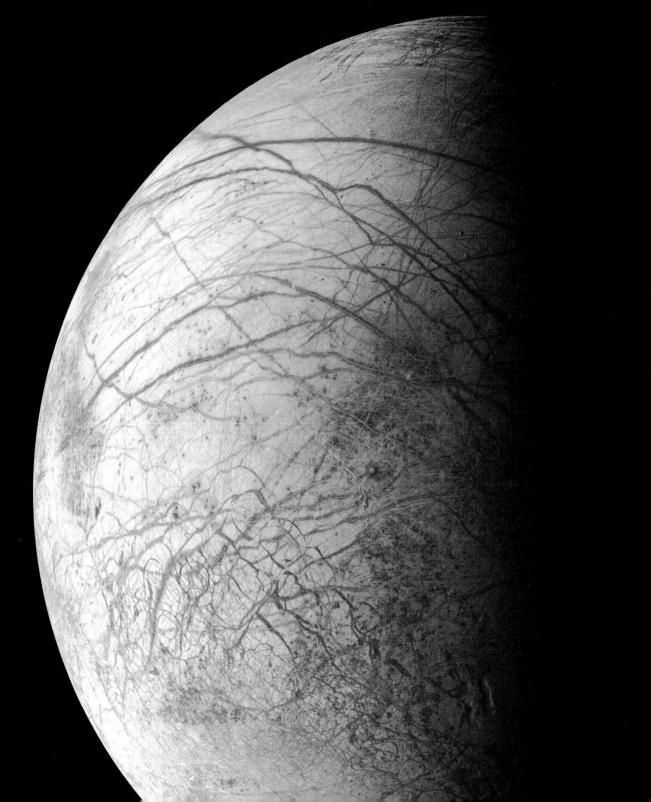

Ganymede

✶ ✶∗ ⭕   ✶

Galileo Galilei (1564–1642), **Pierre-Simon Laplace** (1749–1827)

The third of **Galileo**'s newly discovered moons of 1610 is Ganymede, named after the mythological prince, cupbearer of the gods, and lover of Zeus (Ganymede is the only moon of Jupiter named after a male figure). Ganymede takes a little more than seven days to make one orbit of Jupiter. When Galileo and other astronomers worked out the precise orbits of **Io**, **Europa**, and Ganymede, they noticed something very interesting: for every single orbit of Ganymede, Europa orbits exactly twice, and Io orbits exactly four times. The satellites are described by astronomers as being in resonances.

Discovery of the 4:2:1 orbital resonances of the three inner Galilean satellites spurred a mini revolution of sorts among mathematicians and physicists trying to understand and explain how those resonances came to be. Key explanations were worked out by the French mathematician and astronomer Pierre-Simon Laplace, and in honor of his achievements these kinds of three-body situations are now called Laplace resonances. Orbital resonances have been found to generate gaps in the **Main Asteroid Belt** and in **Saturn's Rings**; even some newly discovered **Extrasolar Planets** are in orbital resonances, too.

Ganymede has also now been closely studied by many robotic spacecraft, which have revealed it to be the largest moon in the solar system, at 3,275 miles (5,270 kilometers) in diameter—even larger than the planet **Mercury**. Its density of 1.9 grams per cubic centimeter implies that it is made of a significantly higher relative percentage of ice than Io and Europa. Brighter, more ice-rich grooves and ridges on Ganymede's surface—potentially the result of past tectonic activity—appear younger than the darker, more heavily cratered terrains. Ganymede is the only moon in the solar system with its own (weak) magnetic field, suggesting that the interior has differentiated into a crust, mantle, and molten iron core. Perhaps most interestingly, the magnetic readings, presence of salty surface minerals, and evidence for past eruption of water into some grooves and ridges could all be consistent with a deep underground layer of liquid water—yet another ocean?—inside Ganymede.

SEE ALSO Galileo's *Starry Messenger* (1610), Io (1610), Europa (1610), Callisto (1610), Saturn Has Rings (1659), Lagrange Points (1772), Kirkwood Gaps (1857), An Ocean on Ganymede? (2000).

TOP: *Galileo's January 8, 1610, drawing of Jupiter's four bright satellites, with Ganymede's position on that night highlighted by the red arrow.* RIGHT: *The NASA Voyager 2 spacecraft's 1996 mosaic of Ganymede, the largest moon in the solar system. The brighter zones are areas of extensive tectonic deformation.*

Callisto

Galileo Galilei (1564–1642)

The farthest of the Galilean satellites from Jupiter is Callisto, named after another nymph and lover of Zeus from Greek mythology. Like the other major moons of Jupiter, it was discovered in early 1610, when Galileo first trained his **Astronomical Telescope** on the giant planet. Callisto is the farthest from Jupiter of the four major satellites, taking almost 17 days to complete one orbit. Perhaps because of this it does not participate in the resonances that characterize the orbits of **Io**, **Europa**, and **Ganymede**.

For more than 350 years after their discovery, it was impossible for astronomers to learn more about Callisto and the other Galilean satellites. Starting in the 1960s, however, it became possible to use **Spectroscopy** from ground-based telescopes to determine their surface compositions. Callisto, Ganymede, and Europa were found to have surfaces dominated by water ice, and Io was discovered to be dry, with colors and spectra dominated by the presence of sulfur. Spectroscopy from more recent space missions has further revealed the presence of ices of carbon dioxide and sulfur dioxide on Callisto and Ganymede, as well as hydrated sulfate salts on all three icy moons.

These missions have also allowed the determination of Callisto's diameter (2,995 miles [4,820 kilometers]—25 percent larger than our **Moon**) and density (1.8 grams per cubic centimeter, mostly icy with some rock), and the mapping of its surface features. Callisto is the most heavily cratered of the Galilean satellites, preserving giant impact basins from the **Late Heavy Bombardment** of the early solar system and suggesting that it has had the least internal activity or resurfacing history of the four. And yet, data from NASA's *Galileo* Jupiter Orbiter suggest that there could be a layer of liquid water— an ocean of sorts—deep beneath the scarred, icy crust. Callisto does not get the tidal flexing and heating that the other moons get from their resonant orbital interactions or proximity to Jupiter, however, so the heat source driving Callisto's possible subsurface ocean is a puzzle waiting to be solved.

SEE ALSO Late Heavy Bombardment (c. 4.1 Billion BCE), First Astronomical Telescopes (1608), Io (1610), Europa (1610), Ganymede (1610), Birth of Spectroscopy (1814), An Ocean on Europa? (1979), *Galileo* Orbits Jupiter (1995), An Ocean on Ganymede? (2000).

TOP: *Galileo's January 8, 1610, drawing of Jupiter's four bright satellites, with Callisto's position on that night highlighted by the red arrow.* RIGHT: *NASA's Galileo Jupiter orbiter spacecraft acquired this "full Callisto" photo of the heavily cratered surface of Jupiter's fourth Galilean satellite in 2001.*

Orion Nebula "Discovered"

Nicolas-Claude Fabri de Peiresc (1580–1637), **Christiaan Huygens** (1629–1695), **Charles Messier** (1730–1817)

The universe is full of atoms and molecules that sometimes form into gas, dust, and rock in strange and beautiful forms. One of the ways that astronomers can search for and catalogue the universe is by finding places where the gas and dust are so warm and dense that they are literally glowing in the dark. These places are interstellar clouds—the wispy, nebulous, filamentary remains of dead and dying stars that are, at the same time, the cocoons and nurseries of new young stellar embryos.

When stars die in nova or supernova events, they eject their outer layers of hydrogen, helium, and other materials into deep space. These ejected remains can often be ionized by energy from nearby stars, or from the explosion that ejected them. Ionized gases in interstellar clouds emit light and heat, allowing them to be observed and characterized by astronomers using spectrometers and large telescopes.

The most famous interstellar cloud in the sky is the Orion Nebula, a fuzzy patch of light visible to the naked eye just below the three stars in Orion's belt. While there is some archaeological evidence that the nebula was noticed by ancient Mayan astronomers, it was French astronomer Nicolas-Claude Fabri de Peiresc who first recorded observing the Orion Nebula in 1610, though he never reported it. It was later "rediscovered" by Christiaan Huygens in 1656 and Charles Messier in 1769.

Modern observations have revealed that the Orion Nebula is the closest interstellar cloud to us, at "only" 1,340 light-years away. Its composition is surprisingly complex and includes hydrogen, carbon monoxide, water, ammonia, formaldehyde, and simple precursors to amino acids.

Interstellar clouds are also places where new stars are born, as pockets and clumps of gas and dust slowly accrete and collapse under their own gravity. It is likely that our own solar system, and perhaps other solar systems in our local neighborhood, were formed from a giant solar nebula cloud of gas and dust, just as the Orion Nebula was formed from previous generations of local stars.

SEE ALSO Solar Nebula (c. 5 Billion BCE), Andromeda Sighted (c. 964), "Daytime Star" Observed (1054), Brahe's "Nova Stella" (1572), Birth of Spectroscopy (1814).

Hubble Space Telescope photomosaic of part of the interstellar cloud known as the Orion Nebula. The full nebula is about 24 light-years across and may contain about 2,000 times the mass of the Sun—enough gas and dust to eventually form more than 1,000 new stars.

Three Laws of Planetary Motion

Johannes Kepler (1571–1630)

While there is significant overlap, astronomers today can be generally characterized as either observationalists, those who primarily collect data from telescopes or space missions, or theorists, those who primarily try to develop models or theories to explain existing observations. Most astronomers (and astrologers) from antiquity through the Middle Ages were observationalists who dabbled in theory. Theoretical astronomy had been primarily considered to be the realm of philosophers, not physicists.

The Renaissance German mathematician, astrologer, and astronomer Johannes Kepler changed that paradigm and arguably became the world's first theoretical astrophysicist. Kepler worked with data from others—most notably **Tycho Brahe** and **Galileo**—in his quest to develop a single unifying model of the cosmos. A deeply religious man, Kepler believed that God had designed the universe in an elegant geometric plan, and that the plan could be reasoned out through careful observations.

Kepler believed in **Copernicus**'s heliocentric cosmology, and also believed that a Sun-centered solar system was entirely consistent with biblical writings. Kepler's book

Astronomia Nova (*New Astronomy*; 1609) described the orbits of Mars and the other planets as elliptical, not circular (first law), and asserted that the planets change speed in a way that allows them to sweep out equal areas in equal time as they orbit (second law). Later, in *Harmonices Mundi* (*Harmony of the Worlds*; 1619) he showed that a planet's orbital period squared is proportional to its average distance from the Sun cubed ($P^2 \propto a^3$; third law). Through Kepler's patience and persistence, the harmony that he sought among the worlds was finally revealed.

Kepler's laws were not widely appreciated until observationalists had verified their precise timing predictions during eclipses and planetary transits (the predictions were right), and ultimately until Isaac Newton found in 1687 that Kepler had discovered the natural consequences of a universal law of gravitation.

SEE ALSO Copernicus's *De Revolutionibus* (1543), Brahe's "Nova Stella" (1572), Galileo's *Starry Messenger* (1610), Newton's Laws of Gravity and Motion (1687).

LEFT: *A 1610 portrait of Kepler by an unknown artist.* RIGHT: *Johannes Kepler struggled to find divine perfection in the orbits of the known planets by trying to match them with the shapes of the so-called perfect solids (i.e., cube, tetrahedron [pyramid], octahedron, icosahedron, and dodecahedron) in this illustration from his book* Mysterium Cosmographicum *(1596).*

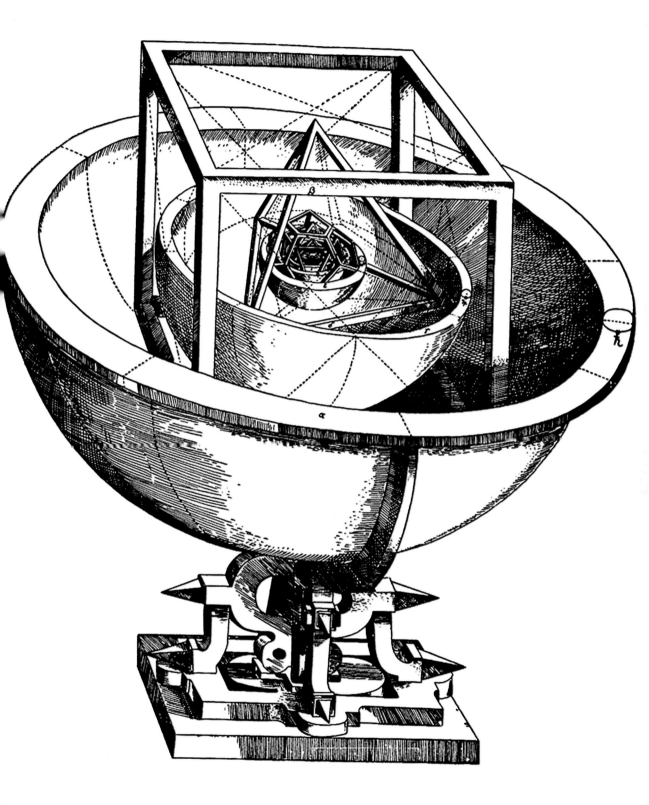

Venus Transits the Sun

Abū ʿAlī ibn Sīnā (c. 980–1037), **Johannes Kepler** (1571–1630), **Jeremiah Horrocks** (1618–1641)

Astronomers describe a transit as an event in which one celestial body passes in front of another from the vantage point of the observer. A Solar Eclipse, for example, is a transit of the Moon across the disk of the Sun. **Galileo** and other astronomers used transits of the Galilean satellites across the disk of Jupiter to refine orbital predictions and to derive an observer's longitude on the Earth.

Johannes Kepler realized that, occasionally, Venus transits the Sun from the vantage point of observers on Earth; if such a transit could be observed from different places, parallax and trigonometry could be used to determine the distance between the Earth and the Sun (known as an astronomical unit, or AU). Estimates of the AU from Aristarchus, Ptolemy, and Arabic astronomers estimated the Sun to be about 20 times farther than the Moon (or about 5 million miles [8 million kilometers] away). But Venus transits are rare events (occurring less than once a century) because the orbit of Venus is tilted by a few degrees relative to the orbit of Earth.

The Persian astronomer Abū ʿAlī ibn Sīnā observed a Venus transit in 1032. Kepler predicted a transit for 1631 and a near miss for 1639. The 1631 event was not visible from Europe, but English astronomer Jeremiah Horrocks used revised calculations to predict and successfully record the Venus transit of December 4, 1639, using the data to estimate the AU as about 60 million miles (96 million kilometers). Although his calculation is 35 percent less than the actual value (93 million miles [150 million kilometers]), Horrocks instantly made the solar system about 250 times bigger than anyone had previously measured.

Captain James Cook went to Tahiti as part of a worldwide campaign to observe the 1769 Venus transit, which resulted in a much more accurate estimate of the AU. Modern astronomers, and millions in the general public, observed two recent Venus transits of the Sun on June 8, 2004, and June 5, 2012 (the next one isn't until 2117!). NASA's Mars Exploration Rovers have observed transits of **Phobos** and **Deimos** from the surface of Mars, and **Extrasolar Planet** hunters have detected transits in other solar systems using ground-based telescopes and satellites such as NASA's *Kepler* **Mission**.

SEE ALSO Mayan Astronomy (c. 1000), Galileo's *Starry Messenger* (1610), Three Laws of Planetary Motion (1619), Speed of Light (1676), Deimos (1877), Phobos (1877), First Extrasolar Planets (1992), The *Kepler* Mission (2009).

A sequence of photographs of the Venus transit of June 5–6, 2012, from NASA's SDO spacecraft. The faint ring around the planet comes from the scattering of sunlight through the Venusian atmosphere.

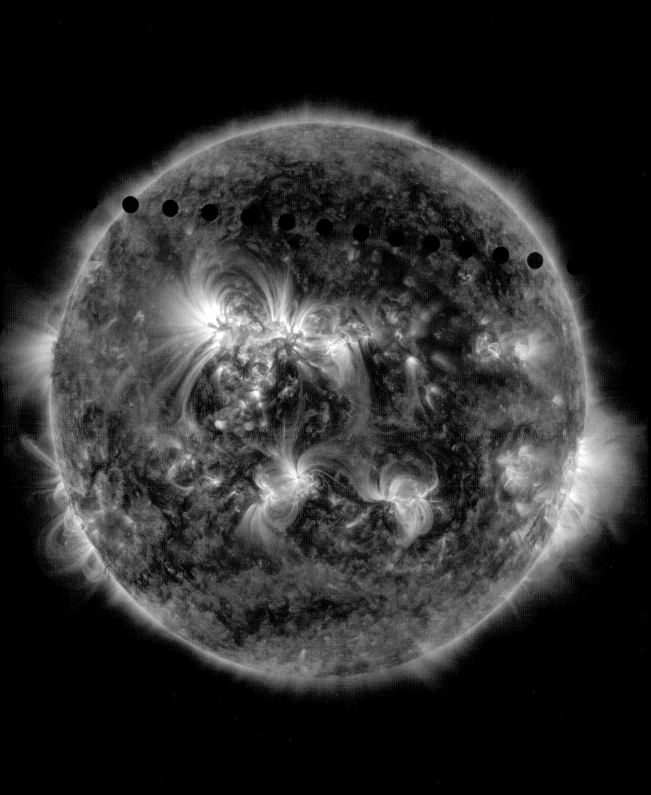

Mizar-Alcor Sextuple System

Giovanni Riccioli (1598–1671)

Early star catalogs of Hipparchus, **Ptolemy**, and al-Sūfī noted many bright stars that appear very close to each other in the sky. Among the most famous are the pair of bright bluish stars second from the end of the Big Dipper's handle, which Arabic astronomers named Mizar and Alcor and often called "the horse and rider." The two stars are about one-fifth of a degree apart and provide a test of a person's naked-eye visual acuity. Can you separate the horse and rider?

There was no way for early astronomers to know whether such pairs of stars were really related to each other or were just coincidentally close. With the advent of the telescope, it became possible to resolve what were eventually recognized as truly multiple-star systems. Galileo appears to have discovered that Mizar is actually a double star, with a fainter companion (Mizar B) only 14 seconds of arc away (one second of arc is ⅓ 600th of a degree), requiring a telescope to find. But Galileo did not publish his discovery, perhaps because Mizar A and B did not show the parallax shifts that he expected would prove **Copernicus**'s heliocentric theory (he and others thought that the stars were much closer to Earth than they truly are). As a result, the "discovery" of Mizar as the first known double star is often credited to Italian astronomer Giovanni Riccioli, who published his data on Mizar A and B in 1650.

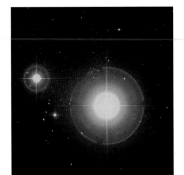

The Mizar system turns out to be even more interesting than that, however. In 1889 astronomers discovered that Mizar A is a spectroscopic binary, itself a double star with a companion that can only be detected by analyzing the merged spectrum of the pair. Then in 1908 Mizar B was also found to be a spectroscopic binary. Mizar is a double-double system of four stars. Mizar's naked-eye companion Alcor was also found to be a double in 2009, and there is now evidence that Mizar and Alcor may indeed be gravitationally bound—which would make the Mizar-Alcor system a sextuplet of six related, co-orbiting stars! This seems amazing to us as residents of a single-star system, but it is estimated that perhaps 60 percent of all stars in the galaxy are members of multiple star systems.

SEE ALSO Stellar Magnitude (c. 150 BCE), Ptolemy's *Almagest* (c. 150 BCE), Andromeda Sighted (c. 964), Copernicus's *De Revolutionibus* (1543), Galileo's *Starry Messenger* (1610).

LEFT: *Binary stars Mizar (brighter) and Alcor, the star system in the middle of the Big Dipper's handle.* RIGHT: *Science fiction has long recognized the possibility of double sunsets in planets around binary star systems, like the one depicted here on Tatooine from the film* Star Wars: A New Hope.

Titan

1655

Christiaan Huygens (1629–1695)

Galileo's *Starry Messenger* of 1610, announcing the discovery of four moons of Jupiter, the phases of Venus, and the presence of mountains and valleys on the Moon, generated an enormous amount of excitement and instigated a "space race" of sorts among seventeenth-century astronomers. If such discoveries could be made with Galileo's relatively simple spyglass, what new wonders would await the even larger telescopes? Soon, bigger telescopes began reaching further into the skies.

Saturn orbits the Sun at nearly twice the distance of Jupiter, and so sunlight there is more than three times less intense than it is at Jupiter. Maybe it's no surprise, therefore, that it took until 1655 for telescopes to become sensitive enough to detect the faint amount of sunlight reflected off a moon in orbit around Saturn. That discovery was made by the Dutch astronomer Christiaan Huygens, using a telescope of his own design. Huygens called the new moon simply Saturni Luna, Latin for "Saturn's moon." It wasn't until 1847 that it was named Titan, as part of a Greek mythology theme for naming the seven then-known satellites of Saturn.

Modern observations of Titan by the *Voyager* and *Cassini* space missions have revealed it to be a strange and unique world. Titan is the second-largest moon in the solar system and, at 3,200 miles (5,152 kilometers) across, it is larger than the planet **Mercury**. Titan's density of 1.9 grams per cubic centimeter implies an icy and rocky interior. It is the only moon with a thick atmosphere—a dense mixture of nitrogen and methane smog that shrouds the surface from view. Because Titan has a temperature of 90 kelvins and a surface pressure about 50 percent higher than Earth's, hydrocarbons created by sunlight in its nitrogen-methane atmosphere are expected to exist as liquids. Indeed, the *Cassini* orbiter's radar mapper found rivers and lakes of liquid ethane or propane on Titan.

Titan's environment is one of sluggish organic chemistry in the absence of oxygen. It is an astrobiology hot spot—the best place in the solar system to study what the early Earth may have been like before life made our atmosphere oxygen-rich. In 2005, the first probe ever to land on a moon of another world was sent to Titan. The successful mission was fittingly named *Huygens*.

SEE ALSO First Astronomical Telescopes (1608), Galileo's *Starry Messenger* (1610), *Huygens* Lands on Titan (2005).

Natural-color view of the haze-shrouded disk of Titan, photographed in 2009 from NASA's Cassini Saturn orbiter. The icy moon Tethys is in the distance beyond Titan.

Saturn Has Rings

Christiaan Huygens (1629–1695), **Giovanni Domenico Cassini** (1625–1712), **James Clerk Maxwell** (1831–1879)

Among the wonders that **Galileo** was the first to glimpse by telescope in 1610 was the planet Saturn. Through his modest astronomical telescope, the planet appeared to be a round disk with two bright blobs on both sides that he referred to as "ears." The nature of these features, which came and went over the years, remained an unresolved puzzle to Galileo for the rest of his life.

In 1659, the Dutch astronomer Christiaan Huygens trained his more powerful telescope on Saturn and became the first person to recognize the "ears" as a disk, or "thin, flat ring," surrounding Saturn. In 1675 the Italian-French mathematician and astronomer Giovanni Domenico Cassini discovered a dark gap in the rings (now called the Cassini Division) and suggested that the rings are actually a series of many narrower, separate rings. Astronomers and mathematicians considered the rings to be solid disks until the Scottish physicist James Clerk Maxwell hypothesized that the rings must instead be made of huge numbers of individual particles because solid rings would be ripped apart by gravitational and centripetal forces.

Maxwell's hypothesis was confirmed by the *Voyager 1* and *2* flybys of (and through) Saturn's rings in the 1980s, which, along with the *Cassini* Saturn Orbiter, have revealed the rings to be an intricate structure of thousands of separate ringlets composed of countless dust- to house-size "particles" of nearly pure water ice, with impurities of silicate dust and possibly some simple organic molecules. The main rings are 174,000 miles (280,000 kilometers) wide but, astonishingly, are less than about 328 feet (100 meters) thick. The "gaps" in the rings aren't really gaps but are areas where ring particles have been greatly depleted by gravitational interactions with small moons that orbit within the rings. Planetary scientists debate the origin and age of Saturn's rings. Are they primordial or "young," perhaps formed only a few hundred million years ago from the catastrophic disruption of an icy moon?

SEE ALSO Saturn (c. 4.5 Billion BCE), First Astronomical Telescopes (1608), Galileo's *Starry Messenger* (1610), Ganymede (1610), Kirkwood Gaps (1857), *Voyager* Saturn Encounters (1980, 1981), *Cassini* Explores Saturn (2004).

LEFT: *A 1659 drawing of Saturn and its rings from Christiaan Huygens's* Systema Saturnia. RIGHT: *Hubble Space Telescope composite of photos of Saturn obtained from 1996 (bottom) to 2000 (top), as our view of the tilt of the planet's rings changed from nearly edge-on to much more open.*

Great Red Spot

Robert Hooke (1635–1703), Giovanni Domenico Cassini (1625–1712)

When the seventeenth-century scientists Robert Hooke and Giovanni Domenico Cassini trained their early astronomical telescopes on the planet **Jupiter**, they were the first to notice and track a circular, reddish blotch in the planet's southern hemisphere. Little did they imagine that they were tracking an enormous, hurricane-like storm system, more than twice the size of the Earth, that would continue churning for nearly 350 more years, and perhaps longer.

Astronomers and planetary scientists have studied Jupiter's Great Red Spot in much more detail from modern telescopes and space missions. To atmospheric scientists, the Great Red Spot is now known to be a persistent, counterclockwise rotating atmospheric vortex. Time-lapse photography of the storm's rotation shows that it takes about 6 Earth days (about 14 Jupiter days) for the storm to spin once. Wind speeds along the edge of the storm zone, where it interacts with other belts and zones in Jupiter's atmosphere, peak around 270 miles per hour (430 kilometers per hour).

The Great Red Spot is cooler than the surrounding parts of Jupiter's atmosphere because its storm clouds are about 6 miles (10 kilometers) higher than the surrounding clouds. If we could somehow fly through that part of Jupiter's atmosphere, we would see something like a giant, slowly rotating thunderhead cloud rising up above the haze. Strong jet-stream winds to the north and the south of the Great Red Spot appear to provide the energy to keep the storm confined to the same latitude. While the spot's size has decreased somewhat in the past few decades, no one knows how much longer it will continue to rage.

Why the Great Red Spot is red is somewhat of a mystery. Various hypotheses are that the color is caused by atmospheric gases or aerosols containing sulfur, phosphorus, or organic molecules. Actually, the spot's color has been observed by astronomers to change over the past few decades, from reddish to brownish, yellowish, and even whitish. The effort to understand the origin of the color of the Great Red Spot, as well as the colors of Jupiter's other lovely atmospheric patterns, is an active area of planetary science research. The mystery of the spot's origin and future only enhances the Van Gogh–like beauty of its colorful, swirling clouds.

SEE ALSO Jupiter (c. 4.5 Billion BCE), First Astronomical Telescopes (1608), Galileo's *Starry Messenger* (1610), *Pioneer 10* at Jupiter (1973), *Galileo* Orbits Jupiter (1995).

Photograph of Jupiter's Great Red Spot from the Voyager 2 *space probe. For scale, more than two Earths could fit across the width of this storm system.*

Globular Clusters

Johann Ihle (c. 1627–1699)

Stars form from the gravitational collapse of immense clouds of gas and dust. Astronomers have found that these clouds are often so large that many stars can form from a single cloud or closely associated collections of clouds, leading to the creation of multiple star systems and particular concentrations of enhanced star formation, such as spiral galaxy arms. Some interstellar clouds, especially in the early universe, appear to have been so massive that they led to the formation of literally hundreds of thousands of stars in relatively close proximity. When gravitational interactions between these nearby stars pull them together into a spherical mass, all orbiting a common center of gravity (and perhaps a black hole), the resulting collection of stars is called a globular cluster.

The first reported observation of a globular cluster was in 1665, from telescopic observations by a German post office official and avid amateur astronomer named Johann Ihle. Ihle observed a dense cluster of stars that is now known in the catalog of astronomer Charles Messier as M22, or the Sagittarius Cluster. M22 is visible to the naked eye as a faint 5th **Magnitude** smudge; Ihle and other seventeenth-century astronomers were able to use their astronomical telescopes to reveal the smudge to be the collected light of a swarm of countless numbers of closely packed stars.

More than 150 bright, densely packed globular clusters like M22 have since been discovered orbiting the center of our **Milky Way** galaxy as part of a semispherical halo of stars and star clusters that are older than the typical stars found in the galactic disk. Other galaxies have been found to have halos of globular clusters as well— apparently, halo formation is an important early stage in the formation of galaxies in general. Globular cluster halos extend so far out from the centers of many galaxies that astronomers speculate that some galaxies actually trade star clusters when they pass by or interact with each other gravitationally.

Stars in globular clusters interact with their close neighbors much more often than stars in "normal" space, and, as a result, astronomers do not believe that globular clusters would be good places to find stable, habitable planets. However, much still remains unknown about these ancient stellar swarms.

SEE ALSO First Stars (c. 13.5 Billion BCE), Milky Way (c. 13.3 Billion BCE), Solar Nebula (c. 5 Billion BCE), Stellar Magnitude (c. 150 BCE).

Hubble Space Telescope photo of the globular star cluster known as NGC 6093, or M80 in Messier's catalog. The cluster is about 28,000 light-years away and contains hundreds of thousands of stars, all bound together by their mutual gravitational attraction.

Iapetus

Giovanni Domenico Cassini (1625–1712)

The discovery of Jovian moons in 1610 and a moon around Saturn in 1655 set off a frenzy of moon hunting by late-seventeenth-century astronomers. The solar system's sixth new moon was discovered in 1671 by the prolific Italian-French mathematician and astronomer Giovanni Domenico Cassini. Strangely, Cassini could only see the new moon when it was on the western side of Saturn. In 1705, Cassini finally observed the satellite on the other side of Saturn, when it was more than six times dimmer!

In 1847 the moon was named after the mythologic Greek Titan Iapetus. Cassini and others had surmised that, like our own **Moon**, Iapetus is tidally locked with the same face pointed to its parent planet. Thus, half of Iapetus always faces forward in its direction of motion (the dark leading side), and half always faces backward (the bright trailing side).

The strange two-toned surface of Iapetus was confirmed by closer encounters

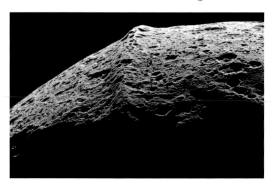

by the *Voyager* missions in the 1980s, which measured the diameter of the moon to be about 930 miles (1,500 kilometers) and its density to be around 1.1 grams per cubic centimeter—consistent with an ice-dominated composition (ice has a density of 1.0 gram per cubic centimeter). More detailed studies and a very close flyby by the *Cassini* orbiter mission have revealed that Iapetus has an ancient, heavily cratered surface and an equatorial ridge that gives the moon a walnut-like shape.

Cassini orbiter observations have helped to solve some of the mysteries of Iapetus's two-toned surface. Because the leading hemisphere is darker, it is slightly warmer than the trailing hemisphere. Ice evaporates from solid to vapor there, leaving behind a lag or residue of darker silicate or organic contaminants that were in the ice; the icy vapor recondenses as cleaner, brighter ice on the colder trailing hemisphere. But mysteries remain—such as the reason the leading side got darker in the first place. Neighboring moon **Phoebe**, the source of a faint outer ring of Saturn, may be the culprit: the leading side of Iapetus may have darkened early on while plowing into Phoebe's ring dust.

SEE ALSO Io (1610), Europa (1610), Ganymede (1610), Callisto (1610), Titan (1655), Phoebe (1899).

LEFT: *A ridge 12.4 miles (20 kilometers) high of unknown origin circles much of the equatorial region of Iapetus.*
RIGHT: *A natural-color photo of the boundary between bright and dark material in the north polar region of Iapetus, taken in December 2004 by NASA's Cassini Saturn orbiter spacecraft.*

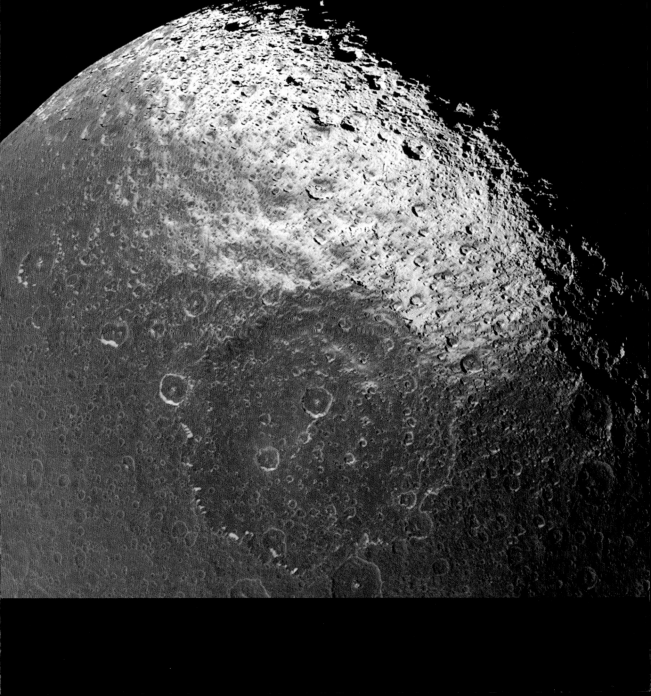

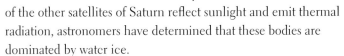

Rhea

Giovanni Domenico Cassini (1625–1712)

The Italian astronomer Giovanni Domenico Cassini was a professor at the University of Bologna in his early career, then moved to France in 1671 to become director of the Paris Observatory under King Louis XIV. Cassini's career in Paris was instantly and phenomenally successful, and it included the discovery of **Iapetus**, the first new solar system moon discovered in more than 15 years. The new French citizen Cassini was hard at work again in 1672, with the discovery of another new moon of Saturn, this one eventually named Rhea (RHEE-uh) after the mother of the gods in Greek mythology.

Rhea orbits 50 percent closer to Saturn than Titan; its discovery closer in to the glare of Saturn's disk is a testament to Cassini's prowess as an observer and the general improvement in telescope and optics technology in the late seventeenth century. Still, little was learned about Rhea other than its orbital period (four and a half days) and other basic orbital characteristics until the advent of telescope-based spectroscopy methods in the 1970s. From detailed studies of the way the surfaces of Rhea and most

of the other satellites of Saturn reflect sunlight and emit thermal radiation, astronomers have determined that these bodies are dominated by water ice.

It was not too surprising, then, that the *Voyager 1* and 2 flybys through the Saturn system in 1980 and 1982 revealed Rhea to have a very bright surface, with a reflectivity higher than 50 percent, consistent with an icy composition. More recent spectroscopy studies of Rhea from the *Cassini* Saturn orbiter have confirmed Rhea's icy nature.

Rhea's ancient surface is also almost completed saturated with circular impact craters and basins, created when high-speed asteroid, comets, or other moonlets crashed into Rhea over the 4.5-billion-year history of the solar system. There is some evidence for very recent impacts on Rhea, however, including fresh-looking craters with bright rays. Some of these impacts may be coming from a diffuse equatorial halo of dust- to boulder-size debris detected in some *Cassini* mission data. If further data and analysis confirm it, Rhea may be the solar system's only moon with its own ring system.

SEE ALSO Iapetus (1671), *Voyager* Saturn Encounters (1980, 1981).

LEFT: A Cassini *orbiter photo of Rhea from November 2005.* RIGHT: *NASA's* Cassini *Saturn orbiter took this stunning photograph of Rhea silhouetted against the rings of Saturn in March 2010.*

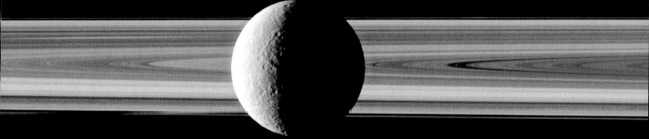

Speed of Light

Ole Christensen Røemer (1644–1710), **Christiaan Huygens** (1629–1695)

The nature of light has been debated throughout history. The Greek philosophers Aristotle, Euclid, and Ptolemy believed light was emitted from the eyes; because we see distant objects like stars instantly when we open our eyes, they believed that light must travel at an infinite speed. Eleventh-century scientists like the Muslim physicist al-Haytham and the Persian scholar al-Bīrūnī proposed that light had a finite speed, however, based on some early experiments in optics. The debate continued into the seventeenth century, with luminaries like Kepler favoring an infinite speed and **Galileo** favoring a finite speed. What was needed was a definitive measurement.

The first good estimate of the speed of light was made by the Danish astronomer Ole Christensen Røemer, who expanded on Galileo's original idea of using the moons of Jupiter as astronomical clocks of sorts, not for measuring longitude but for measuring light time. Røemer observed hundreds of eclipses of Jupiter's moon Io, carefully timing its disappearance into or reappearance from Jupiter's shadow. He noticed that the

predicted eclipse times differed from the observed times in a systematic way: they occurred about 11 minutes earlier when Earth was closest to Jupiter, and 11 minutes later when Earth was farthest away. Røemer deduced that the difference was because of the finite speed of light. With the help of his Dutch colleague Christiaan Huygens, the speed of light from Røemer's eclipse data was estimated to be about 137,000 miles (220,000 kilometers) per second. While the actual value was later determined to be about 35 percent greater, based on the more definitive eighteenth-century stellar aberration measurements by the English astronomer James Bradley and the nineteenth-century Michelson-Morely Experiment, Røemer's insight and years of careful observing enabled the first good estimate of the speed of light.

SEE ALSO Galileo's *Starry Messenger* (1610), End of the Ether (1887).

LEFT. *Part of Ole Røemer's observation log book with recorded times for some of the Io eclipses that he used to estimate the speed of light.* RIGHT: *Hubble Space Telescope false-color infrared photo from March 2004 showing Io (center) and Ganymede (right) transiting across Jupiter's disk. The shadows of Ganymede and Io are at left and the shadow of Callisto (itself not in the image) is at upper right.*

Halley's Comet

Edmond Halley (1656–1742)

Comets are occasional, often dramatic interlopers among the otherwise clockwork regularity of the solar system. Early **Chinese Astronomers** recorded them as "broom stars" with long tails. Isaac Newton suspected that at least some of them orbit the Sun, though he didn't pursue proving the idea.

But the idea was soon picked up by Newton's friend and colleague Edmond Halley, an English astronomer, geophysicist, and mathematician. Halley observed a bright comet that appeared in 1682; later, using historical records and **Newton's Laws** to calculate its orbit, he proposed that it was the same comet that had been seen in 1531 and 1607. These spacings of approximately 76 years enabled him to predict that the comet would return again in 1758. It did, and while Halley did not live to see it return, astronomers named it Halley's comet (now just "Halley") in his honor.

More modern telescopic and photographic studies of the comet were made during its 1835 and 1910 passes, and the 1986 Soviet *Vega* and European *Giotto* probes flew close by Halley's central nucleus, revealing it to be surprisingly small (approximately $9 \times 5 \times 5$ miles [$15 \times 8 \times 8$ kilometers]), peanut-shaped, rugged, porous (with a density of approximately 0.6 grams per cubic centimeter), and dark as coal. Bright jets of ices composed of water, carbon monoxide, and carbon dioxide sublime from the surface and interior, releasing dust and organic molecules and creating a long tail.

Halley was the first discovered periodic comet. Nearly 500 other "short-period" (less than 200 years) comets are now known; most come from the **Kuiper Belt,** beyond Neptune, but some, including Halley, may have started as long-period comets from the distant **Oort Cloud**. Astronomers have found that more than 20 bright comets recorded between 240 BCE and 1682 CE were all actually previous apparitions of Halley's comet. Its next pass will be in the summer of 2061.

SEE ALSO Chinese Observe "Guest Star" (185), Newton's Laws of Gravity and Motion (1687), Öpik-Oort Cloud (1932), Kuiper Belt Objects (1992), "Great Comet" Hale-Bopp (1997).

LEFT: *The nucleus of Halley's comet, photographed by the European Space Agency Giotto spacecraft on March 13, 1986, from inside the coma of gas and dust surrounding the comet.* RIGHT: *Halley's comet, in a negative photo from Mauna Kea Observatory on March 5, 1986, just after its closest approach to the Sun during its latest 76-year journey.*

1684

Tethys

Giovanni Domenico Cassini (1625–1712)

Having discovered Saturn's moons **Iapetus** and **Rhea** in 1671 and 1672, the Italian-French astronomer Giovanni Domenico Cassini continued hunting for fainter moons using ever more sophisticated telescopes at the Paris Observatory. In 1684 he discovered a fourth moon of Saturn using a "tubeless," or aerial, telescope 100 feet (30 meters) long.

This fourth new moon was eventually named after the Greek Titan and sea goddess Tethys. Cassini found the moon to be in an almost perfectly circular orbit close to Saturn's equatorial plane (like all Saturn's major moons except Iapetus, which orbits at an approximate 15-degree tilt), able to circle the planet in just under two days.

Images of Tethys from the *Voyager* flybys of the Saturn system in 1980 and 1982 showed the moon to be about 670 miles (1,080 kilometers) in diameter (about 30 percent the size of Earth's Moon) and to have a highly reflective and heavily cratered surface. The high reflectivity and density of 0.97 grams per cubic centimeter are consistent with Tethys being primarily made of water ice. More detailed studies of the moon by the *Cassini*

Saturn orbiter confirmed the icy nature of the surface.

Voyager and *Cassini* images have revealed some dramatic icy landscapes on Tethys, including a huge impact basin—250 miles (400 kilometers) wide—named Odysseus and an enormous valley named Ithaca Chasma, 62 miles (100 kilometers) wide and 1.8–3 miles (3–5 kilometers) deep, which extends about 75 percent of the way around the moon. Ithaca Chasma is itself heavily cratered and ancient, and may have been formed when the crust of an early, warmer Tethys (which perhaps had once been an ocean moon, warmed by tidal heating, like **Europa** and **Ganymede**) slowly cooled, expanded, and cracked.

Curiously, astronomers using ground-based telescopes in the 1980s discovered that Tethys is the "parent" to two smaller moons, Telesto and Calypso, which share the same orbit but, like **Jupiter's Trojan Asteroids**, are, in so-called **Lagrange Points,** 60 degrees ahead of and behind Tethys, respectively.

SEE ALSO Europa (1610), Ganymede (1610), Iapetus (1671), Rhea (1672), Dione (1684), Lagrange Points (1772), Jupiter's Trojan Asteroids (1906), *Voyager* Saturn Encounters (1980, 1981).

LEFT: *A 2009 Cassini color composite of Tethys captured through infrared, green, and ultraviolet filters.* RIGHT: *Tethys photographed by the NASA Cassini orbiter spacecraft in September 2005, showing part of the enormous rift 1.8–3 miles (3–5 kilometers) deep in the icy crust called Ithaca Chasma.*

160

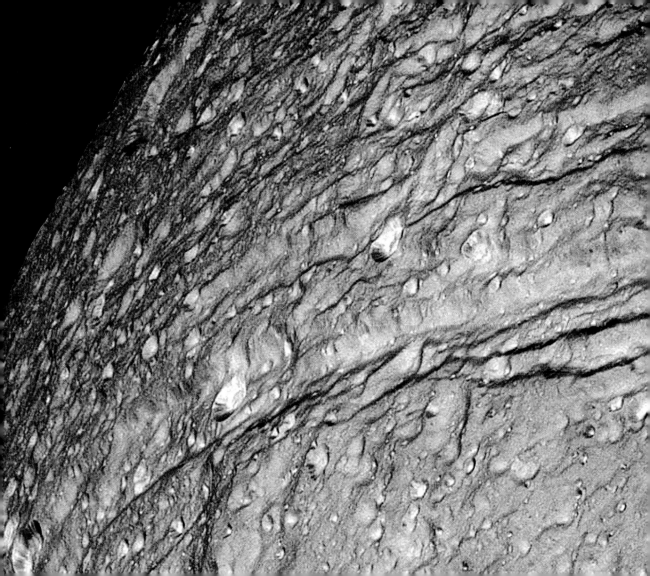

Dione

Giovanni Domenico Cassini (1625–1712), **John Herschel** (1792–1871), **William Herschel** (1738–1822)

Astronomer Giovanni Domenico Cassini had discovered Saturn's moons **Iapetus, Rhea, and then Tethys** using what were at the time some of the world's most powerful telescopes. He discovered the last of his four Saturnian moons in 1684 using the Paris Observatory's aerial (open tube) telescope—at 136 feet (41 meters), about as long a design as was possible to build at the time.

Cassini named the four moons that he discovered Sidera Lodoicea ("the stars of Louis") to honor the French king Louis XIV, who was his benefactor as well as the observatory's patron. But, Cassini's proposed names didn't sit well with fellow astronomers; instead, in 1847 a new scheme by the English astronomer John Herschel named the seven then-known moons of Saturn, including two discovered by his father, William Herschel, after Greek Titans. Cassini's fourth new moon was thus named Dione, after the mother of Aphrodite.

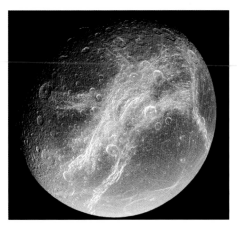

Modern exploration of Dione from the *Voyager* and *Cassini* space missions have revealed it to be an ice-covered world about 695 miles (1,120 kilometers) in diameter, similar in size to Tethys. Dione's surface is ancient and heavily cratered, but some regions have fewer craters than others. This suggests that parts of Dione may have been resurfaced early in the moon's history, perhaps by "volcanoes" of liquid water erupting from a once-warmer interior and depositing younger "ice lava" floes on top of older parts of the surface. Dione has a higher density than most of Saturn's other moons (about 1.5 grams per cubic centimeter), meaning that it has a more rocky interior; radioactive heating from that rocky material could have provided the heat to melt water in the interior and drive early water-ice volcanism.

Bright, wispy streaks seen in *Voyager* photos of Dione's trailing side (opposite the direction of orbital motion) were revealed in *Cassini* imaging to be icy cliffs hundreds of meters high and formed by some not-yet-understood tectonic process.

SEE ALSO Iapetus (1671), Rhea (1672), Tethys (1684), Mimas (1789), Enceladus (1789).

LEFT: *Bright ice cliffs make streaks on Dione's trailing hemisphere in this July 2006 Cassini photo.* RIGHT: *Icy Dione seems to float above the incredibly thin rings of Saturn in this October 2005 photo from NASA's Cassini Saturn orbiter. The shadow of the rings can be seen on Saturn in the background.*

Zodiacal Light

Giovanni Domenico Cassini (1625–1712), **Nicolas Fatio de Duillier** (1664–1753)

Seasoned and amateur sky watchers alike know that the best way to view the full splendor of the night sky is to go out on a clear, moonless night to a dark, isolated observing site far from city lights. Even then, however, the sky is not completely dark. Besides the thousands of visible stars and the diffuse glow of the **Milky Way** galaxy, another faint glow can often be observed, especially in the west after sunset or the east before sunrise. This glow is called the zodiacal light because it appears as a whitish band or faint spire from the horizon away from the Sun, roughly along a line that follows the path of the constellations of the zodiac. This line is also called the ecliptic, the plane of the Sun's equator and the orbits of most of the planets in the solar system.

Islamic astronomers referred to the zodiacal light as the "false dawn," and understanding its behavior and influence on the determination of true sunrise or sunset was an important part of determining the timing of daily prayers. Many Renaissance astronomers, including Giovanni Domenico Cassini, who observed the "luminous streak" in 1683, believed it was just an extension of the Sun's atmosphere. It remained a puzzle why it was confined to being bright only along the Sun's equatorial plane.

The first to propose the explanation that turned out to be right was the Swiss mathematician Nicolas Fatio de Duillier, who worked under Cassini at the Paris Observatory and who proposed that the glow was caused by particles reflecting the Sun's light. Fatio's hypothesis was shown to be correct by modern spectroscopy and space probes that showed that the zodiacal light is the result of sunlight reflecting off of interplanetary dust grains that measure from a few hundredths to a few hundred microns in size (the width of a human hair is about 100 microns). Because these grains slowly spiral in to the Sun because of the absorption of sunlight (an effect called Poynting-Robertson drag, discovered by early-twentieth-century physicists), there must be a continuous source to resupply them. Astronomers believe that dust from comets and occasional collisions between asteroids—most of which travel in or near the ecliptic—provide that continuous source of cosmic dust.

SEE ALSO Milky Way (c. 13.3 Billion BCE), Main Asteroid Belt (c. 4.5 Billion BCE), Halley's Comet (1682).

A December 2009 photo of the white glow of the zodiacal light seen from Cerro Paranal, Chile, near the site of the European Southern Observatory's Very Large Telescope (VLT).

Origin of Tides

Isaac Newton (1643–1727)

Coastal communities and seafaring civilizations have been tuned in to the twice daily rise and fall of the sea—the tides—throughout human history. Babylonian and Greek astronomers recognized the relationship of the height of the tides with the position of the **Moon** in its orbit, and thought they were connected through the same almost spiritual forces that governed the motions of the planets. Early Arabic astronomers thought the tides were mediated by sea temperature changes. Seeking support for a heliocentrism, Galileo proposed that tides come from the sloshing of the oceans during the Earth's motion around the Sun.

The first to propose the correct origin of the tides—linking the Earth, Moon, and Sun—was the English mathematician, physicist, and astronomer Isaac Newton. Among other things, Newton had been working on a generalized theory to explain Kepler's **Laws of Planetary Motion**, and by 1686 had developed the basic outline for new theories of universal **Gravitation and Laws of Motion**. Newton hypothesized that both the Moon and the Sun exert a strong gravitational attraction to the Earth (and vice versa). His breakthrough discovery, since verified and enhanced by space-age observations, was that gravitational forces alone (not the Earth's spin or orbital motion), acting on the Earth's thin fluid ocean "shell," are almost entirely responsible for the tides.

The Moon's gravitational attraction raises a deep water ocean tide of about 20 inches (50 centimeters); the Sun's tidal effect is about half that. Tidal heights can be up to 10 times larger in shallow water, and the tides at any particular coastal location are a strong function of the positions of the Sun and Moon as well as the local seafloor depth and coastline shape. The solid parts of the Earth and Moon bulge in response to gravitational tidal attraction as well, with an amplitude typically about half of that of the oceanic tides. Solid and liquid deformation dissipate energy in the Earth–Moon system through what is called tidal friction. As a result, the Earth's spin is slowing down by a few milliseconds per century, and the Moon is slowly receding from the Earth by about 13 feet (4 meters) per century.

SEE ALSO Three Laws of Planetary Motion (1619), Newton's Laws of Gravity and Motion (1687).

The Moon, Earth, and Sun are all connected gravitationally, a link made by Sir Isaac Newton. Each exerts strong gravitational attractive forces on the other in proportion to their masses and inversely proportional to the square of their distances. The fluid nature of Earth's oceans allows these forces to be expressed as tides.

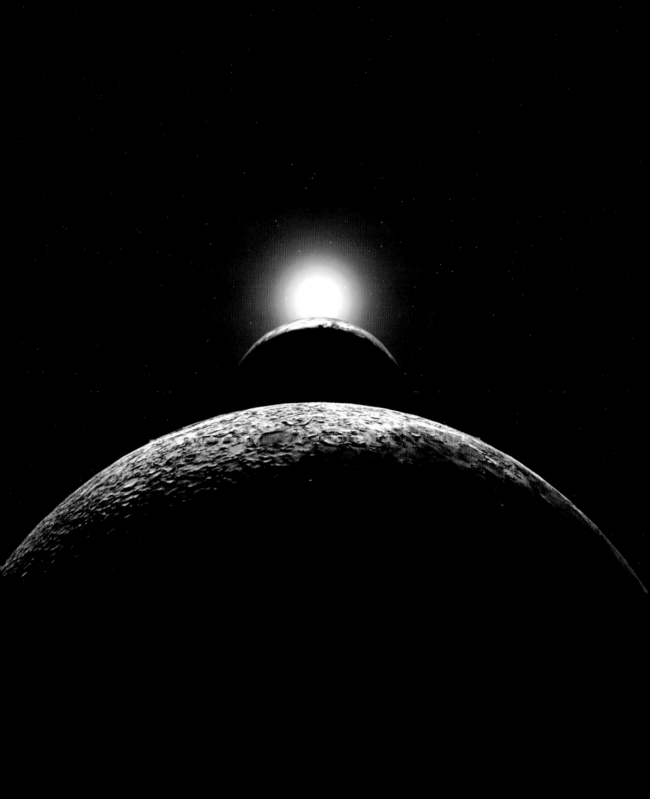

Newton's Laws of Gravity and Motion

Isaac Newton (1643–1727)

The scientific revolution begun by **Aristarchus,** who proposed removing the Earth from its cosmological position of centrality, was carried on for 2,000 years by scientific rebels such as Aryabhata, al-Bīrūnī, Nīlakantha, Copernicus, Tycho, Kepler, and Galileo. This revolution culminated—decisively—in the work of the Englishman Isaac Newton. Newton was a mathematician, physicist, astronomer, philosopher, and theologian, and is widely regarded as one of the most influential scientists in all of human history.

Newton was both an experimentalist and a theorist, and he excelled in both realms. He developed new concepts and tools in optics, including the first astronomical telescope to use mirrors instead of lenses, a design that bears his name. In the theoretical realm—using basic principles of then-modern physics and essentially inventing the new mathematical field of calculus as he went along—Newton discovered that Kepler's **Laws of Planetary Motion** were the natural consequence of a force that exists between any two masses and that decreases as the square of the distance between them (known as $1/r^2$ behavior). He called this force gravitas (Latin for "weight"). We now call it gravity, and the $1/r^2$ behavior is known as Newton's law of universal gravity.

Newton built on that foundation to derive his three famous laws of motion: (1) bodies at rest or in motion remain at rest or in motion unless acted on by an external force; (2) a body of mass (m) subjected to a force (F) will accelerate (a) at a rate according to $F = ma$; and (3) the mutual forces of action and reaction between two bodies are equal and opposite. Newton published these transformational theories in 1687 in a book called *Philosophiae Naturalis Principia Mathematica*, known and revered today simply as the *Principia*. Newton's laws of gravity and motion destroyed any remaining shreds of geocentrism and were the definitive solutions to planetary orbits for more than 200 years, until **Albert Einstein** showed them to be a subset of an even bigger theory known as general relativity. In one of the most famously quoted examples of scientific humility, Newton once wrote, "If I have seen further, it is by standing on the shoulders of giants." Indeed.

SEE ALSO Sun-Centered Cosmos (c. 280 BCE), *Aryabhatiya* (c. 500), Early Arabic Astronomy (c. 825), Early Calculus (c. 1500), Copernicus's *De Revolutionibus* (1543), Brahe's "Nova Stella" (1572), First Astronomical Telescopes (1608), Galileo's Starry Messenger (1610), Three Laws of Planetary Motion (1619), Einstein's "Miracle Year" (1905).

LEFT: *A replica of one of Newton's reflecting telescopes, built in 1672.* RIGHT: *A vintage engraving of Isaac Newton from 1856.*

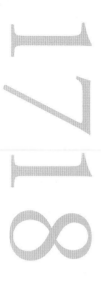
Proper Motion of Stars

Edmond Halley (1656–1742)

The British astronomer Edmond Halley is most famous, of course, for discovering the periodic return of **Halley's Comet**. But he also studied other aspects of astronomy, including comparing the positions of stars relative to their positions recorded by earlier astronomers, in order to identify the ones closest to the Sun and to potentially determine their absolute distances.

For most of the history of astronomy, the stars were assumed to be fixed denizens of a crystalline or otherwise solid celestial sphere that rotated around the Earth (to some) or that appeared to rotate as the Earth spun underneath (to others). The occasional appearance of supernovae or comets ("guest stars") cast doubt on the concept of a fixed celestial sphere but did not disprove it.

Halley provided that proof, by carefully comparing the positions of bright stars in 1718 to the positions recorded by Hipparchus in the second century BCE. Three bright stars—Sirius, Arcturus, and Aldebaran—had moved significantly over 1,850 years relative to background stars. Halley calculated what he dubbed the "proper motion" of these stars—that is, the motion belonging to the star proper, rather than perceived motion that would be due to parallax. Stars with the largest proper motion are relatively closer to the Sun; indeed, the three stars above are only about 9, 37, and 65 light-years (ly) away. Stars closer to the Sun, such as Proxima Centauri (4.3 ly) and Barnard's star (6 ly; discovered in 1916 by the American astronomer E. E. Barnard), have even larger proper motions—Barnard's star has the largest known, moving more than 10 arc seconds (0.003°) per year.

Halley and other pioneers of stellar astrometry (positional measurements) have helped us understand that the sphere of stars that we see overhead at night is just the projection of what is actually an incomprehensibly enormous three-dimensional volume of space. Everything is in motion relative to everything else, and, as **Edwin Hubble** would discover in the twentieth century, the entire volume is actually increasing over time. The universe is dynamic, if we only watch patiently enough.

SEE ALSO Stellar Magnitude (c. 150 BCE), Ptolemy's *Almagest* (c. 150), Halley's Comet (1682), Hubble's Law (1929).

Hubble Space Telescope images of the proper motion of a nearby (200 light-years away) neutron star from 1996 to 1999.

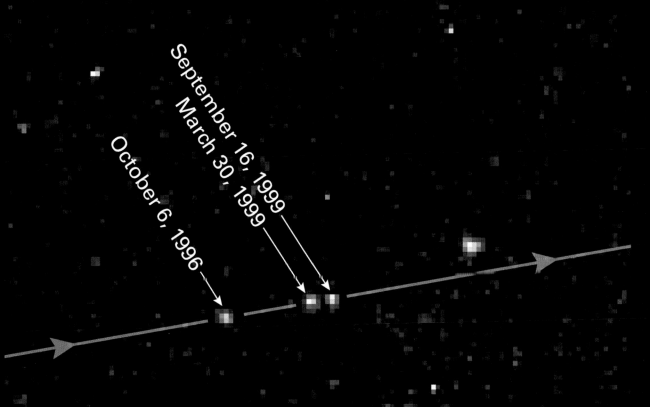

September 16, 1999
March 30, 1999
October 6, 1996

Neutron Star RX J185635-3754
Hubble Space Telescope • WFPC2

Celestial Navigation

Tycho Brahe (1546–1601), Isaac Newton (1643–1727), John Bird (1709–1776)

Prior to the invention of the first **Astronomical Telescopes**, ancient through medieval astronomers had to use naked-eye instruments such as armillary spheres, celestial globes, and astrolabes to establish the altitude and azimuth of celestial bodies (what astronomers call their declination and right ascension, respectively), as well as the relative distances between objects in the sky. The desire for higher-precision measurements meant that circular instruments like the astrolabe would have to be built in very large and cumbersome sizes. Astronomers and instrument makers thus began using only semicircular segments to decrease instrument size while maintaining precision.

Measuring instruments made from a quarter of a circle—quadrants—began appearing in the mid-sixteenth century. Tycho Brahe invented a large pedestal-mounted or framed version spanning one-sixth of a circle and called it a sextant. These devices continued to provide unsurpassed positional measurement accuracy even after the invention and proliferation of the telescope. One-eighth circle versions—octants—appeared in the eighteenth century.

Framed quadrants, sextants, and octants were fine instruments on land, but they were impractical at sea, where more precision was needed. Miniaturization, flexibility, and simplicity of operation on a moving platform were required. Isaac Newton had proposed a modification to the quadrant using two mirrors (like his reflecting telescope design), and in 1757 a portable version of the sextant was built by the English instrument maker John Bird that incorporated Newton's double-mirror design. Many modern sextants essentially mimic this eighteenth-century design, with the addition of higher-powered optics and the incorporation of composite materials. Even in our computerized and global-positioning-system era, it is heartening that many mariners today are still required to know how to perform the basics of celestial navigation using a sextant.

SEE ALSO Astronomy in China (c. 2100 BCE), Western Astrology (c. 400 BCE), Large Medieval Observatories (1260), Brahe's "Nova Stella" (1572), First Astronomical Telescopes (1608).

LEFT: A portable sextant from around 1890 used by the US Coast and Geodetic Survey and based on the original sextant design by John Bird. RIGHT: Polish astronomer Johannes Hevelius and his wife, Elisabeth, using an astronomical sextant in 1673.

Planetary Nebulae

Charles Messier (1730–1817), William Herschel (1738–1822)

The advent of telescopes in the seventeenth century, and significant improvements in their size and ability to detect faint objects in the eighteenth century, enabled exciting new discoveries by astronomers during the Enlightenment. Not just fainter stars, but entire new classes of objects became visible. The largest class comprised fuzzy, extended patches of light that resembled faint, fixed clouds among the stars. Astronomers called them nebulae (singular: nebula)—Latin for "clouds."

A number of bright nebulae had been recorded by ancient and medieval astronomers, such as the **Orion Nebula** and a nebula in **Andromeda** later recognized by twentieth-century astronomers to be a separate galaxy. Edmond Halley published a short list of nebulae in 1715. The crown prince of eighteenth-century nebular observations, however, was the French astronomer Charles Messier, who developed a definitive early catalog of more than one hundred nebulae.

One kind of nebula that Messier noted was typified by an object that he discovered

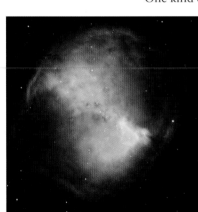

in 1764 and catalogued as object M27. With a somewhat circular, fuzzy appearance, M27 and similar objects looked a lot like the giant planets in telescopes of the day, causing the English astronomer William Herschel to dub them "planetary nebulae."

The choice of name stuck but was unfortunate, as modern spectroscopy observations show these objects to have nothing to do with planets at all. Rather, planetary nebulae have since been revealed to be enormous, expanding shells of hot, glowing gases ejected from red giant stars near the end of their stellar lifetimes. Typical planetary nebulae can be more than a light-year across, and they glow from the emission of ionized carbon, oxygen, nitrogen, and other elements heavier than hydrogen and helium created during the collapse of their precursor stars. As such, planetary nebulae are part of a cosmic recycling program that turns hydrogen and helium into heavier elements and, eventually, life.

SEE ALSO Chinese Observe "Guest Star" (185), Andromeda Sighted (c. 964), "Daytime Star" Observed (1054), Mira Variables (1596), Orion Nebula "Discovered" (1610), Messier Catalog (1771), End of the Sun (5–7 Billion).

LEFT: *The Dumbbell Nebula (Messier's M27), photographed in 1998 from the European Southern Observatory in Chile.* RIGHT: *Hubble Space Telescope photo from 1994 of the Cat's Eye Nebula (NGC 6543), a spectacular example of a multishelled planetary nebula created as a dying star sheds its outer layers.*

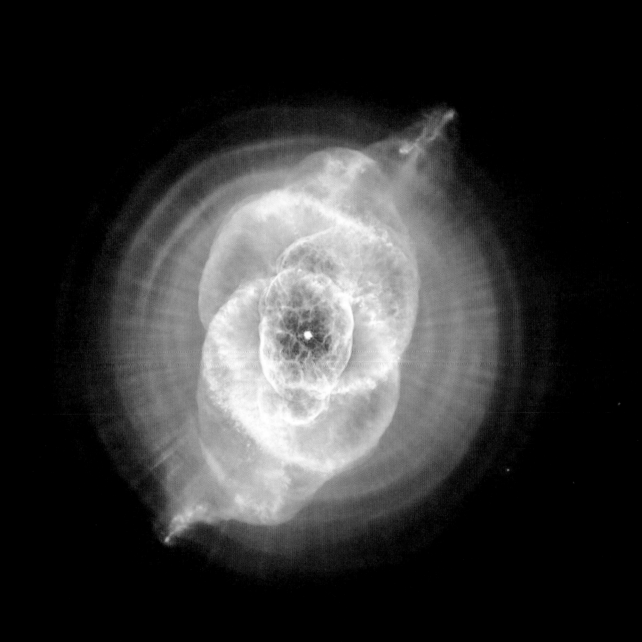

Messier Catalog

Charles Messier (1730–1817), Pierre Méchain (1744–1804)

Frenchman Charles Messier had a lifelong interest in astronomy that was stimulated by witnessing exciting celestial events when he was a child—such as the great comet of 1744 and the solar eclipse of 1748. His passion for observing led him to take a job in Paris working as a depot clerk for the official astronomer of the French navy, a job that entailed the use of an observatory on the roof of the Hôtel de Cluny.

Messier had ample time to observe, and a good dark-sky observing location in pre–Industrial Revolution Paris. His early passion appears to have been comet hunting. He was among the first astronomers to verify the predicted return of Halley's comet in 1758–1759; during the search he discovered another comet as well as a comet-like smudge of light in the constellation Taurus. Unlike a comet, however, this smudge didn't move relative to the stars. He made a note of it.

As he continued hunting comets over the next decade, he continued encountering one fuzzy, cloud-like nebula after another, giving each a designation of the letter *M* followed by a number. Some, like M22, which had been discovered in 1665, he could barely resolve as a circular cluster of huge numbers of closely packed stars (later called **Globular Clusters**); others, like M31, had an elongated shape and had been previously recognized as the great nebula in **Andromeda**. Many were objects that Messier himself was the first person to discover and describe. He had collected so many that by 1771 he had amassed a collection of 45 objects, which he published as his "Catalogue des nébuleuses et des amas d'étoiles" ("Catalog of Nebulae and Clusters of Stars") in the journal of the French Academy of Sciences. By 1781 he and his colleague Pierre Méchain, a future director of the Paris Observatory, had amassed even more; their final catalog contained 103 objects.

Twentieth-century astronomers discovered seven more objects that Messier and Méchain observed after 1781, bringing the total number of Messier objects—now recognized as star clusters, **Planetary Nebulae**, molecular clouds, and galaxies—to its modern value of 110. Some amateur astronomers and astronomy clubs try to observe all 110 objects during springtime "Messier marathons." Who knows—such events might help inspire another child's passion for astronomy.

SEE ALSO Andromeda Sighted (c. 964), Globular Clusters (1665), Planetary Nebulae (1764).

A compilation of all 110 official Messier objects, gathered from observations by the Students for the Exploration and Development of Space (SEDS) and the Paris Observatory at Meudon.

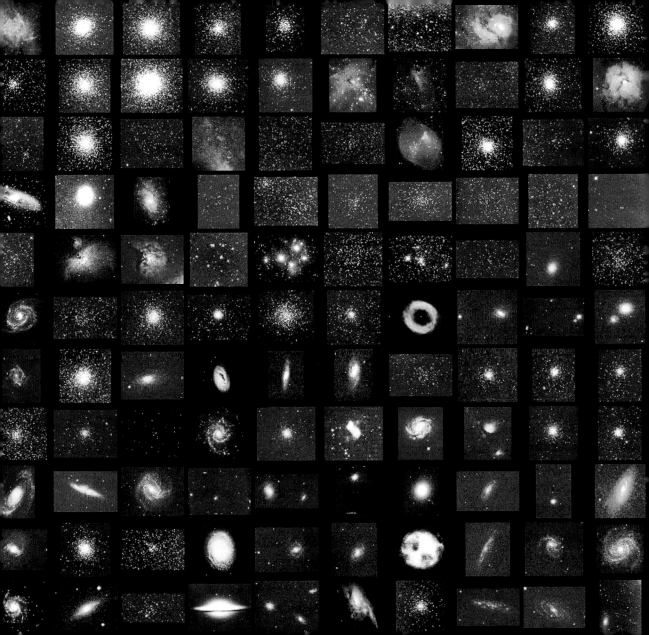

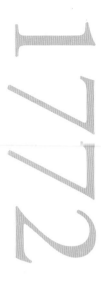

Lagrange Points

Isaac Newton (1643–1727), **Joseph-Louis Lagrange** (1736–1813), **Pierre-Simon Laplace** (1749–1827)

With his 1687 publication of the *Principia*, **Isaac Newton** had expanded Kepler's **Laws of Planetary Motion** into laws of gravitation and motion in general, thus establishing a reliable framework within which astronomers, physicists, and mathematicians could begin to understand the detailed intricacies of not only how planets, moons, and comets moved in the sky but why they did so. Among the most challenging applications of Newton's laws was the so-called three-body problem. Newton's equations could be easily applied to understand the gravitational effects that governed the motions of two masses, like the Earth and the Moon, or the Sun and the Earth, but the calculations got pretty thorny when theorists introduced the gravitational effects of a third body.

The first to successfully solve one particular case of the three-body problem was the French mathematician Joseph-Louis Lagrange, who predicted in 1772 that there should be five special places in any three-body system that has two large masses and one small mass where the gravitational forces among the bodies would be balanced. These places are now called Lagrange points. Lagrange's prediction was verified with the twentieth-century discovery of the **Trojan Asteroids**, which are "trapped" in the L4 and L5 Lagrange points of the Sun–Jupiter system.

The French mathematician Pierre-Simon Laplace coined the phrase "celestial mechanics" in 1799 to describe the physics and math of studying complex solar system motions. Lagrange's solution of the three-body problem turned out to be relatively "easy" because it assumed one small object was interacting with two large, massive objects, one of which was in an essentially circular orbit around the other. Celestial mechanicians expanded on Lagrange's discovery to eventually tackle more general three-body problems; indeed, twenty-first-century researchers often use high-speed computers to solve even more complicated N-body problems—those in which the Newtonian motions of huge numbers of individual bodies (be they planets, moon, and asteroids; ring particles; or even grains of cosmic dust) are tracked by computer in order to understand, explain, and predict very complex motions in our solar system.

SEE ALSO Ganymede (1610), Three Laws of Planetary Motion (1619), Newton's Laws of Gravity and Motion (1687), Jupiter's Trojan Asteroids (1906).

A contour map of the gravitational forces (white lines) in the Sun-Earth-Moon system and the five special Lagrange points (L1 through L5) where those forces are balanced. The image is not to scale (L2 is only about 932,000 miles [1.5 million kilometers] from Earth; the Sun is 100 times that in the other direction).

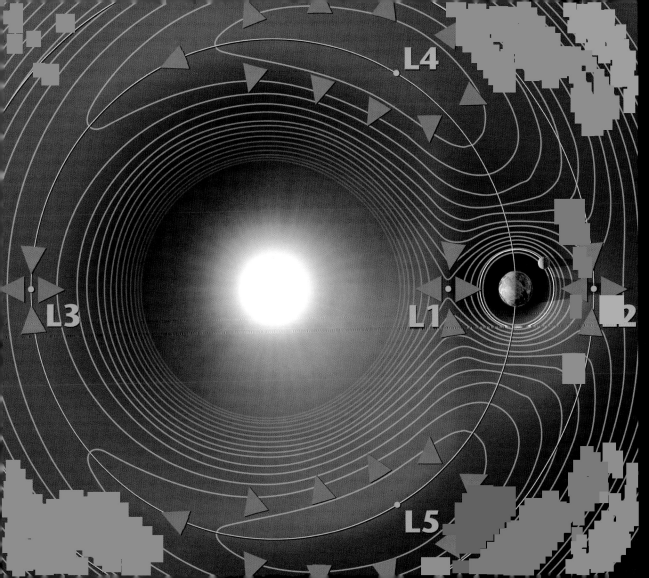

Discovery of Uranus

William Herschel (1738–1822)

Over the entire span of human history no one had ever discovered a planet before the year 1781. The five planets besides Earth that are visible to the naked eye had been known since prehistoric times. With the invention of the **Astronomical Telescope**, Galileo, Huygens, and Cassini discovered new moons, but no new planets.

The honor of being the first person to discover an entirely new planet went to the English musician and astronomer William Herschel. Herschel's musical skills led to an interest in astronomy and optics, and he began making his own telescopes after befriending the English Astronomer Royal Nevil Maskelyne.

In March of 1781 Herschel observed a 6th magnitude extended (not starlike) object—his first reaction was that he had discovered a comet. He tracked and reported its position, and other astronomers were able to verify by 1783 that it was indeed a new planet orbiting at 19 astronomical units, well beyond Saturn. Herschel wasn't the first to spot the planet, but as the first to recognize it for what it really was, he got to name it; he chose Georgium Sidus ("George's Star"), after King George III, but, like previous celestial names inspired by patronages, it didn't stick. Astronomers later named it Uranus (pronounced YUR-uh-nus), after the Greek god of the sky.

Modern telescopic observations and the 1986 **Voyager 2** space probe flyby have revealed **Uranus** to be an ice giant planet (density: 1.3 grams per cubic centimeter) 4 times the size and 15 times the mass of Earth. The atmosphere is much less active than Jupiter's and Saturn's but is still rich in hydrogen and helium; it is colored bluish-green by methane and covers an Earth-size icy, rocky core. Uranus is tilted on its side and thus seems to roll rather than spin on its axis. The planet has 27 known moons, including 5 large icy worlds (2 discovered by Herschel). In 1977 the thin, dark **Rings of Uranus** were discovered by watching starlight wink on and off as it passed by the planet. We passed through the plane of those rings in 2007; astronomers are now busily studying springtime on the seventh planet.

SEE ALSO Uranus (c. 4.5 Billion BCE), Titania (1787), Oberon (1787), Ariel (1851), Umbriel (1851), Miranda (1948), Uranian Rings Discovered (1977), *Voyager 2* at Uranus (1986).

LEFT: *Herschel's 7-foot (2.1-meter) telescope, used to discover Uranus in 1781.* RIGHT: Keck II *composite image of Uranus taken on May 28, 2007.*

Titania

William Herschel (1738–1822), **Caroline Herschel** (1750–1848), **John Herschel** (1792–1871)

After **Discovering Uranus** in 1781, English astronomer William Herschel continued to observe the planet with successively larger telescopes. Better technology and careful observations, as well as assistance from his sister Caroline Herschel, led to the discovery of new moons around both Uranus and Saturn.

Herschel discovered two moons orbiting Uranus on the same night in early 1787. By monitoring their motions and assuming (as for the then-known satellites of Jupiter and Saturn) that the moons orbit in or close to the planet's equatorial plane, Herschel and other astronomers were able to quickly deduce that the tilt of the Uranian spin axis—what astronomers call the obliquity—was far different from that of other planets. Uranus is tilted on its side, with an obliquity of about 98 degrees. A planet's obliquity is the reason for the seasons. Earth's obliquity of 23.5 degrees causes our familiar winters, springs, summers, and falls as well as the six-month-long polar days and nights in the Arctic and Antarctic. On Uranus and its moons, the entire northern and southern hemispheres experience polar days and polar nights that are 42 Earth years long! An extreme tilt leads to extreme seasons.

Herschel did not name the two moons of Uranus that he discovered. Instead, his son, John Herschel, also an accomplished astronomer, named these and the other Uranian moons much later, choosing as a theme characters from the works of William Shakespeare and Alexander Pope. He named the brighter of the two moons of Uranus Titania, after the queen of the fairies in Shakespeare's *A Midsummer Night's Dream*.

The *Voyager 2* encounter with Uranus in 1986 revealed Titania to be a heavily cratered world about 980 miles (1,577 kilometers) across (about half the size of our Moon). Having discovered that its density is about 1.7 grams per cubic centimeter, astronomers now deduce that the interior is both rocky and icy. **Spectroscopy** from recent telescope observations reveals both water and carbon dioxide ices on the surface, just as on Jupiter's icy moons **Callisto** and **Ganymede**. Perhaps the most puzzling features on Titania are the large canyons and scarps (chasma and rupes, in planetary nomenclature) that crosscut the surface. These huge cliffs may be frozen remnants of a once-liquid crust and interior that expanded and fractured dramatically when it froze.

SEE ALSO Discovery of Uranus (1781), Oberon (1787), Ariel (1851), Umbriel (1851), Miranda (1948), Birth of Spectroscopy (1814), *Voyager 2* at Uranus (1986).

Voyager 2's highest-resolution photo of Titania, acquired on January 24, 1986. The large crater on the right is called Gertrude, and the Messina Chasma, 932 miles (1,500 kilometers) long, is the bright streak at center.

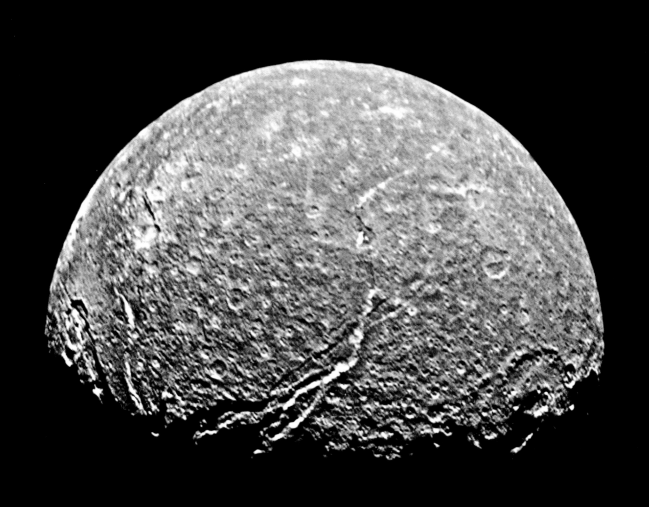

Oberon

William Herschel (1738–1822), **John Herschel** (1792–1871), **William Lassell** (1799–1880)

English astronomer William Herschel discovered two new moons of **Uranus** on the same night in early 1787. Both are very faint (about 14th magnitude, or around 1,500 times dimmer than Uranus itself) and difficult to observe close to the glare of the planet.

Herschel's fellow English astronomer William Lassell discovered two more moons of Uranus in 1851, and he decided to follow the convention of referring to his and Herschel's moons as Uranus I through Uranus IV. In 1852, however, Lassell asked Herschel's son, the astronomer John Herschel, to suggest proper names for these new, distant worlds. John named the dimmer of his father's two new moons Oberon, after the king of the fairies in Shakespeare's *A Midsummer Night's Dream*—a nice counterpoint to his father's other discovery, named after the fairy queen **Titania**.

Oberon is the most distant of the five large moons in the Uranian system, with an orbital period of about 13.5 days. Modern spectroscopy from telescopes reveals the

surface to be primarily composed of water ice, darkened and reddened most likely by simple organic molecules like those produced from the exposure of methane or other carbon-bearing ices to the Sun's high-energy radiation. Oberon is the reddest (though not very red) of any of its fellow moons.

Everything we know about the geology and interior of this small world comes from the brief *Voyager 2* fly-through of the Uranus system in 1986. Oberon is 945 miles (1,520 kilometers) across—almost the same as Titania—and has a density of about 1.6 grams per cubic centimeter, slightly lower than Titania's but still indicating a significantly rocky interior. Oberon is the most heavily cratered of all the Uranian moons—cratered almost to the point of saturation (new craters erasing old ones). Still, some deep canyons (chasma) cut the icy crust, perhaps indicating a warmer, more active ancient history.

SEE ALSO Discovery of Uranus (1781), Titania (1787), Ariel (1851), Umbriel (1851), Miranda (1948), *Voyager 2* at Uranus (1986).

LEFT: *Voyager 2's highest-resolution photo of Oberon, showing the moon's natural, slightly reddish color.* RIGHT: *Engraving by James Godby of William Herschel against the background of the constellation Gemini, where he discovered the planet Uranus in 1781.*

Fr. Rehberg del ad viv Windsor 1814.   James Godby sculp.

DR. HERSCHEL,

Member of the Royal Society of London, Imp. Academies of St. Petersburg and Vienna, National Institut at Paris,

Royal Academies of Berlin, Stockholm, &c. &c.   Born at Hanover, the 15th of November, 1738.

The background represents part of the Constellation of Gemini, with a telescopic aspect of the Georgium Sidus, as it was discovered by Dr Herschell at Bath, the 13th of March 1781, in consequence of which he was soon after introduced to the most gracious patronage of His Majesty King George III.

Dedicated to the Rt. Honble. SIR JOSEPH BANKS, Bar. Presdt. of the Royal Society,

by his very respectful and obedient Servant, Frederick Rehberg, Member of the Royal Academy of the fine Arts at Berlin, &c.

Enceladus

William Herschel (1738–1822)

Not content with aiming his larger and larger telescopes only at newly discovered Uranus, English astronomer William Herschel also set his sights on searching for new moons around Saturn. In August 1709 he was rewarded with the discovery of Saturn's sixth moon; nearly 60 years later, Herschel's son, John, named the moon Enceladus (en-CELL-a-dus), after a mythic Greek giant.

Several centuries of telescope observations of Enceladus revealed little more about the moon than its orbital period (1.4 days), its extreme brightness (it reflects nearly 100 percent of the sunlight from its surface), and its apparent association with Saturn's E ring. Spectroscopy of Enceladus showed evidence for very fine-grained water ice.

The *Voyager* **Flybys** of 1980 and 1982 and the *Cassini* **Saturn Orbiter** mission beginning in 2004 have revealed the true nature of this strange and interesting little world. It is only about 310 miles (500 kilometers) across but still exhibits some of the most varied geologic terrains in the solar system. Some of Enceladus is heavily cratered and ancient,

but other areas show almost no craters and thus must be geologically very young. Some places have deep troughs, steep ridges, or sharp grooves, suggesting that significant tectonic forces were at work.

But the most exciting discovery has been finding active plumes of water vapor coming from Enceladus's south polar region. The *Cassini* probe was flown through these plumes and showed them to contain minor amounts of nitrogen, methane, carbon dioxide, and even propane, ethane, and acetylene. This indicates that Enceladus is venting ices and organic molecules into space as though it were a giant comet rather than a planetary moon.

What is powering all this activity on (and in) Enceladus? Its density of 1.6 grams per cubic centimeter implies some rocky material, so it could be radioactive heating. Also, like **Io**, **Europa**, and **Ganymede**, it is being tidally heated because of an orbital resonance with **Dione**. *Cassini* data provide evidence that a salty subsurface liquid-water layer—an ocean—may lurk under that icy crust. Astrobiologists are taking note.

SEE ALSO Europa (1610), Io (1610), Ganymede (1610), Saturn Has Rings (1659), Dione (1684), Mimas (1789), *Voyager* Saturn Encounters (1980, 1981), *Cassini* Explores Saturn (2004).

LEFT: Cassini *image of a crescent Enceladus and its plumes and jets of water vapor.* RIGHT: NASA Cassini *orbiter photo of Enceladus, showing heavily and lightly cratered plains, troughs, ridges, and the blue-ice "tiger stripes" near the south pole, where active water-ice vents and geysers were discovered.*

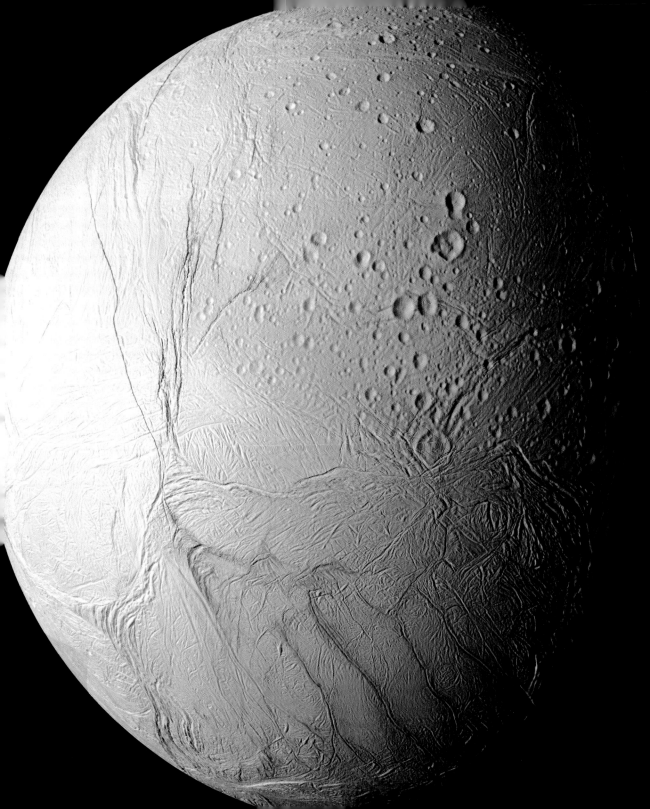

Mimas

William Herschel (1738–1822), **Caroline Herschel** (1750–1848), **John Herschel** (1792–1871)

Shortly after discovering **Enceladus**, William Herschel struck Saturnian gold again, spotting the seventh known moon of the ringed planet in September of 1789. At the time, this newest one was the closest to Saturn itself, with an orbital period of less than a day. William's son, John Herschel, would later name this moon Mimas, after a giant in Greek Titan mythology.

Herschel's ability to detect faint moons close to bright planets was made possible by his innovative telescope designs, which could collect unprecedented numbers of photons from distant sources. For example, his discoveries of both Mimas and Enceladus were made using a 40-foot (12-meter) reflecting telescope with a primary mirror 4 feet (1.2 meters) in diameter—the largest telescope in the world at the time.

Telescopic spectroscopy of Mimas showed that it has a water ice–dominated surface, like all the other Saturnian moons except for Titan. Little else was known

about Mimas until the *Voyager* and *Cassini* spacecraft encounters obtained close-up images and other data. The images revealed Mimas to be a small world, only about 250 miles (400 kilometers) in diameter (less than one-eighth the size of our Moon), with a density of just under 1.2 grams per cubic centimeter, indicating a primarily icy interior. Astronomers believe Mimas is close to the smallest possible astronomical body that can form into a spherical shape from its own self-gravity. Actually, Mimas is noticeably oblate (about 10 percent wider at the equator than at the poles) because of the strong gravitational pull of nearby Saturn.

Mimas is heavily cratered, but not uniformly so. That and the presence of canyons and fractures indicate that some areas might have been resurfaced in the distant past by tectonic forces or cryovolcanism (ice volcanoes). The most distinctive feature on Mimas is a single enormous crater, almost a third of the moon's diameter across, that gives Mimas its distinctive "Death Star" look.

SEE ALSO Saturn Has Rings (1659), Iapetus (1671), Rhea (1672), Dione (1684), Iapetus (1671), Enceladus (1789), *Voyager* Saturn Encounters (1980, 1981), *Cassini* Explores Saturn (2004).

LEFT: *Herschel's famous 40-foot (12-meter) telescope, used to discover both Mimas and Enceladus.* RIGHT: *NASA Cassini Saturn orbiter photo of Mimas taken in October 2010. The large (80-mile [130-kilometer] wide) impact crater that dominates this satellite's hemisphere is named Herschel, after the moon's discoverer.*

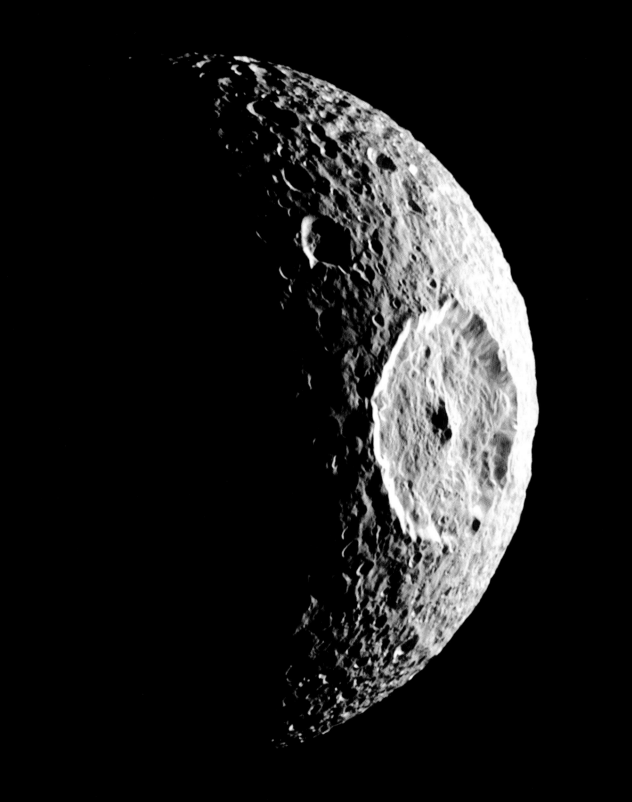

Meteorites Come from Space

Ernst Chladni (1756–1827), Jean-Baptiste Biot (1774–1862)

We take it for granted that rocks sometimes fall from the sky, but for much of human history such an idea was regarded as simply crazy. Many ancient and indigenous cultures were aware of special stones with unique magnetic properties or high concentrations of iron metal, but the fact that these were extraterrestrial samples of **Main Belt** and near-Earth asteroids was not deduced until the late eighteenth and early nineteenth centuries.

The German physicist Ernst Chladni proposed in 1794 that some of these special rocks, including a large metal-rich sample that he studied called the Pallas Iron, found in 1772 near the Russian city of Krasnoyarsk, were debris from outer space. The idea was dubbed preposterous, as many scientists believed these rocks to be volcanic or produced by lightning strikes. By the early nineteenth century it was possible to make detailed laboratory measurements of these rocks. In 1803, the French physicist and mathematician Jean-Baptiste Biot proved Chladni's hypothesis by showing that the chemical compositions of rocks found by the thousands near the town of L'Aigle shortly after a spectacular shower of shooting stars were unlike any known Earth rocks. Such rocks became known as meteorites, and the scientific field of meteoritics was born.

Scientists have now collected more than 30,000 meteorites from around the world—many from desolate deserts or Antarctic snow fields, where it is relatively easy to notice an odd rocky interloper that fell from above. The vast majority of rocks that fall to Earth (86 percent) are made of simple silicate minerals and tiny spherical grains called chondrules, thought to be some of the first materials condensed from the **Solar Nebula** and the building blocks of asteroids and eventually planets. About 8 percent more are silicates without chondrules—samples of igneous rocks from formerly geologically active crusts of large asteroids, the Moon, and Mars—and only about 5 percent are made of iron and iron-nickel (like the rocks originally studied by Chladni and Biot) and are pieces of the cores of ancient, now shattered asteroids and planetesimals that had grown large enough to differentiate into core, mantle, and crust before being destroyed by impacts in the solar system's violent early history.

SEE ALSO Violent Proto-Sun (c. 4.6 Billion BCE), Main Asteroid Belt (c. 4.5 Billion BCE), Arizona Impact (c. 50,000 BCE).

A small (14-ounce [408-gram]) ordinary chondrite meteorite that was found in 2008 on the hard pebbly surface of the Rub 'al-Khali desert near Ash-Sharqīyah, Saudi Arabia. The black surface is a thin fusion crust formed during the rock's short, fiery passage through Earth's atmosphere.

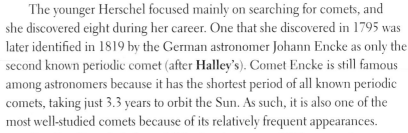

Encke's Comet

Caroline Herschel (1750–1848), Johann Encke (1791–1865)

While there have been significant advances toward establishing gender equity in the practicing of science in the past few decades, the overall history of astronomical observations and discoveries from **antiquity** to modern times has been dominated by men. Among the first female pioneers to break into this old boys' club was the English astronomer Caroline Herschel, the younger sister of William Herschel, the discoverer of Uranus.

Caroline was an accomplished vocalist and often performed in concerts with her brother in their younger years. Her interest in astronomy appears to have paralleled William's, and as he began devoting more of his time to telescope making and observations, she joined him as a constant assistant. She became proficient in astronomical calculations and developed a reputation that exceeded her brother's as a mirror polisher and telescope engineer. In 1782, at the urging of her brother, she began making her own observations.

Caroline Herschel
From an oil painting by Tielemann, 1829
In the possession of Mr. J. Ronald Blunt

The younger Herschel focused mainly on searching for comets, and she discovered eight during her career. One that she discovered in 1795 was later identified in 1819 by the German astronomer Johann Encke as only the second known periodic comet (after **Halley's**). Comet Encke is still famous among astronomers because it has the shortest period of all known periodic comets, taking just 3.3 years to orbit the Sun. As such, it is also one of the most well-studied comets because of its relatively frequent appearances.

Caroline Herschel also published an important 1798 star catalog, and codiscovered **Messier** 110, a companion to the Andromeda galaxy. She was widely recognized as the most accomplished female astronomer up until that time, and was among the first women elected to honorary membership in the Royal Astronomical Society (it would be another 80 years until they let women become full members). In a field long dominated by men, she made a lasting impression as a skilled observer and a role model for women aspiring to scientific careers.

SEE ALSO Halley's Comet (1682), Messier Catalog (1771).

LEFT: *An 1829 portrait of Caroline Herschel.* RIGHT: *Infrared image of Comet Encke and its rocky trail of debris (long diagonal band), acquired by the NASA Spitzer Space Telescope in 2005. The more horizontal band is from dust and gas emitted by jets on the comet's small rocky and icy nucleus.*

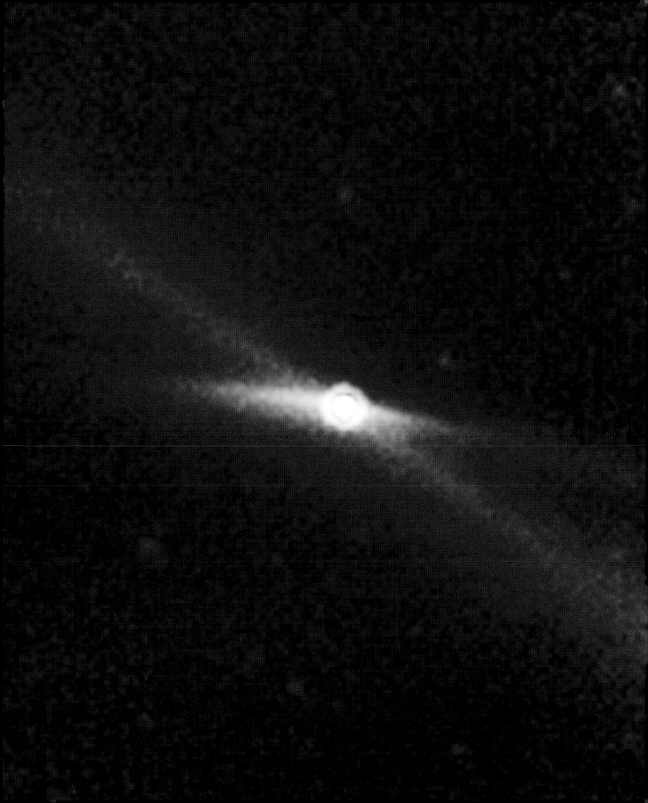

Ceres

Giuseppe Piazzi (1746–1826)

The discoveries of Uranus and a dozen new solar system moons in the late eighteenth century spurred astronomers into even more detailed cataloging and searching of the skies. Among the most meticulous and methodical of those sky mappers was the Italian priest, mathematician, and astronomer Giuseppe Piazzi of the Palermo Astronomical Observatory. Between 1789 and 1803 Father Piazzi oversaw the creation of the Palermo star catalog, a survey of nearly eight thousand stars.

As part of his catalog work, on January 1, 1801, Piazzi noticed a dim (8th magnitude) star near the head of the constellation Cetus that hadn't appeared in any previous catalogs. He observed that it moved against the fixed stars over the following nights, and thought that he might have discovered a comet. But the lack of any comet-like coma or tail led him to suspect that it might be "something better."

Indeed, follow-on observations by Piazzi and others led to the discovery late in 1801 that the starlike object was instead orbiting the Sun as though it were a planet, between Mars and Jupiter (near 2.7 astronomical units). Piazzi named it Ceres Ferdinandea, after the Roman goddess of plants and his patron, King Ferdinand III of Sicily. Astronomers weren't sure how to classify Ceres—too small and dim to be a planet but clearly not a star, comet, or moon. In 1802 the astronomer William Herschel coined the term *asteroid* (meaning "starlike") to describe Ceres and the newly discovered Pallas.

Modern telescopic observations have shown that Ceres is the largest of the **Main Belt Asteroids**, with a diameter of around 590 miles (950 kilometers) and a density around 2 grams per cubic centimeter, suggesting the possible presence of an ice-rich interior. Because Ceres is large enough to have been formed into a round shape by its own gravity, astronomers today classify it as a dwarf planet (like **Pluto**). We know little else about this world but will learn much more when the NASA *Dawn* mission encounters Ceres in 2015.

As of mid-2012, astronomers have determined the orbits of more than 500,000 asteroids, and that number is expected to grow much larger as more detailed surveys discover fainter objects. The solar system is teeming with minor planets!

SEE ALSO Main Asteroid Belt (c. 4.5 Billion BCE), Vesta (1807), Discovery of Pluto (1930), Demotion of Pluto (2006).

The dwarf planet Ceres, photographed from the Hubble Space Telescope in December 2003.

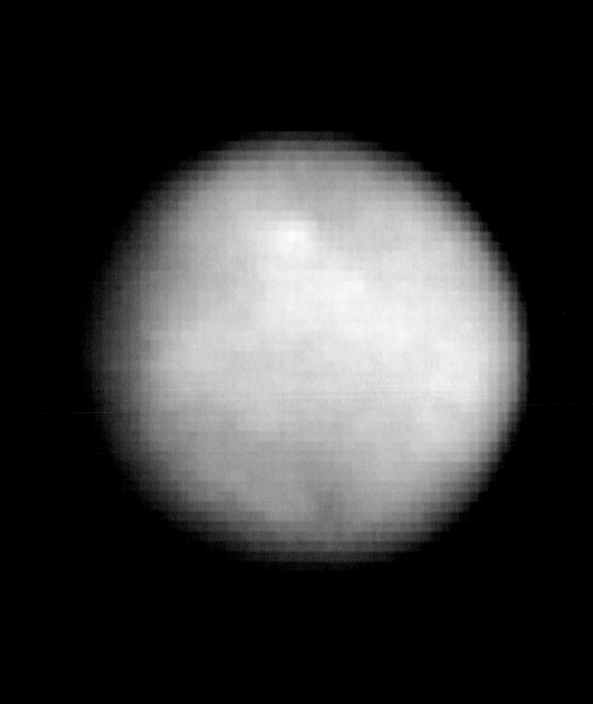

Vesta

Heinrich Wilhelm Olbers (1758–1840), Carl Friedrich Gauss (1777–1855)

The discovery of **Ceres** in 1801 was quickly followed by the discovery of a second asteroid, Pallas, in 1802, and then a third, Juno, in 1804. The German astronomer Heinrich Wilhelm Olbers was directly involved in two of those discoveries, having recovered Ceres near its predicted position in late 1801, and having discovered and named Pallas in 1802. Physician by day and astronomer by night, Olbers proposed that minor planets such as Ceres and Pallas might be the fragmented remains of a now-destroyed planet that must have existed between Mars and Jupiter.

Olbers set about searching for even more potential planetary fragments, and in 1807 he discovered the fourth known asteroid. Its orbit was calculated by the mathematician Carl Friedrich Gauss, who named it Vesta after the Roman virgin goddess of the hearth. Gauss found that, like Ceres, Pallas, and Juno, Vesta orbits in what is now known as the **Main Asteroid Belt**, ranging from about 2.1 to 3.3 astronomical units (AU) and centered around 2.7 AU. Astronomers now know that Ceres, Pallas, Juno, and Vesta comprise about 50 percent of the mass of the entire main asteroid belt; it is perhaps no surprise, then, that they were commonly referred to as new planets in early-nineteenth-century astronomy textbooks. Today we might call them dwarf planets.

Vesta is the brightest asteroid in the main asteroid belt. Ground-based observations, the Hubble Space Telescope, and NASA's **Dawn Mission** reveal it to have a diameter of about 330 miles (530 kilometers), a rocky density of 3.4 grams per cubic centimeter, and a squashed shape due to two gigantic impact basins that gouged out the asteroid's south pole. Spectroscopic data indicate that the asteroid appears to have been once molten, and differentiated into a core, mantle, and volcanic crust, much like the terrestrial planets. Meteorites from the classes known as Howardites, Eucrites, and Diogenites are likely samples of Vesta, perhaps blasted off by the giant south polar impact events.

The *Dawn* mission provided much-needed details about Vesta's geology, composition, and history during its year-long orbital mission around Vesta in 2011–2012, and has revealed the asteroid to be a rare surviving example of an ancient transitional object known as a protoplanet—part asteroid, part planet. Vesta is an important link in understanding how terrestrial planets like the Earth are formed.

SEE ALSO Main Asteroid Belt (c. 4.5 Billion BCE), Discovery of Ceres (1801), *Dawn* at Vesta (2011).

The many faces of Vesta, photographed repeatedly over the course of its fast 5.3-hour rotation by the Hubble Space Telescope in May 2007. A large crater at the south pole gives Vesta its eggy shape.

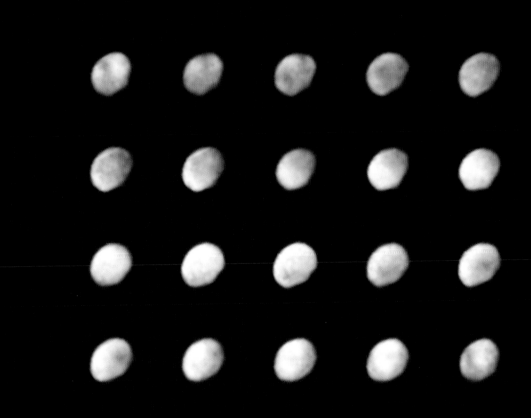

Birth of Spectroscopy

Isaac Newton (1643–1727), **William Hyde Wollaston** (1766–1828), **Joseph von Fraunhofer** (1787–1826)

In 1672, experiments by **Isaac Newton** showed that sunlight is not white or yellow but instead is composed of many colors of light that can be separated into a spectrum because they refract slightly differently when passing through an object, such as a prism. Newton's experiments were widely repeated and expanded by others, including fellow English scientist William Hyde Wollaston, who, in 1802, was the first to observe that some parts of the Sun's spectrum showed mysterious dark lines.

Scientists needed a tool, a method, for understanding these dark lines and exactly where they occurred in the solar spectrum. In 1814 the German optician Joseph von Fraunhofer developed that tool, called a spectroscope, which was a specially designed prism that could be used to measure the positions or wavelengths of the lines in an experimental technique known as spectroscopy. He observed more than five hundred dark lines in the solar spectrum with his spectroscope—astronomers still call these Fraunhofer lines. In 1821 he constructed a higher-resolution spectroscope using a diffraction grating instead of a prism, and founded stellar spectroscopy by discovering that bright stars such as Sirius also have spectral lines, and they are different from the Sun's.

By the mid-nineteenth century, physicists and astronomers were able to reproduce these kinds of lines in the laboratory by filtering light through various gases, thus discovering that the lines are caused by different kinds of atomic elements absorbing different, very narrow and specific, wavelengths of light. Spectroscopy instantly became the primary way to measure the atomic and molecular composition of distant light sources such as the Sun, planetary atmospheres, stars, or nebulae without having to touch the object directly—all that was needed was a telescope and some kind of spectral-line measuring device, or spectrometer. Indeed, spectroscopy from ground- and space-based telescopes and from orbiting and landed space missions to other worlds continues to be an important part of modern astronomy and space exploration.

SEE ALSO Newton's Laws of Gravity and Motion (1687), Speed of Light (1676), Helium (1868), Pickering's "Harvard Computers" (1901).

A series of high-resolution visible light spectra of the Sun showing Fraunhofer lines, from the McMath-Pierce Solar Facility at Kitt Peak National Observatory in Arizona. Wavelengths increase from left to right in each row, starting with violet at lower left and ending with red at upper right.

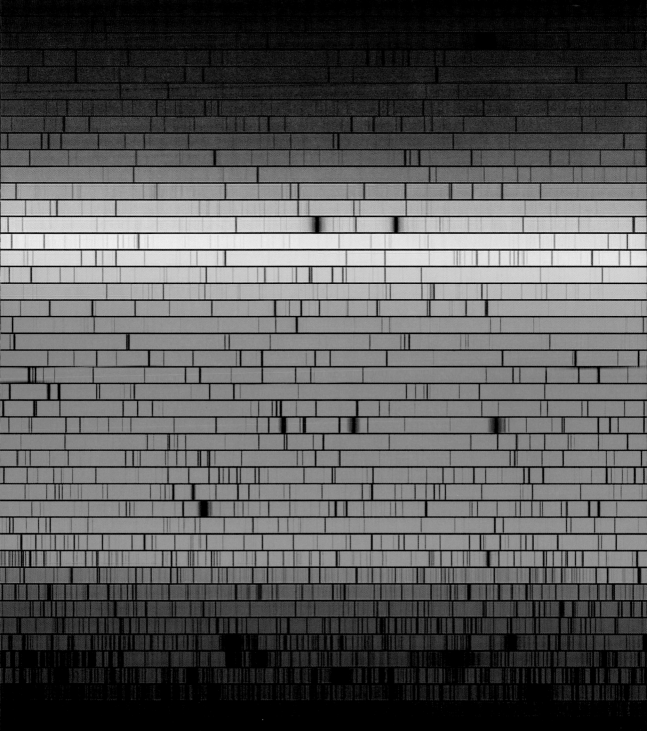

Stellar Parallax

Friedrich Wilhelm Bessel (1784–1846), **Georg Wilhelm Struve** (1793–1864), **Thomas Henderson** (1798–1844)

Parallax is the apparent motion of an object from the changing perspective of the observer. Hold a finger in front of your face and blink back and forth between your left and right eyes; your finger will appear to move because of the small change in viewing perspective—that's parallax.

It is possible to do a similar experiment using a much larger change in viewing perspective—the diameter of the Earth's orbit—to search for parallax motions among the stars and thus to determine their distances from us. Going back to Aristarchus's time and even up to Tycho's, the inability of astronomers to observe parallax for any stars was used as evidence that the Earth was motionless at the center of the universe. Once Copernicus's heliocentric view was established, the continued inability to measure stellar parallax was cited as evidence that the stars must be extremely far away. But how far?

In 1838 the race to finally measure a star's parallax was won by the German mathematician and expert stellar cartographer Friedrich Wilhelm Bessel, who used observations six months apart (when Earth was on opposite sides of the Sun) to measure a parallax of 0.314 arc seconds (0.000087 degrees) for the star 61 Cygni. Knowing that the viewing perspective had changed by 2 astronomical units (186 million miles [300 million kilometers]), Bessel estimated the distance to 61 Cygni as about 58 trillion miles (93 trillion kilometers), or nearly 10 years' distance at the speed of light.

Bessel's estimate was very close to the modern accepted value of 11.4 light-years. Similar parallax discoveries were also reported in 1838 by the astronomers Georg Wilhelm Struve and Thomas Henderson, which led to estimates of distances to the stars Vega (25 light-years) and Alpha Centauri (4.3 light-years). Alpha Centauri also has one of the largest observed **Proper Motions** in the sky, and is now known to be the star system closest to the sun—at nearly 272,000 times the Earth's average orbital distance. Even the closest stars, it turns out, are immensely far away.

SEE ALSO Sun-Centered Cosmos (c. 280 BCE), Mizar-Alcor Sextuple System (1650), Proper Motion of Stars (1718), White Dwarfs (1862).

LEFT: *The astronomer Friedrich Bessel is depicted in this 1898 engraving.* RIGHT: *A 3-D map of the 50 star systems within 16 light-years of the Sun (center). Each grid box is 1 light-year. 61 Cygni is the yellowish star just above the Sun. The Alpha Centauri system (closest) is just below and to the left of the Sun.*

off

200

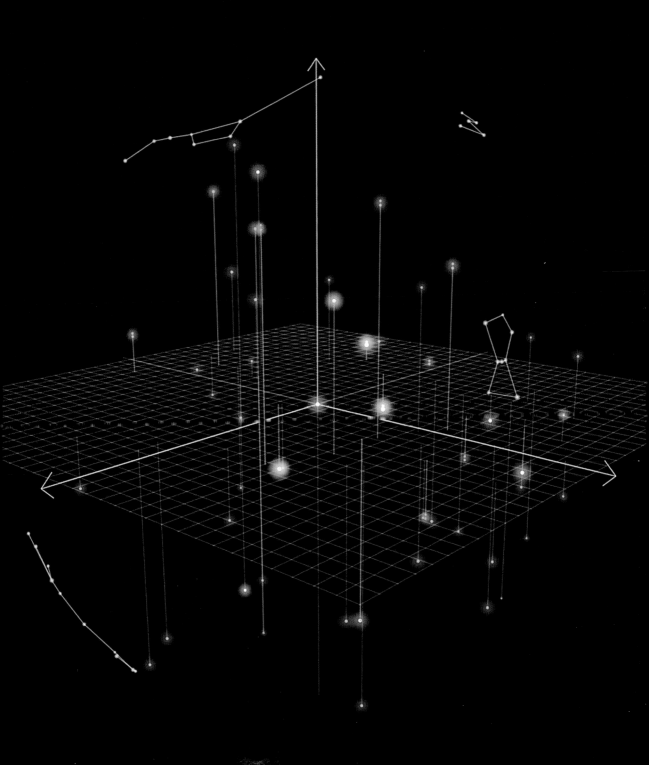

First Astrophotographs

John William Draper (1811–1882), **Henry Draper** (1837–1882)

Most successful pre-nineteenth-century astronomers were also talented artists by necessity, because the only way for them to record their observations was to sketch, draw, or paint what they saw with their eyes. With the invention of photography in 1839 however, the situation quickly changed. Over time, new and improved methods of taking photographs would forever revolutionize astronomy.

Early photographic methods were cumbersome, crude, and dangerous. The daguerreotype process, invented by the French inventors and artists Louis Daguerre and Joseph Niépce, could create relatively sharp images on wet copper plates coated with silver, but the process required photographers to work closely with toxic vapors of mercury, iodine, or bromine. Still, the results were impressive, and the French government made the process publicly available to the world for free very soon after its invention.

The American doctor, chemist, and photographer John William Draper was quick to develop some improvements on Daguerre's process. His scientific interests compelled him to point his equipment skyward, and he took a series of increasingly higher quality daguerreotype plates of the Moon from 1839 through 1843. Draper's lunar plates were the first recorded photos of a recognizable astronomical object; thus he is hailed as the inventor of astrophotography.

The invention of dry plate technology in the 1870s helped astrophotography become an important scientific research tool for the recording of images as well as spectra. John Draper's son Henry Draper continued his father's legacy in astrophotography by mounting his cameras to large telescopes and recording the first photographic spectrum of the star Vega in 1872, as well as acquiring the first photograph of the **Orion Nebula** in 1880.

Film eventually replaced plates, but analog photo technology reached its limit of sensitivity in the mid-twentieth century. Electronic imaging technology (mainly the Charge-Coupled Device) was developed in the 1970s and has now taken over as the photographic medium of choice for both scientific and consumer uses.

SEE ALSO Orion Nebula "Discovered" (1610), Birth of Spectroscopy (1814), Astronomy Goes Digital (1969).

One of the first known astronomical photographs of any kind, from a daguerreotype plate of the nearly full Moon taken by John William Draper in the winter of 1839–1840 from New York City. Draper focused the Moon's image onto the silvered copper plate using a 3-inch (7.6-centimeter) lens, and had to keep the image stable on the same part of the plate for nearly 20 minutes.

Discovery of Neptune

Urbain LeVerrier (1811–1877), **Johann Galle** (1812–1910), **John Couch Adams** (1819–1892)

During the decades following its discovery in 1781, astronomers had carefully tracked the position and refined the orbit of the slow-moving planet Uranus. Several noted slight differences between what **Newton's Law of Gravity** predicted and its actual path across the sky. Two particularly talented theoreticians, the English astronomer John Couch Adams and the French mathematician Urbain LeVerrier, thought that these differences might be caused by the gravitational pull of another, yet unseen planet.

Working independently during 1845 and 1846, LeVerrier and Adams each developed predictions about where this hypothesized "perturber" of Uranus might be found in the sky. Adams convinced colleagues at Cambridge to search for the planet, but it was not recognized as more than a star by the observers, who had a lot of sky to cover.

In contrast, LeVerrier's prediction was over a narrower region of the sky, and it took only a few hours on the night of September 24–25, 1846, for his colleague, the German astronomer Johann Galle, to find it from Berlin Observatory and (over several more nights) confirm its identity as the eighth planet. The discovery was hailed as a triumph of Newtonian physics, and the French mathematician and politician François Arago proclaimed that LeVerrier had discovered a planet "with the point of his pen." LeVerrier and Galle were jointly credited (by a gracious Adams, among others) with the discovery of the new planet, and LeVerrier chose to name it **Neptune**, after the Roman god of the sea. Its discovery doubled the size of the solar system yet again.

With an average orbital distance of 30 astronomical units and an orbital period of nearly 165 years, Neptune's small apparent diameter and very slow motion explain why several earlier astronomers, including Galileo, had observed Neptune but, like Adams's colleagues at Cambridge, had not recognized it as more than just another bluish star. Neptune has been studied up close by just one space mission, the *Voyager 2* flyby in August 1989, providing scientists with a fleeting glimpse of a beautiful and stormy blue world in the far outer reaches of our solar system.

SEE ALSO Neptune (c. 4.5 Billion BCE), Newton's Laws of Gravity and Motion (1687), Discovery of Uranus (1781), Triton (1846), *Voyager 2* at Neptune (1989).

A nearly "full Neptune" photographed by the Voyager 2 *space probe as it approached the planet. The Great Dark Spot seen here near the center of the planet's disk is thought to be a large swirling storm system similar to Jupiter's Great Red Spot.*

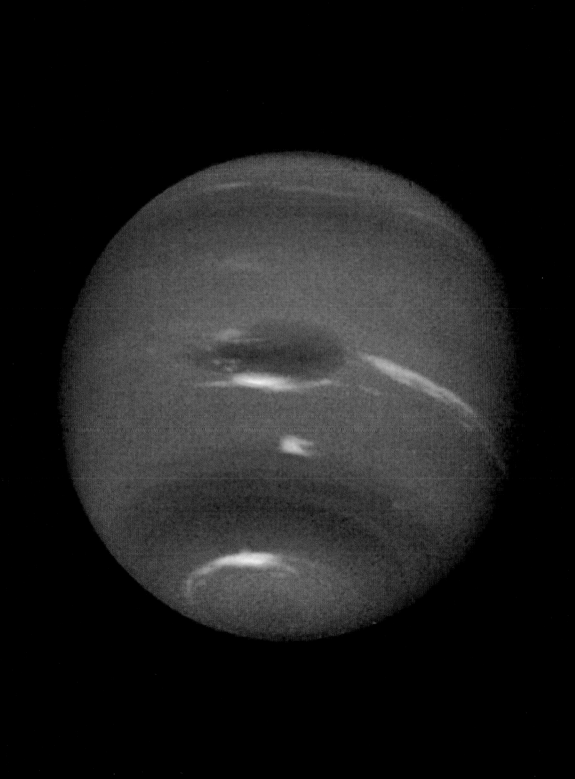

Triton

John Herschel (1792–1871), **William Lassell** (1799–1880)

The **Discovery of Neptune** in the autumn of 1846 provided a new target for moon hunters—a challenge because of Neptune's great distance from the Sun. It was the British merchant and amateur-turned-professional astronomer William Lassell who discovered Neptune's moon just 17 days after Neptune itself was discovered. Lassell had made his money as a beer brewer, and used it to build his own telescope mounts, grinding his own mirrors and assembling a Newtonian reflector 24 inches (61 centimeters) in diameter that was, at the time, the largest functioning telescope in the world. When John Herschel heard about the discovery of Neptune, he suggested to Lassell that he use his telescope to search for moons around the new planet.

It didn't take long—only eight days of searching—for Lassell to find a moon, and it was a strange one indeed. Tracking its orbit revealed that it was orbiting backward— retrograde—compared to all of the other known moons of the solar system. Further, its orbit is highly tilted to the plane of Neptune's own orbit, so much so that each of its poles is sometimes pointed almost right at the Sun, like highly tilted Uranus. Lassell didn't name his discovery; astronomers later agreed on Triton, the Greek sea god and son of Poseidon (the Greek equivalent of Neptune).

Little else was known about distant Triton before the **Voyager 2** encounter in 1989, which revealed it to be large (1,678 miles [2,700 kilometers] in diameter), bright (70–80 percent reflective), and with an icy-rocky density of 2.1 grams per cubic centimeter. Most surprising, a very thin nitrogen atmosphere and gently erupting geysers were found, along with a geologically young surface (few craters) that appears to be continually resurfaced through icy cryovolcanism.

Triton's strange landforms have more recently been found to be composed of nitrogen, water, and carbon dioxide ices only 30 to 40 degrees above absolute zero at the surface but perhaps warmed by radioactive heat in the interior. Triton appears to be a twin of **Pluto**, and many astronomers believe that it may have formed in the **Kuiper Belt** and was captured—somehow—by Neptune into its tilted, backward orbit.

SEE ALSO Pluto and the Kuiper Belt (c. 4.5 Billion BCE), Mimas (1789), Discovery of Neptune (1846), Discovery of Pluto (1930), *Voyager 2* at Neptune (1989).

Voyager 2 *false-color image of Neptune's large moon Triton, photographed on August 25, 1989. This view shows Triton's bright southern hemisphere polar cap of nitrogen and methane ices, crisscrossed with ridges and fractures and dotted with streaks from active geyser-like eruptions of nitrogen gas.*

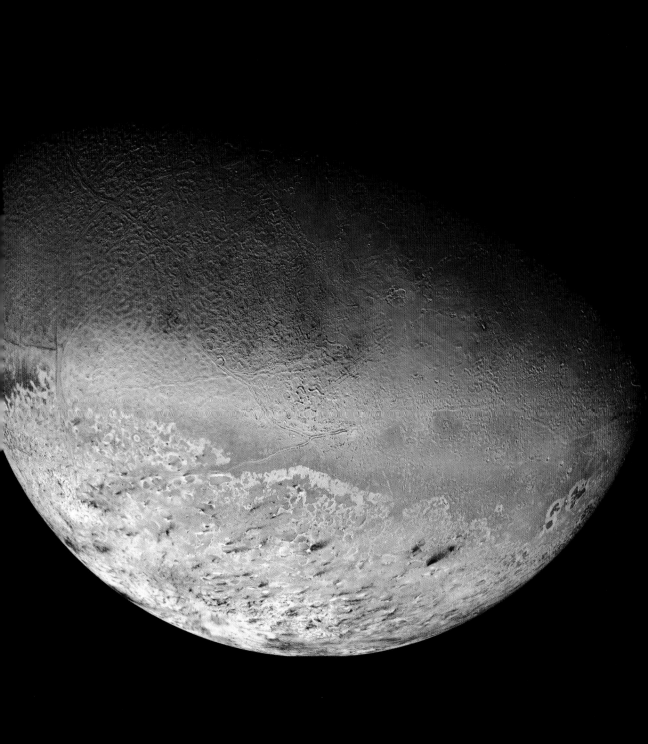

Miss Mitchell's Comet

Maria Mitchell (1818–1889)

Even though a pioneering path for women interested in science and astronomy had been established around the turn of the eighteenth century by the English astronomer Caroline Herschel, extremely few women were able to follow that path successfully for some time. Indeed, it would be more than another half century until the first female scientist made a professional academic career in astronomy.

That woman was Maria Mitchell, a native of Nantucket and a skilled observational astronomer and educator. Mitchell helped teach at a school that her father founded on the island and eventually became a librarian—soaking up the scientific and literary works of the time while continuing to enhance her experience observing the night sky. In 1847, using her father's small observatory, she discovered a faint comet only visible via telescope. The comet was informally dubbed "Miss Mitchell's comet."

Mitchell was widely recognized for her astronomical work. She traveled to Europe to meet other astronomers and receive awards, and was eventually offered a job computing the positions of Venus for a nautical almanac publisher. She became the first female member of the American Academy of Arts and Sciences and the first woman elected to the American Association for the Advancement of Science.

As much as she must have loved astronomical research, she apparently never lost the passion for teaching that she had developed on Nantucket. When the wealthy New York businessman Matthew Vassar offered her a teaching position in 1865 at the new women's university he founded, she took the offer and became not only Vassar College's first faculty member but the world's first female professor of astronomy. She became the director of the Vassar College Observatory and used the facility as a teaching tool for astronomy students, while at the same time conducting her own research on sunspots and the changing appearances of Jupiter, Saturn, and their moons. She taught on the Vassar faculty for 23 years, and trained scores of women who went on to pursue scientific careers. After her death, the Maria Mitchell Society was formed to preserve her legacy, and in 1908 the society opened the Maria Mitchell Observatory on Nantucket, a facility still dedicated to astronomical teaching and research.

SEE ALSO Encke's Comet (1795).

Nantucket Historical Association photograph of comet hunter Maria Mitchell from around 1865, about the same time she became the world's first female professor of astronomy.

Doppler Shift of Light

Christian Doppler (1803–1853), **Armand Hippolyte Fizeau** (1819–1896), **Vesto Slipher** (1875–1969), **Edwin Hubble** (1889–1953)

Most of us are familiar with the dramatic change in the sound of an ambulance or a train whistle or a race car as it approaches us and then speeds past. As the vehicle recedes, its sound changes to a distinctly lower pitch or frequency than when it was approaching. This change is known as the Doppler effect, named after the Austrian physicist Christian Doppler, who first proposed in 1842 the idea that the observed frequency of any kind of wave should depend on the relative difference in speeds between the wave's source and the observer.

In 1845 Doppler's hypothesis was verified for sound waves in some clever experiments by the Dutch meteorologist C. H. D. Buys Ballot, who hired musicians to play notes on a moving train and then had stationary observers report the pitches that they heard as the musicians approached and receded. In 1848, the French physicist Armand Hippolyte Fizeau showed how Doppler's hypothesis applied to light waves by noting slight changes in frequency or shifts of absorption lines in the spectra of stars.

Astronomers call these frequency changes Doppler shifts, and the size and direction of the frequency change can be used to determine the speed at which astronomical bodies are approaching or receding from each other. Objects approaching us have their spectra shifted to higher frequencies, or shorter (bluer) wavelengths; conversely, objects receding from us have spectra that are red-shifted. In the 1860s the first accurate relative stellar velocities were measured, and in the 1870s it became possible to detect the Doppler shift of the stars caused by the Earth's annual motion around the Sun. In the early twentieth century, the American astronomer Vesto Slipher made observations showing that most of the known nebulae (such as those in **Messier**'s list) were red-shifted—or moving away from us. Soon after, **Edwin Hubble**, another American, showed that many of these nebulae were actually other galaxies, enormously far from the Milky Way. Hubble's work led directly to the concept of an expanding universe and the **Big Bang** theory.

SEE ALSO Big Bang (c. 13.7 Billion BCE), Messier Catalog (1771), Birth of Spectroscopy (1814), Hubble's Law (1929).

A graphical representation of the Doppler effect: waves are being emitted by a source moving here from right to left. To the observer, waves in front of the source are compressed to a higher frequency (shorter wavelength, or bluer); waves behind the source have a longer wavelength (redder).

Hyperion

William Bond (1789–1859), **William Lassell** (1799–1880), **George Bond** (1825–1865)

No new moons around Jupiter, Saturn, or Uranus were discovered for more than a half century after the 1789 discovery of Mimas by William Herschel, but the eighth known satellite of Saturn was independently discovered by two groups within days of each other in late 1848. The American father-and-son astronomer team of William Bond and George Bond apparently observed the new moon first, but the English astronomer William Lassell was the first to announce his observations. All three astronomers are credited with the discovery. Just one year earlier, John Herschel had published a compendium of new observations in which he had proposed a Greek Titans naming scheme for the moons of Saturn; Lassell had endorsed the scheme and suggested that the newest moon be named Hyperion, after the Greek Titan and brother of Cronus, the Greek equivalent of Saturn.

Little was known about Hyperion, except that it traveled in a relatively eccentric orbit at an average distance of about 25 times the radius of Saturn (well outside the planet's ring system), until the *Voyager* flybys in the 1980s and then the **Cassini** **Saturn Orbiter** mission's encounters with the moon in the early 2000s. Hyperion turns out to have been the first, and largest, irregular (nonspherical) moon discovered by telescope. Space mission images show it to be an elongated body with dimensions around 204 x 162 x 133 miles (328 x 260 x 214 kilometers), and more recent spectroscopy from telescopes shows it to have a surface composed of dark, reddish, "dirty" water ice, perhaps similar to that on the dark side of **Iapetus**. Hyperion has an extremely weird, spongy surface, characterized by tons of deep impact craters and sharp-edged pits and cliffs. One whole side of the moon is scarred by a giant impact crater 75 miles (120 kilometers) wide and 6 miles (10 kilometers) deep, suggesting that Hyperion is only a fragment of a larger satellite broken apart by a giant impact. Hyperion's density is only about 0.56 grams per cubic centimeter, suggesting that it is made mostly of very porous water ice. Indeed, astronomers think that a remarkable 40–50 percent of Hyperion's interior is empty space between icy fragments.

SEE ALSO Iapetus (1671), Mimas (1789), Phoebe (1898–1899), *Cassini* Explores Saturn (2004).

NASA Cassini Saturn orbiter enhanced color photo of the irregular, almost spongy-looking moon Hyperion, taken in September 2005. The long axis of the satellite is about 205 miles (330 kilometers) across; the impact crater Helios (75 miles [120 kilometers] wide) dominates most of this hemisphere of the satellite.

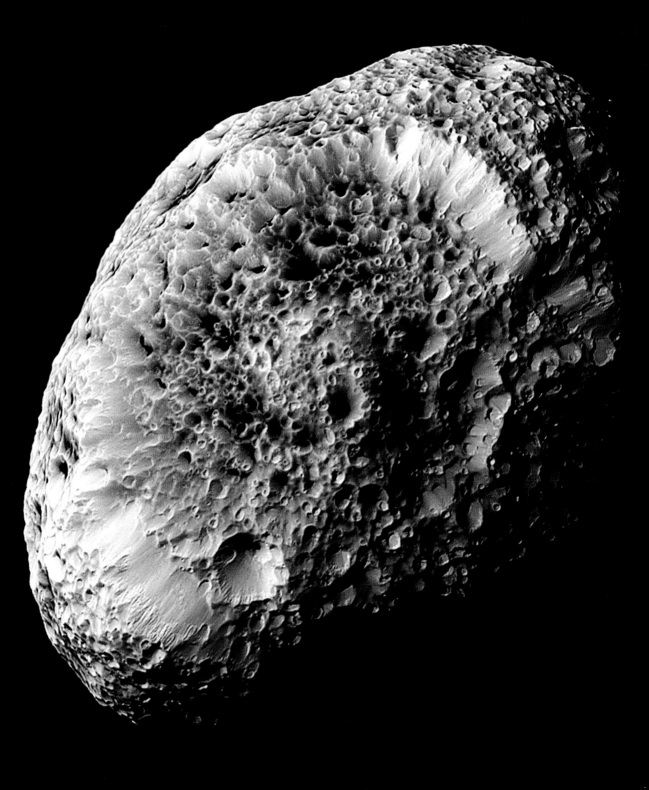

Foucault's Pendulum

Jean Bernard Léon Foucault (1819–1868)

Our space-age perspective on our planet makes it now a matter of common knowledge that the Earth spins. But imagine for a moment going back in time to an era when there were no satellites or space probes or fancy computerized planetarium programs, and try to convince someone that the Earth is actually spinning. It's not intuitive—the Sun and sky appear to move, not the Earth! If the Earth were spinning as fast as it would need to be spinning to rotate once per day (about 1,000 miles per hour at the equator), wouldn't we all get flung off into outer space? Even today, it's hard to prove to someone that the Earth spins. What is needed is a simple and repeatable experiment that provides a physical demonstration of the Earth's rotation.

Although a number of such experiments have been proposed and conducted, by far the most famous is the one first performed in 1851 by the French physicist Jean Bernard Léon Foucault. Foucault (pronounced foo-KOH), like any good physicist, understood **Newton's Laws** and exploited the first law (bodies at rest or in motion remain at rest or in motion unless acted on by an external force) for his experiment. He constructed a long, heavy, stable pendulum using a ball, or bob, made from lead-coated brass that was suspended on a wire 220 feet (67 meters) long from the ceiling to the floor of the Panthéon in Paris. Foucault knew that in the absence of any other forces, once he started the pendulum swinging, it would continue to swing in that same plane—that is, it would remain in the same inertial reference frame relative to the "fixed" stars, rather than to the Earth. By setting up hour markers like those on a sundial (or small obstacles for the bob to knock over), and by compensating for friction in the wire or from the bob's motion through the air, it became easy to demonstrate that the room (indeed, the whole Earth) was slowly rotating relative to the plane of the pendulum's swing. The following year Foucault perfected a gyroscope based on similar principles.

Foucault's pendulum became a nineteenth-century sensation because of its simplicity; hundreds can still be found around the world in universities, museums, and science centers.

SEE ALSO Newton's Laws of Gravity and Motion (1687).

A large Foucault pendulum from the Príncipe Felipe Science Museum of the City of Arts and Sciences in Valencia, Spain. In this setup, a little ball and stick model is tipped over roughly every 30 minutes because the Earth is spinning underneath the inertially fixed plane of the pendulum's swing.

Ariel

William Lassell (1799–1880)

The British industrialist and amateur astronomer William Lassell had gained significant notoriety and fame with his discovery of Neptune's large moon **Triton** in 1846 and his codiscovery of Saturn's small moon **Hyperion** in 1848. These discoveries, made from a relatively modest (and often cloudy) observatory location named Starfield near his home in Liverpool, England, were made possible by his development of new telescope mounting methods (he perfected the equatorial mount design still in use today for many large telescopes) and new mirror fabrication and polishing techniques.

Lassell's countryman, William Herschel, had discovered Uranus's moons **Titania** and **Oberon** in 1787 using a lens 18.5 inches (47 centimeters) in diameter. But when Lassell trained his 24-inch (61-centimeter) Newtonian reflector—still the largest routinely functioning telescope in the world at that time—on the Uranian system in 1851, it was with an expectation that even dimmer moons, perhaps even closer to Uranus, might be found. His patience paid off, and he discovered two new moons around the seventh planet. In 1852 his colleague John Herschel named the innermost of these Ariel, after fictional characters who appeared in works by both Pope and Shakespeare.

The *Voyager 2* space probe's flyby through the Uranus system in January 1986 enabled high-resolution images to be taken of Ariel's sunlit southern hemisphere (the moons, like Uranus itself, were oriented in a sort of pole-on, bull's-eye pattern pointed at the Sun at the time of the *Voyager* flyby). The data revealed Ariel to be about 721 miles (1,160 kilometers) across (about one-third the size of the Earth's Moon) and to have a density of about 1.7 grams per cubic centimeter, suggesting a mixed icy and rocky interior. Long networks of complex extensional ridges and wide, flat-floored canyons point to significant past geologic activity on Ariel, perhaps driven by tidal interactions with Uranus or its other large moons.

More recent ground-based telescope observations have been able to identify both water and carbon dioxide ice on Ariel's surface. The carbon dioxide appears to be concentrated on the hemisphere that receives the most intense bombardment from Uranus's magnetic field, suggesting that the interaction between a magnetic field and an icy surface can yield interesting chemistry in the outer solar system.

SEE ALSO Titania (1787), Oberon (1787), Triton (1846), Hyperion (1848), Umbriel (1851).

Voyager 2 colorized mosaic of the closest-approach photographs of Ariel, taken on January 24, 1986. Complex ridges and canyons indicate a once-active geologic surface on this small, icy world.

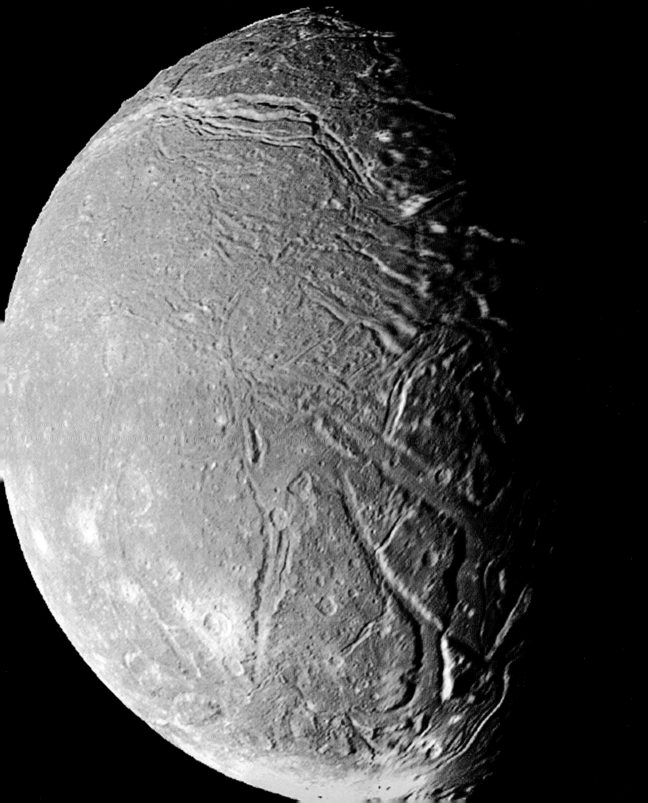

Umbriel

William Lassell (1799–1880)

The English astronomer William Lassell made two of his four solar system moon discoveries on the night of October 24, 1851, spotting two satellites orbiting closer in to Uranus than either of the two already known, Titania and Oberon. Adopting the theme of characters from classical British fiction, one of the new moons was named **Ariel** by Lassell's colleague John Herschel, and the second was named Umbriel, after the "dusky melancholy sprite" in Alexander Pope's *The Rape of the Lock*.

Lassell was not trained as an academic astronomer; rather, he made his mark in business (as a brewer) and used some of his fortune to indulge his passion for telescopes and observational astronomy. He appears to have had significant skills as a designer and mechanical engineer as well. Not content with simply building larger mirrors for his Newtonian reflector telescopes, he made significant advances in perfecting the "speculum metal" alloys that went into the mirror blocks themselves—usually combinations of arsenic, copper, and tin. In addition, he developed new methods and

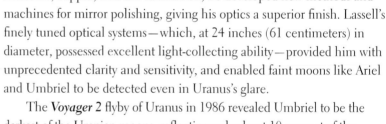

machines for mirror polishing, giving his optics a superior finish. Lassell's finely tuned optical systems—which, at 24 inches (61 centimeters) in diameter, possessed excellent light-collecting ability—provided him with unprecedented clarity and sensitivity, and enabled faint moons like Ariel and Umbriel to be detected even in Uranus's glare.

The *Voyager 2* flyby of Uranus in 1986 revealed Umbriel to be the darkest of the Uranian moons, reflecting only about 10 percent of the incident sunlight (compared to about 25 percent for Ariel, for example) except for some brighter, possibly fresher patches of material in a few craters. *Voyager* found only a heavily cratered surface on Umbriel, with no good examples of more complex geology. Umbriel is about the same size as Ariel, so it's not obvious why it lacks ridges and canyons. Perhaps the moon's lower density (1.4 grams per cubic centimeter) and less rocky interior is the key. Only about 20 percent of Umbriel was imaged at high resolution, however. Perhaps some future space mission to Uranus will complete the geologic reconnaissance of these worlds.

SEE ALSO Titania (1787), Oberon (1787), Triton (1846), Hyperion (1848), Ariel (1851), *Voyager 2* at Uranus (1986).

LEFT: *An image of Umbriel taken during the 1986 Voyager 2 flyby through the Uranian system.* RIGHT: *An undated portrait of English astronomer William Lassell.*

Kirkwood Gaps

Daniel Kirkwood (1814–1895)

The discoveries of the asteroids **Ceres**, Pallas, Juno, and **Vesta** in the early nineteenth century opened up a new area of solar system study—so-called minor planet research. For more than 33 years after the discovery of Vesta in 1807, however, no new asteroids were discovered. Improvements in telescope sensitivity by the middle of the century led to a sort of mini explosion in the size of the known population, though, and by 1857 there were 50 known asteroids, all orbiting in what would be known as the **Main Asteroid Belt**, between Mars and Jupiter.

This asteroid population drew the attention of the American mathematician Daniel Kirkwood. Kirkwood had gained some fame in 1846 for proposing a sort of Kepler's law for planetary spin rates versus distance from the Sun that initially seemed to be correct. Although the idea was eventually rejected once better data became available, he continued to pursue other solar system dynamical studies. When he examined the orbital properties of the 50 or so asteroids known as of 1857 he noticed a remarkable thing: instead of a uniform or random or even bell-shaped distribution of asteroid distances from the Sun, they appeared to bunch up into clusters, with many orbiting at certain distances and none orbiting at other distances.

Kirkwood found that the places in the main belt with no asteroids are places where an object orbits a number of times that can be expressed as a simple integer relative to Jupiter's orbit. For example, no asteroids were found near 2.25 astronomical units, where an object would orbit the Sun exactly three times for every single Jupiter orbit. He correctly proposed that the resonance between Jupiter and any objects in such places would give the objects a little extra gravitational tug from Jupiter that would cause them to eventually be nudged away from that area, or even out of the main belt entirely.

More than 150 years of data have verified the prediction, and astronomers call these "gaps" in the main belt the Kirkwood gaps, in honor of their discoverer. The gap positions are also known as mean-motion resonances. Other such gaps have been found at the 5:2, 7:3, and other mean-motion resonances, as well as in the **Rings of Saturn**, where they form by mean-motion resonances among embedded moons.

SEE ALSO Main Asteroid Belt (c. 4.5 Billion BCE), Ganymede (1610), Saturn Has Rings (1659), Ceres (1801), Vesta (1807).

A histogram of the number of known asteroids (vertical axis; peak is about 15,000) versus their average distance from the Sun in astronomical units, or AU (1 AU = average distance between the Earth and the Sun).

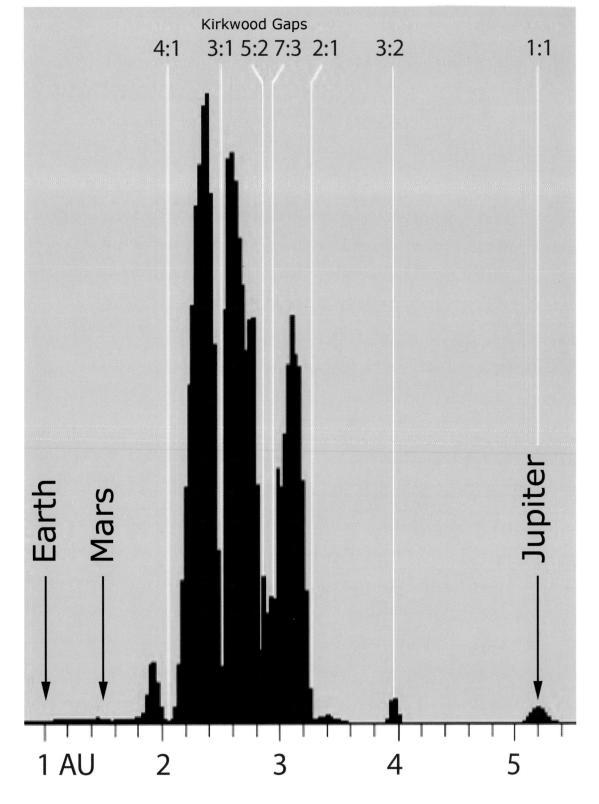

Solar Flares

Richard Carrington (1826–1875)

The **Sun** is the most massive, energetic, and important (to us, at least) object in the solar system, and so it should not be surprising that many nineteenth-century astronomers chose to train their increasingly powerful telescopes on our nearby star in order to study its inner workings. By using proper filters or projecting the solar disk onto a wall or screen, astronomers could measure and monitor features like sunspots on the Sun's visible "surface," the photosphere. Sunspots had been studied for centuries, with observations going back to the early seventeenth century. Improvements in telescopes and observing methods allowed them to be studied in ever-more exquisite detail.

One of the most noted and prolific observers of sunspots was the English amateur astronomer Richard Carrington. On September 1, 1859, Carrington observed an intense brightening near a particularly dense cluster of sunspots. The event lasted only a few minutes. The next day, however, saw reports worldwide of intense auroral activity and major disruptions to telegraphs and other electrical systems.

What Carrington had witnessed was the first recorded example of a solar flare—an enormous explosion in the Sun's atmosphere that can hurl high-energy particles at enormous speeds out into the solar system. The dramatic effects from this solar "wind" crashing into the Earth's protective magnetic field is known as a solar storm. Many such flares and storms have been observed since then, but visual records and ice core data indicate that the 1859 event was not only the first but also the largest in recorded history—perhaps a once-in-a-millennium mega flare.

Carrington's scientific observations established a connection between the Sun's activity and the Earth's environment, and led to intense interest in the study of space weather—the interaction of the solar wind with all the planets. Today's armada of Earth-orbiting satellites, in particular, represents billions of dollars of technology and infrastructure that is highly vulnerable to disruption by solar flares and their ensuing storms, which is just one reason why NASA and other space agencies are highly motivated to continue Carrington's important work of predicting, monitoring, and understanding the effects of space weather.

SEE ALSO Birth of the Sun (c. 4.5 Billion BCE), Astronomy in China (c. 2100 BCE), Mira Variables (1596).

A spectacular solar prominence eruption captured in time-lapse frames on March 30, 2010, in the extreme ultraviolet light of ionized helium by the NASA Solar Dynamics Observatory satellite. For scale, hundreds of Earths would fit into the loop in the top frame.

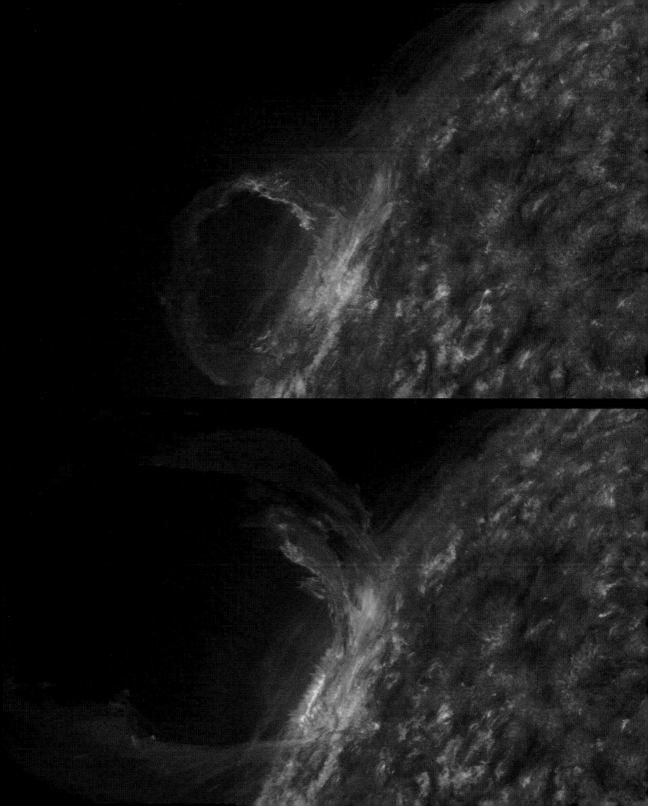

Search for Vulcan

Urbain LeVerrier (1811–1877)

The French mathematician Urbain LeVerrier was still basking in the glory of his 1846 mathematical **"Discovery" of Neptune** when he took on his next theoretical challenge. Astronomical observations of the rapid motion of **Mercury** (which orbits the Sun in just 88 days) over many decades had revealed some inconsistencies in its motion, resulting in inaccurate predictions of solar transits and other observational phenomena related to the planet. **Kepler's and Newton's Laws of Motion** worked incredibly well for the other planets, moons, and asteroids of the solar system except Mercury. What was going on?

LeVerrier guessed that the source of Mercury's orbital inconsistencies was the same as the source of the orbital inconsistencies seen for Uranus more than a decade earlier: another planet must be exerting an occasional gravitational tug on Mercury, perturbing its orbit. Another planet was waiting to be discovered! LeVerrier did some calculations and came up with a prediction for the unseen planet's location, which he figured must be very close to the Sun, with an orbital period of only about 20 days. He even named it: Vulcan, after the Roman god of fire. He announced his "discovery" to the French Academy of Sciences in 1859, and the search for Vulcan was on.

Mercury is exceedingly hard to observe with a telescope because it never gets more than about 20 degrees away from the Sun in the sky, so astronomers are always battling the Sun's glare to observe the planet. The search for Vulcan was even more challenging, as it was predicted to never get farther than about 8 degrees from the Sun. Still, professionals and amateurs searched. LeVerrier pursued a few reports of small objects in roughly the right place transiting the Sun, or reports of objects of about the appropriate brightness being observed during eclipses, but none of the reports could be verified by follow-up observations. LeVerrier died in 1877 thinking that Vulcan was still out there, waiting to be found.

Astronomers never did find Vulcan, because Mercury's orbital motions were eventually found to be an effect of **Einstein**'s theory of general relativity and the curvature of space-time that close to the Sun. Still, LeVerrier's search lives on in a way; modern astronomers are now searching for a hypothesized population of small asteroids interior to Mercury's orbit called, fittingly, vulcanoids.

SEE ALSO Three Laws of Planetary Motion (1619), Newton's Laws of Gravity and Motion (1687), Discovery of Neptune (1846), Einstein's "Miracle Year" (1905).

Artist's impression of a possible vulcanoid asteroid orbiting close in to the Sun.

White Dwarfs

Friedrich Wilhelm Bessel (1784–1846), **Alvan Clark** (1804–1887), **Alvan Graham Clark** (1832–1897)

The second half of the nineteenth century saw the design, manufacture, and operation of larger and higher-quality telescopes by skilled astronomers and instrument makers such as William Lassell in England and Alvan Clark in America. Clark's specialty was designing and grinding large glass refracting lenses that could be combined to yield high angular resolution and achromatic performance—free of the colored rainbow and halo artifacts that plagued many large lenses in previous refractor designs. Alvan Clark and Sons, telescope makers of Massachusetts, gained a worldwide reputation for high-quality instruments, many of which are still in operation today.

One of those Clark sons was Alvan Graham Clark, who often tested the company's new lenses on his own astronomical research. On one such occasion, observing on January 31, 1862, from the outskirts of Boston, the younger Clark trained a new 18.5-inch (47-centimeter) refracting telescope on the bright, nearby star Sirius. In 1844 the German mathematician Friedrich Wilhelm Bessel had predicted that the bright stars Sirius and Procyon had **Proper Motion** changes that were caused by unseen companions. The superbly clear night and the excellent quality of the lenses enabled Clark to discover that faint companion to Sirius, now known as Sirius B.

By the early twentieth century, Sirius B was found to have a spectrum similar to Sirius itself, even though it is much fainter. It and a few other visually dim stars were soon recognized to be part of a new class of small, hot stars dubbed white dwarfs. White dwarfs are now known to be a common end result of low-mass, Sun-like stars that have run out of hydrogen fuel but are too small to have exploded into supernovas.

By analyzing their orbits and using **Kepler's Laws**, white dwarfs like Sirius B have been found to be extremely dense—packing anywhere from 0.5 to 1.3 times the mass of the Sun into a volume the size of the Earth, for a density of more than 1,000,000 grams per cubic centimeter! As such, white dwarf stars are members of a special club of compact objects, also including **Neutron Stars** and **Black Holes**, that have the highest known densities of anything in the universe.

SEE ALSO Three Laws of Planetary Motion (1619), Mizar-Alcor Sextuple System (1650), Proper Motion of Stars (1718), Stellar Parallax (1838), Neutron Stars (1933), Black Holes (1965).

NASA Chandra Observatory X-ray image of the nearby star Sirius and its white dwarf companion, Sirius B. Sirius B is the brighter star in this image; with a surface temperature of 25,000 kelvins, it is a prodigious emitter of X-rays. In visible-wavelength light, Sirius B is actually 10,000 times dimmer than Sirius.

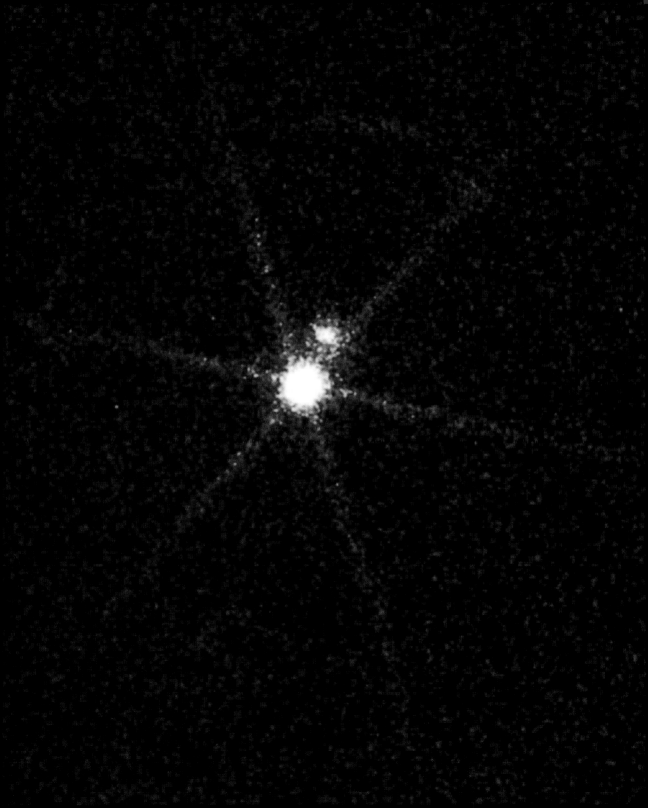

Source of the Leonid Meteors

Urbain LeVerrier (1811–1877)

Go out on a clear, moonless night to a dark-sky location away from the city, park yourself down on a blanket or in a nice reclining chair, and look up. Once your eyes have adapted to the dim light of the stars, it won't be long until—out of the corner of your eye—you spot a short, bright streak of light zipping across the sky. You've just seen a meteor, a tiny bit of rock or ice from space entering the Earth's atmosphere and burning up from the friction. Often called by the misnomer "shooting stars," a few such meteors per hour can be seen on any typical clear night. Once in a while, though, usually around the same time each year, careful and lucky observers can spot dozens or perhaps hundreds of meteors per hour—a meteor shower. And, extremely rarely, the night sky can be briefly lit by a storm of thousands and thousands of meteors per hour—a cosmic fireworks show that can rival any New Year's Eve or Independence Day celebration.

For millennia, such showers or storms of meteors were viewed as ominous. It wasn't until the late 1860s that astronomers were able to piece together some important clues and determine the origin of these cosmic spectacles: meteor showers are related to comets.

The key to solving the mystery was the discovery in 1866, independently in France and America, of a new short-period (33-year) comet named Tempel-Tuttle after its discoverers. Other astronomers, including the French mathematician and **Neptune** discoverer Urbain LeVerrier, realized that Tempel-Tuttle's orbit was remarkably similar to the orbits of meteors often seen in the mid-November shower, which are known as the Leonids. This allowed astronomers to accurately predict the next big Leonid shower, which occurred around the turn of the twentieth century, and thus to prove that meteor showers and storms happen when the Earth passes through patches of icy and rocky debris by a comet that passed through the same part of space some time before.

In addition to the Leonids in mid-November, you can also enjoy watching the fiery demise of tiny bits of comets during the Perseid shower in mid-August (from comet Swift-Tuttle), the Orionids in late October (from **Halley's Comet**), and more than a dozen other increased meteor activity events throughout the year.

SEE ALSO Halley's Comet (1682), Discovery of Neptune (1846).

An 1888 artist's depiction of a spectacular storm of tens to hundreds of thousands of Leonid meteors per hour. This particular event was widely viewed across North America on November 12–13, 1833; similarly large numbers of meteors were reported during storms in 1866 and again in 1966.

Helium

Jules Janssen (1824–1907), Norman Lockyer (1836–1920)

A total **Eclipse of the Sun**—when the new Moon completely covers the **Sun**'s disk and casts its shadow onto a small part of the Earth—is a spectacular and inspirational sight. For most of human history such events were treated as portents of fear, doom, and change. By the nineteenth century, however, scientists were treating accurately predicted eclipses as grand natural experiments—chances to study the Sun's atmosphere under special circumstances.

The French astronomer Jules Janssen mounted a special expedition to India to observe the August 18, 1868, total solar eclipse and to take **spectroscopic** observations of the solar corona. His data revealed a new, unidentified Fraunhofer-like emission line in the Sun's spectrum. A few months later, the British astronomer Norman Lockyer developed a method to acquire spectra of the Sun's atmosphere without an eclipse, and he observed the same new spectral line. Lockyer named the new element helium, after the Greek word for the Sun; both he and Janssen are credited with the discovery.

By the end of the nineteenth century, helium (He) was discovered on Earth as well, as a gas associated with radioactive uranium mineral deposits. Scientists began studying the new gas in detail, learning about its extraordinary properties. For example, helium liquefies at 4 degrees above absolute zero, and at temperatures very close to absolute zero it becomes a superfluid—an almost frictionless, zero-viscosity material.

More than a century of experiments and observations have revealed helium to have a relatively simple and stable structure, with a nucleus most commonly having two protons and two neutrons (He-4), but rarely with only one neutron (the isotope He-3). Because of its simplicity, helium is the second-most abundant element in the universe, and most of it likely formed in the **Big Bang**. Some helium forms even today during decay of radioactive elements such as uranium.

It is perhaps a humbling scientific lesson worth remembering that Janssen and Lockyer, taking the time to search for the unexpected, enabled the discovery of a colorless, odorless, nontoxic, inert material that we've since discovered constitutes nearly a quarter of the observable mass of our entire galaxy.

SEE ALSO Big Bang (c. 13.7 Billion BCE), First Stars (c. 13.5 Billion BCE), Birth of the Sun (c. 4.6 Billion BCE), Birth of Spectroscopy (1814), North American Solar Eclipse (2017).

This enhanced and processed photograph shows spectacular details in the solar corona during the August 1, 2008, total solar eclipse over central Asia.

Deimos

Asaph Hall (1829–1907)

By the mid- to late nineteenth century, the use and refinement of the telescope had allowed astronomers to discover two new planets and more than a dozen new moons in the outer solar system. Except for the discovery of a small number of main belt asteroids, however, few discoveries were made in the inner solar system.

The American astronomer and US Naval Observatory professor Asaph Hall was surprised to learn that few astronomers had made serious attempts to search for any moons around **Mars**. To be sure, it was a difficult problem because of the intense glare from the Red Planet itself, but Hall knew that he had unique access to an excellent instrument: a 26-inch (66-centimeter) Alvan Clark and Sons refractor that was, at the time, the largest refracting telescope in the world.

Mars and the Earth pass each other approximately every 26 months during what is called an opposition (Mars is opposite the Sun). Hall took advantage of the excellent opposition of 1877 and the high image quality of the Clark refractor to search for satellites of Mars. On August 11, 1877, he spotted a faint object near and moving with the planet; on subsequent nights he was able to verify that it was a moon in orbit around the planet, and that it was accompanied by a second moon even closer to Mars. A colleague suggested naming the moons after the Greek sons of Mars, Dread (Deimos) and Fear (**Phobos**), and the names stuck.

Hall and others realized that Deimos must be very small, but little more was known until space missions to Mars began observing the satellites as opportunities arose. Deimos has since been discovered to indeed be tiny—9.3 x 7.5 x 6 miles (15 x 12 x 10 kilometers)—irregular, and asteroid-like in shape, and with a cratered but strangely smooth and muted surface texture. With a density of about 1.5 grams per cubic centimeter and no evidence of an icy surface, astronomers speculate that Deimos may be a relatively porous, rocky object with a composition similar to some of the chondritic meteorites, the building blocks of Earth and other planets. It may take a dedicated mission to Deimos, however, to figure out if it really is a captured asteroid.

SEE ALSO Mars (c. 4.5 Billion BCE), Meteorites Come from Space (1794), Search for Vulcan (1859), White Dwarfs (1862), Phobos (1877).

The strange, smooth, reddish surface of the Martian moon Deimos, photographed on February 21, 2009, by the Mars Reconnaissance Orbiter spacecraft. The moon is only about 7.5 miles (12 kilometers) wide.

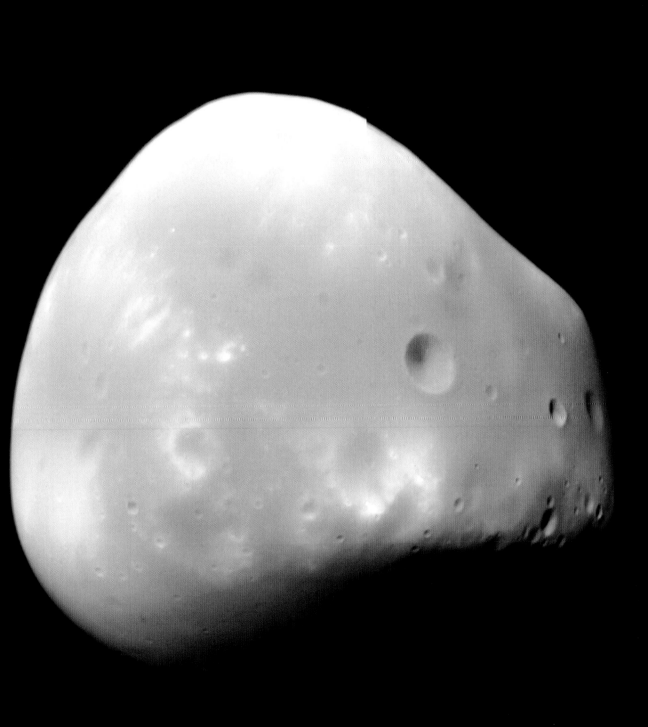

Phobos

Asaph Hall (1829–1907)

After the initial discovery of **Deimos** on August 11, 1877, the American astronomer Asaph Hall continued to scan the skies around **Mars** using the 26-inch (66-centimeter) Alvan Clark and Sons refractor of the US Naval Observatory in Washington, D.C. Hall's efforts were frequently interrupted by fog or bad weather, but his persistence paid off. Not only was he able to confirm that Deimos was in orbit around Mars, but on August 17–18 he was able to peer even closer in to Mars and discover a second faint, small moon, which was eventually named Phobos.

Hall and other astronomers quickly realized that Phobos orbits closer to Mars than any other known moon relative to its primary planet—so close, in fact, that with an orbital period of just over 7.5 hours it actually spins around Mars faster than Mars itself rotates on its axis. This meant that an observer on the surface of Mars would see Phobos rise in the west and set in the east, even though it orbits Mars in the same direction that Mars spins.

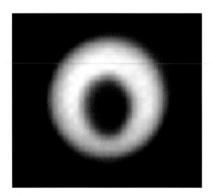

Modern space missions have been able to reveal more about this tiny, slightly reddish, asteroid-like world, but much still remains a mystery. Phobos is irregular in size—17 × 13.7 × 11 miles (27 × 22 × 18 kilometers)—and has a density near 1.9 grams per cubic centimeter, suggesting a porous, rocky, chondritic composition perhaps similar to Deimos's. The surface is heavily cratered, and one large crater (Stickney, named for Hall's wife, Angeline Stickney) is surrounded by a series of deep grooves that cover much of the surface.

Is Phobos an asteroid that Mars somehow captured from the nearby **Main Asteroid Belt**? Or is it perhaps a piece of Mars created from ejected fragments of a giant impact into the Red Planet? Two Soviet robotic probes were sent in 1988 to land on Phobos, but they failed—one en route and the other after attaining Mars orbit for only a few months. A new Russian robotic mission was launched in 2011, but it also failed, shortly after launch. New mission ideas are in the works, but in the meantime it seems that Phobos will continue to guard its secrets.

SEE ALSO Mars (c. 4.5 Billion BCE), Meteorites Come from Space (1794), Deimos (1877).

LEFT: *The silhouette of Phobos as it eclipses (transits) the Sun, photographed from the NASA Mars rover Opportunity on January 21, 2006.* RIGHT: *The reddish Martian moon Phobos, 13 miles (21 kilometers) wide, photographed on March 23, 2008, by the NASA Mars Reconnaissance Orbiter.*

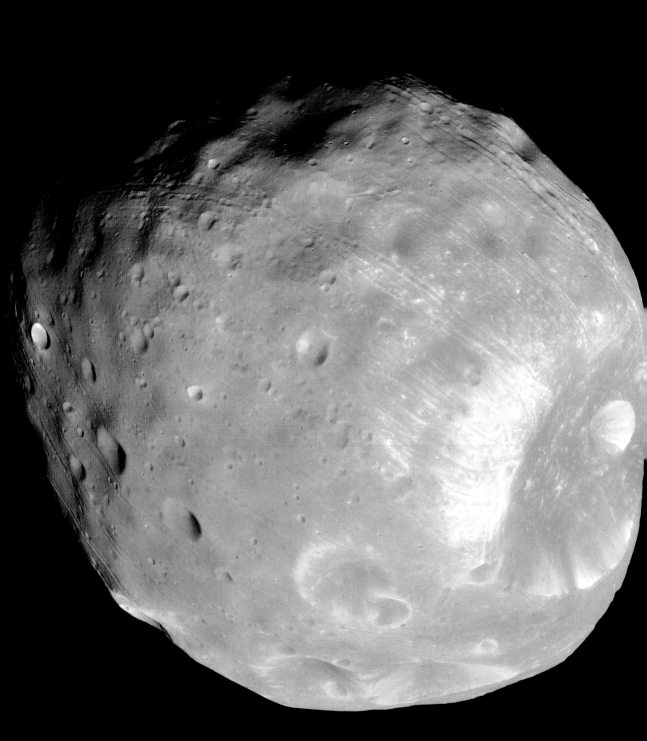

End of the Ether

James Clerk Maxwell (1831–1879), **Albert Michelson** (1852–1931), **Edward Morley** (1838–1923)

The Danish astronomer Ole Christensen Røemer and his Dutch colleague Christiaan Huygens are widely regarded as being the first scientists to prove, in 1676, that light travels at a finite speed. Their initial estimate of the **Speed of Light** was a little low, but by the 1860s physicists such as **Jean Foucault** had come very close to the modern accepted value (186,282 miles [299,792 kilometers] per second).

What was still uncertain to nineteenth-century scientists, however, was whether light required some sort of medium to travel through. Isaac Newton, following on Aristotle, had argued that there might be a "luminiferous aether" (or simply "ether") within which particles of light traveled. In the late 1870s the Scottish physicist James Clerk Maxwell described a way that physicists could test for the existence of the ether, by searching for slight variations in the speed of light depending on whether the Earth was moving toward or away from the ether.

In 1887, the American physicists Albert Michelson and Edward Morley turned Maxwell's ideas into an elegant experimental test for the existence of the ether. The Michelson-Morley experiment split a beam of light into two beams and used mirrors and a small telescope to recombine the beams. By changing the distance between the mirrors, they could cause the beams to add to or subtract from each other until a characteristic interference pattern of light waves—like wave ripples merging together on a pond—emerged in the telescope.

As the apparatus (floating on a pool of liquid mercury) was rotated, if the light beams changed speed because of their interactions with the ether, the scientists would see the very sensitive fringe pattern change. No such change was observed, however, proving the absolutely constant value of the speed of light, and providing definitive evidence that there is no "luminiferous aether." The experiment set the stage for dramatic advances in physics and astronomy in the twentieth century, including **Einstein**'s theories of general and special relativity.

SEE ALSO Speed of Light (1676), Foucault's Pendulum (1851), Einstein's "Miracle Year" (1905).

A cartoon drawing of the setup used in 1887 by Michelson and Morley for their definitive experiment to measure the speed of light in a variety of directions, superimposed upon the pattern of light waves generated by the interaction of the two beams of a helium-neon laser in a modern version of the Michelson-Morely experiment.

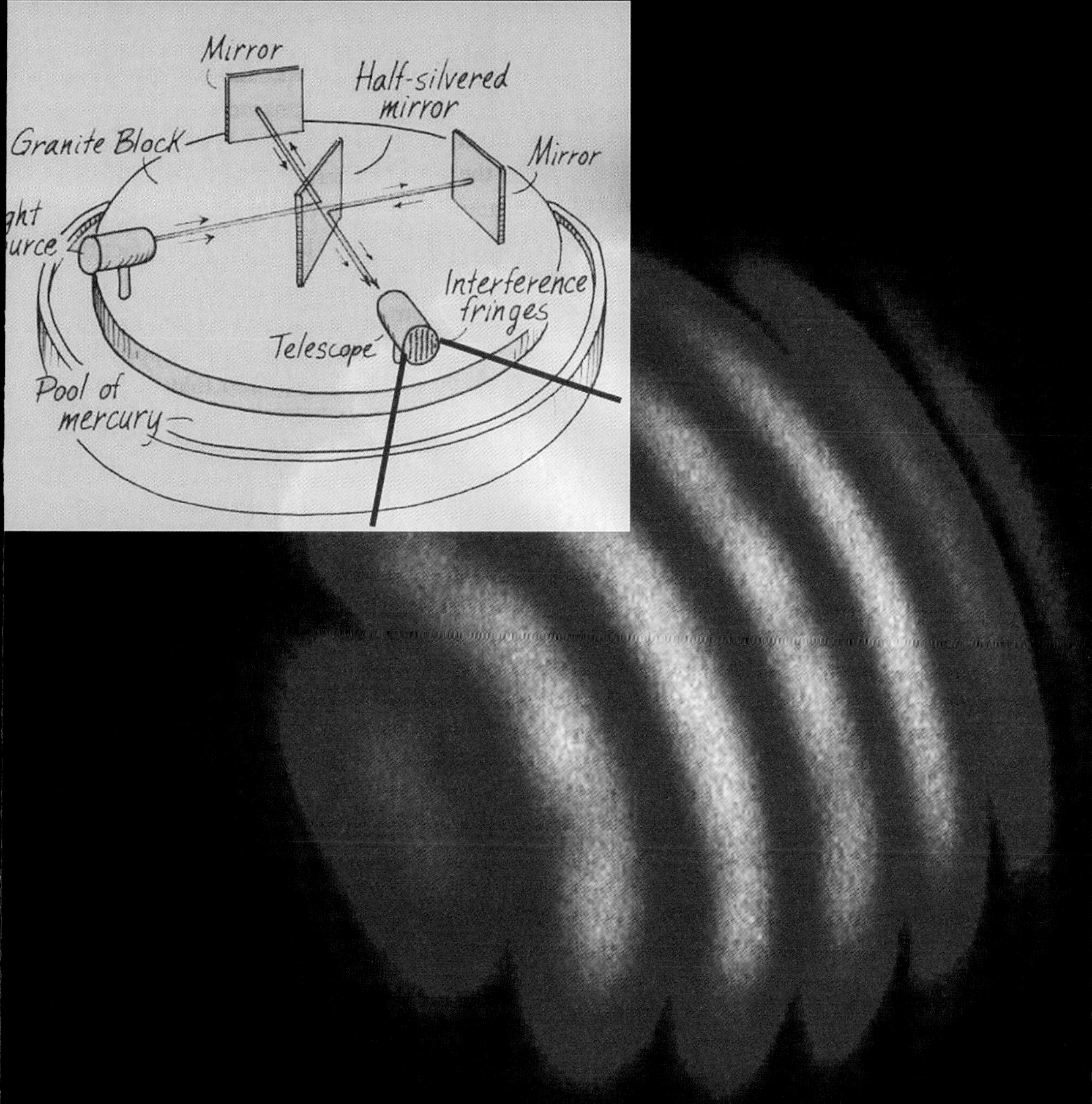

Mirror

Half-silvered mirror

Granite Block

Mirror

Light source

Interference fringes

Telescope

Pool of mercury

Amalthea

Camille Flammarion (1842–1925), Edward Emerson Barnard (1857–1923)

Despite improvements in telescopes that enabled new moons to be discovered around Saturn, Uranus, and Neptune, no new moons were discovered around **Jupiter** during more than 280 years of observations, dating back to the discovery of the four large **Galilean Satellites** in 1610. The American astronomer Edward Emerson Barnard figured that the "king of the planets" likely harbored some additional companions, and so he took it upon himself to use some of his weekly telescope time to search for them.

Barnard was a staff member of the Lick Observatory (named after its patron, the craftsman and California land baron James Lick). Located in the mountains above San Jose, since 1889 the observatory has been the home of the 36-inch (91.4-centimeter) Lick refractor, the largest refracting telescope in the world until 1897, and the third-largest refractor still in use in the world today. With the telescope's fine Alvan Clark and Sons lens craftsmanship at his disposal, Barnard had unique access to one of the best telescopes on the planet to conduct his search.

For three months Barnard patiently scanned the space around Jupiter, and finally, on September 9, 1892, he spotted a faint "star" near Ganymede that appeared to be moving with the planet. Subsequent nights of observing allowed him to track the object's motion and confirm that it was, indeed, a new moon of the giant planet. While Barnard called it simply the fifth satellite, the French astronomer Camille Flammarion suggested its eventual official name—Amalthea—after the mythological Greek nymph who nursed the infant Zeus.

Amalthea remained just a speck of light until the *Voyager* flybys of Jupiter in 1979 and the ***Galileo* Orbiter Mission** in 1995, when it was revealed to be a cratered, irregular moon of some 155 x 90.7 x 79.5 miles (250 x 146 x 128 kilometers) in size. While the moon's composition is unknown, with a density of only about 0.9 grams per cubic centimeter, it must be either very icy, very porous, or both. Amalthea orbits relatively close in to Jupiter and appears to be the source of one of Jupiter's faint, dusty rings. Impacts into Amalthea allow dust and ice to be launched to escape velocity; over time the thin cloud of debris has slowly spread out into a diffuse, gossamer ring.

SEE ALSO Io (1610), Europa (1610), Ganymede (1610), Callisto (1610), White Dwarfs (1862), Jovian Rings (1979), *Galileo* Orbits Jupiter (1995).

Space artist Michael Carroll's painting of the November 2002 Galileo orbiter spacecraft's flyby of the giant planet's fifth satellite, Amalthea.

Star Color = Star Temperature

Gustav Kirchhoff (1824–1887), **Max Planck** (1858–1947), **Wilhelm Wien** (1864–1928)

Physicists made major advances in the second half of the nineteenth century in understanding light and energy. For example, the German physicist Gustav Kirchhoff developed fundamental equations describing the way a hypothetically perfect light-absorbing object called a blackbody should emit energy as electromagnetic radiation at a particular temperature. He found that blackbodies at the temperatures of typical real-world objects radiate a continuous spectrum of energy, from the long-wavelength part of the radio and infrared spectrum to the higher-energy, shorter-wavelength visible and ultraviolet wavelengths.

The German physicist Wilhelm Wien expanded upon those ideas, and in 1893 derived a simple relationship, now called Wien's law, that showed that the peak wavelength of the energy being emitted by an object is inversely proportional to its temperature. That is, hotter objects emit most of their energy at shorter UV and visible wavelengths; cooler objects emit mostly in the infrared. Max Planck, another German physicist, would further expand on these ideas about blackbodies and light and help to create the field of **Quantum Mechanics**.

Astronomers took advantage of this new understanding of light and energy to begin to understand objects that they could observe visually. Specifically, Wien's law helped astronomers deduce the relative temperatures of the stars: Hotter stars should be emitting more of their energy at shorter wavelengths and thus should appear bluer, and cooler stars should be emitting most of their energy at longer wavelengths, with their spectra peaking toward the yellow, orange, and red end of the spectrum. Our **Sun**, for example, is a yellowish star, putting it near the average to slightly cooler-than-average end of the stellar color scale.

The colors of the stars thus became a key observational parameter that could be used to begin to classify them according to temperature and, thus, in the twentieth century, to develop a systematic understanding of their origin, evolution, inner workings, and ultimate fate.

SEE ALSO Birth of Spectroscopy (1814), Quantum Mechanics (1900), Pickering's "Harvard Computers" (1901), Main Sequence (1910), Eddington's Mass-Luminosity Relation (1924).

Hubble Space Telescope photo of part of the globular star cluster Omega Centauri (NGC 5139), a cluster of more than 10 million stars all bound together gravitationally. The wide range of star colors indicates a wide range of temperatures, from blue/white (hottest) to orange/red (coolest).

Milky Way Dark Lanes

Edward Emerson Barnard (1857–1923), **Max Wolf** (1863–1932)

People fortunate enough to live in or at least occasionally visit truly dark, non-light-polluted night skies on moonless nights are treated to a stunning view: the grand **Milky Way** sweeps from horizon to horizon, with bright starry bands and black inky lanes stretching out like a celestial Jackson Pollock painting splashed across a grand cosmic canvas. On such wonderful nights it's easy to understand our ancestors' reverence for the night sky, as well as their need to try to make sense of what they were seeing.

In the late nineteenth century, many of the world's major cities could still be considered "dark sky" observing sites; the electrification of the night sky didn't become truly ubiquitous until sometime after World War II. Thus, the American astronomer E. E. Barnard jumped at the opportunity to move to the University of Chicago in 1895 to gain access to what had just become the world's largest refracting telescope, the giant 40-inch (102-centimeter) lens at Yerkes Observatory. Armed with a great telescope and his newfound interest in the nascent field of **Astrophotography**, Barnard began taking the best data ever acquired of bright star fields and dark, seemingly empty gaps across the Milky Way's grand sweep.

An important collaborator on Barnard's Milky Way studies was the German astronomer and astrophotographer Max Wolf. Wolf was aware that many astronomers were puzzled by the Milky Way's dark lanes—what the English astronomer William Herschel had called holes in the sky. Barnard's photos and Wolf's analysis revealed that these "holes" weren't really holes at all—careful observations could reveal faint embedded stars, or even background stars, that could be used to derive the properties of the dark lanes.

Wolf made a convincing and ultimately accurate argument that the dark areas in the Milky Way are enormous clouds of relatively opaque dust that obscure the otherwise blazing light of the background stars and prevent them from shining through. He noticed that the dark lanes were often associated with pockets of bright nebulosity, potentially from newly formed stars, and deduced that the dark regions might be cosmic cocoons, places where dust and gas are being compressed and thickened and are "about to form new suns." Wolf and Barnard's early speculations about the origin of the dark lanes have turned out to be spot-on.

SEE ALSO Milky Way (c. 13.3 Billion BCE), Solar Nebula (c. 5 Billion BCE), First Astrophotographs (1839).

The glorious spectacle of the Milky Way and its dark, dusty lanes rises over Long's Peak (14,259 feet [4,346 meters] high), in the Rocky Mountain National Park in Colorado.

Greenhouse Effect

Joseph Fourier (1768–1830), Svante Arrhenius (1859–1927)

We often think of our home planet as a natural "Goldilocks" world, not the hellish inferno of closer-to-the-Sun **Venus** or the frozen ice world of farther-from-the-Sun **Mars**. It wasn't until the end of the nineteenth century, however, that scientists realized that **Earth** is a habitable oceanic world only because of the influence of two relatively minor, but critically important, atmospheric gases: water vapor (H_2O) and carbon dioxide (CO_2). Without them, Earth's oceans would freeze solid, and life on our planet would likely be very different, if indeed any life had developed at all.

In the 1820s the French mathematician Joseph Fourier was the first scientist to realize that the Earth's equilibrium temperature—how warm the surface would be if it were heated by sunlight alone—was actually well below freezing. So why are the oceans liquid? Fourier speculated that the atmosphere might act as an insulator, perhaps trapping heat like the panes of glass in a greenhouse. But Fourier wasn't sure.

It was the Swedish physicist and chemist Svante Arrhenius who provided the answer, showing that gases in our atmosphere are indeed warming the surface, by more than 30 degrees—thus keeping our planet above freezing. The specific gases responsible are mostly water and carbon dioxide. These gases are transparent and thus let sunlight reach the surface, but they absorb a large part of the outgoing infrared heat energy emitted by the planet, thereby warming the atmosphere. Even though the warming is different from that in a closed glass box, it's still called the greenhouse effect, partly because of the earlier ideas and experiments discussed by Fourier.

Arrhenius knew that greenhouse warming was a simple—and fortunate—consequence of Earth's natural abundance of water and carbon dioxide, and he speculated that past decreases in carbon dioxide, especially, could explain the ice ages. He was the first to further speculate that future burning of fossil fuels could enhance the carbon dioxide abundance and lead to global warming. The Earth's climate is more complex than Arrhenius envisioned, but still his concern over the role that people might have on changing the Earth's environment has turned out to be prescient.

SEE ALSO Venus (c. 4.5 Billion BCE), Life on Earth (c. 3.8 Billion BCE), Cambrian Explosion (550 Million BCE), Dinosaur-Killing Impact (65 Million BCE), Life on Mars? (1996).

UPPER RIGHT: *Global map of atmospheric water vapor (blue is more) in July 2000.* LOWER RIGHT: *Global map of atmospheric carbon dioxide (red is more) in July 2009. The information comes from the NASA Atmospheric Infrared Sounder satellite.*

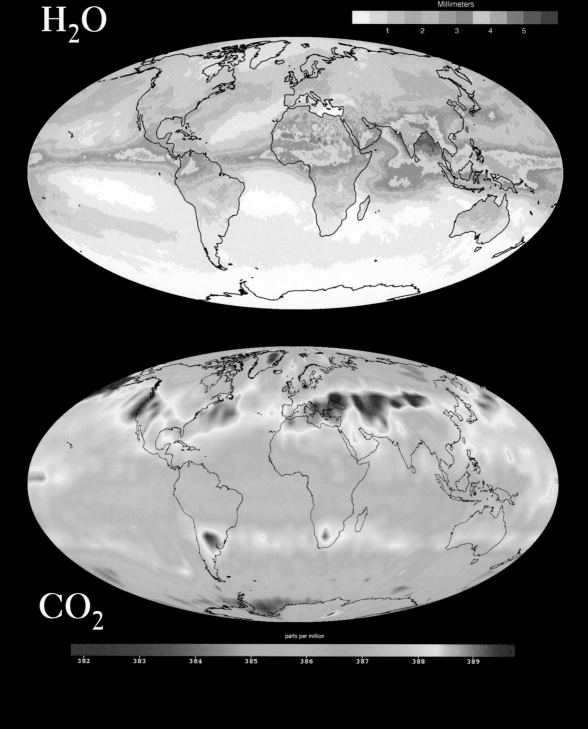

H₂O

Millimeters
1 2 3 4 5

CO₂

parts per million
382 383 384 385 386 387 388 389

Radioactivity

Wilhelm Röntgen (1845–1923), **Henri Becquerel** (1852–1908), **Pierre Curie** (1859–1906), **Marie Skłodowska Curie** (1867–1934)

Late-nineteenth-century physics labs in Europe and America were quite literally abuzz with new discoveries related to electricity and magnetism. The newly created ability to generate and store large voltages and currents for use in a variety of experiments often led to surprises, such as the German physicist Wilhelm Röntgen's 1895 study of high-voltage cathode ray tubes that generated a mysterious new form of radiation, which he dubbed X-rays.

The French physicist Henri Becquerel suspected that the ability of some natural materials to phosphoresce (glow in the dark) might be related to X-rays. He conducted a series of experiments in 1896 to determine if these materials emitted X-rays when exposed to sunlight, and accidentally discovered that one of the materials, uranium salts, spontaneously emitted radiation on its own. He had discovered radioactivity, something altogether different from X-rays.

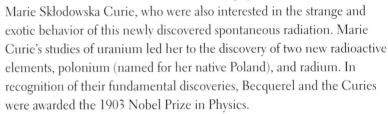

Becquerel went on to collaborate with fellow French physicists Pierre Curie and Marie Skłodowska Curie, who were also interested in the strange and exotic behavior of this newly discovered spontaneous radiation. Marie Curie's studies of uranium led her to the discovery of two new radioactive elements, polonium (named for her native Poland), and radium. In recognition of their fundamental discoveries, Becquerel and the Curies were awarded the 1903 Nobel Prize in Physics.

Over the past century, radioactivity has been exploited as a natural "clock" because radioactive elements release energy and decay into other elements at a predictable rate. Radioactivity has been used to accurately determine the ages of the **Earth**, **Moon**, meteorites, and, by extension, the entire solar system and beyond, using the age and evolution of the Sun as a guide. Because of radioactivity and the pioneering work of scientists like Becquerel and the Curies, we now know, with astonishing precision, that the Earth is 4.54 billion years old, and that the solar system formed 4.567 billion years ago.

SEE ALSO Birth of the Sun (c. 4.6 Billion BCE), Earth (c. 4.5 Billion BCE), Birth of the Moon (c. 4.5 Billion BCE), Mapping the Cosmic Microwave Background (1992), Age of the Universe (2001).

LEFT: *A 1918 photo of Henri Becquerel, discoverer of radioactivity.* RIGHT: *Pierre and Marie Curie, at work studying radioactivity in their laboratory in Paris, sometime before 1907.*

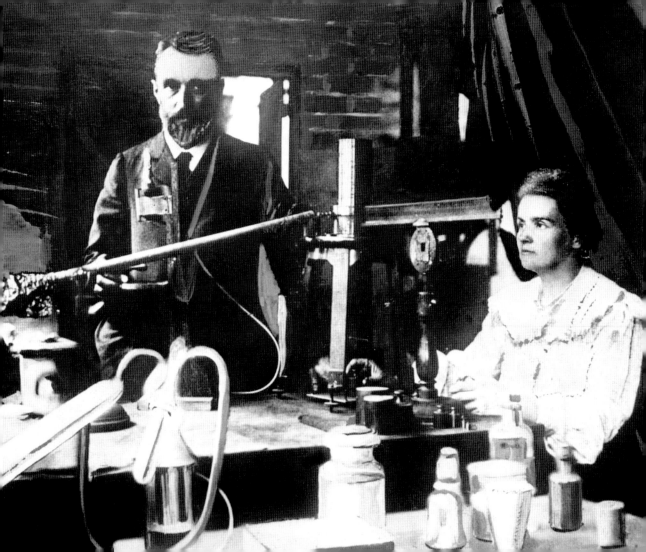

Phoebe

Edward Charles Pickering (1846–1919), **William Henry Pickering** (1858–1938), **DeLisle Stewart** (1870–1941)

By the end of the nineteenth century, most astronomers realized that simply building progressively larger telescopes was not the only way to study progressively fainter objects. Just as important as increasing light-gathering power was increasing the sensitivity of the "detector" used to record that light. Thus, more observatories were converting from the human eye to the photographic plate as their detector of choice.

The Harvard College Observatory (HCO) was no exception; indeed, Edward Charles Pickering, the late-nineteenth-century HCO director, was a pioneer in the use of **Astrophotography** to collect and record high-resolution spectra of stars. Pickering's brother, William Henry Pickering, was also an HCO astronomer. In 1899 he analyzed photographic plates of the sky near **Saturn** taken a year earlier by HCO staff member DeLisle Stewart and discovered a faint new moon orbiting the ringed planet. But this was quite an oddball moon: it orbited backward compared to Saturn's other moons, in a highly eccentric and tilted orbit four times farther from Saturn than the next closest moon (**Iapetus**). W. H. Pickering named the new moon Phoebe, following the theme of naming Saturn's satellites after Titans from Greek mythology. Phoebe was the first moon in our solar system to be discovered using photography rather than the naked eye.

It wasn't until the ***Voyager 2* Saturn Flyby** in 1981 and then especially the ***Cassini* Saturn Orbiter** mission beginning in 2004 that astronomers got more detailed information about Phoebe. It is a relatively large and somewhat spherical moon about 134 miles (220 kilometers) in diameter, with a low reflectivity (approximately 6 percent) and moderately low density (1.6 grams per cubic centimeter). Bright icy patches appear beneath the dark surface layers along steep slopes, and spectroscopy measurements show some to be carbon dioxide ice. Phoebe's composition and strange orbit suggest that it might be a captured **Centaur Object**—an interloper that was somehow diverted inward from the Kuiper belt. Impacts into Phoebe have created an enormous, tilted, dark, and diffuse distant ring of icy, rocky material around Saturn, some of which appears to be responsible for darkening the leading hemisphere of two-toned Iapetus.

SEE ALSO Saturn Has Rings (1659), Iapetus (1671), First Astrophotographs (1839), Hyperion (1848), "Centaur" Asteroids (1920), Pickering's "Harvard Computers" (1901), *Voyager 2* Encounters (1980, 1981), *Cassini* Explores Saturn (2004).

NASA Cassini Saturn orbiter photo of part of the heavily cratered, dark surface of Saturn's outer moon, Phoebe. Bright crater walls reveal icy deposits beneath a layer of darker materials.

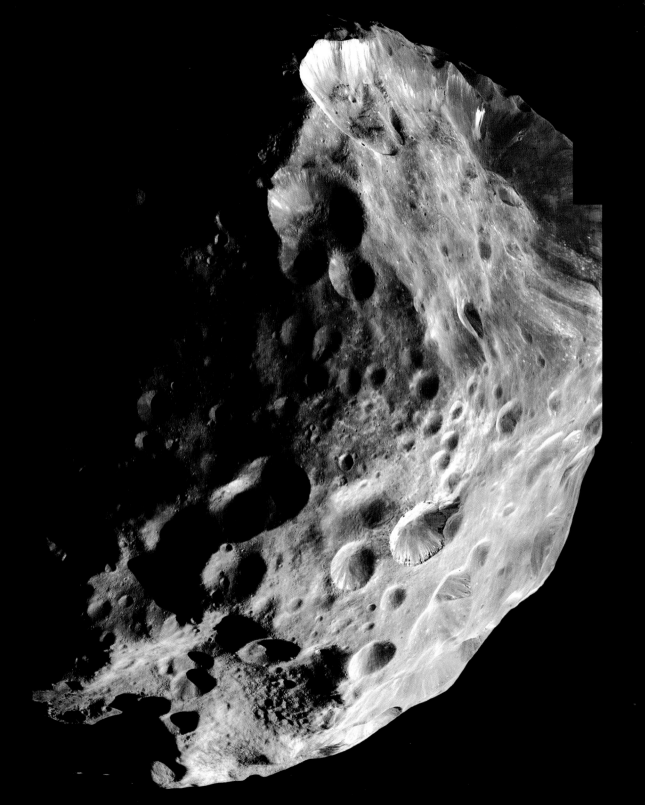

Quantum Mechanics

Max Planck (1858–1947), Albert Einstein (1879–1955)

What is light? The question has perplexed philosophers and physicists for millennia. Aristotle and his followers thought it to be a wavelike disturbance propagating through the air, while followers of Democritus subscribed to a so-called atomist theory, which posited that light existed as particles. Debates about light's wave-particle duality permeated Renaissance physics as well: Isaac Newton believed that only corpuscles (particles) of light could explain its behavior in optics; Christiaan Huygens held just as firmly that light must be wavelike, because it required a medium to travel through and refract. It was upon this confusing stage that late-nineteenth-century physicists began to propose a fundamental paradigm shift in our scientific understanding of the very nature of matter.

The revolution began with a mathematical trick of sorts by the German physicist Max Planck. Planck was trying to understand why objects of a given temperature, whether atoms or molecules or stars, radiate and absorb energy, and why they sometimes produce distinct bright emission lines or dark absorption lines in their spectra. Planck's turn-of-the-century trick was to assume that light (or, equally, radiation or energy) could only be emitted or absorbed by matter in discrete packets called quanta, whose energies depend only on the frequency or wavelength of the light.

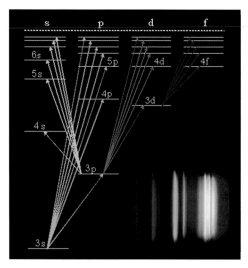

To Planck, the quantization of energy was a simple mathematical assumption needed to solve an equation (like the physics joke "Assume a spherical cow . . ."), not necessarily an expression of any physical reality. But his contemporaries, including the physicist **Albert Einstein**, saw a deeper truth in Planck's work, proposing that light consists of quanta of energy called photons, whose interactions with matter follow wavelike equations. Rather than a conundrum, the wave-particle duality of light became a fundamental tenet of an entirely new branch of physics called quantum mechanics.

SEE ALSO Speed of Light (1676), Birth of Spectroscopy (1814), End of the Ether (1887), Star Color = Star Temperature (1893), Einstein's "Miracle Year" (1905).

LEFT: *Planck's quantization of energy led to a detailed "energy level" theory of electrons orbiting an atomic nucleus, put forth by physicists such as Niels Bohr (1885–1962). In the Bohr model of energy levels in an atom (such as the sodium atom depicted here), electrons gaining and losing energy between levels explain the bright and dark lines in atomic spectra (lower right) at discrete wavelengths.* RIGHT: *Max Planck at his desk in an undated photograph.*

DR. MAX PLANECK

Pickering's "Harvard Computers"

Annie Jump Cannon (1863–1941), Edward Charles Pickering (1846–1919)

Astronomers, like most other scientists, like to be able to group the objects they study into convenient classes or groups, making it easier to compare and contrast their proportion and histories. Coming up with a suitable scheme to classify the stars was particularly important; thousands of them are visible to the naked eye and millions are accessible from the telescope.

The early star catalogs of Hipparchus, Ptolemy, and al-Sūfī recorded the relative magnitudes of the brighter stars and sometimes their relative colors, from bluer to redder. In the 1860s, the Italian astronomer Father Angelo Secchi acquired spectroscopic data for thousands of stars, developing the first stellar classification scheme and dividing the stars into five main classes based on their spectral patterns.

Many astronomers worked on refining and extending Secchi's scheme to classify millions more stars, including Harvard College Observatory director Edward Charles Pickering. Pickering had access to excellent telescopes for his project, and, like other observatory directors of the time, he hired human "computers" to help sift through and analyze the enormous data set (thousands of photographic plates) that was being collected.

Many of these computers were women, hired for little or no pay to perform what many of their male employers regarded as the rather menial and tedious work of measuring stellar spectral lines. Some of these women became quite skilled in stellar spectroscopy and went on to make important contributions to the field. Among the standouts was Annie Jump Cannon, who was able to use her knowledge of the strengths of absorption lines to reorganize and simplify what had been becoming overly complex and competing schemes. Cannon's 1901 class names—OBAFGKM, from bluer and weaker to redder and stronger lines—are still used by astronomers today. Later, her classes would be shown to be directly correlated with **Stellar Temperature** and **Stellar Evolution**.

SEE ALSO Stellar Magnitude (c. 150 BCE), Birth of Spectroscopy (1814), Star Color = Star Temperature (1893), Main Sequence (1910).

LEFT: *Portrait of Annie Jump Cannon from 1922.* RIGHT: *Schematic representation of the so-called Harvard classification scheme developed by Annie Jump Cannon in 1901, grouping the stars by weakest lines (O) to strongest (M) based on their spectra.*

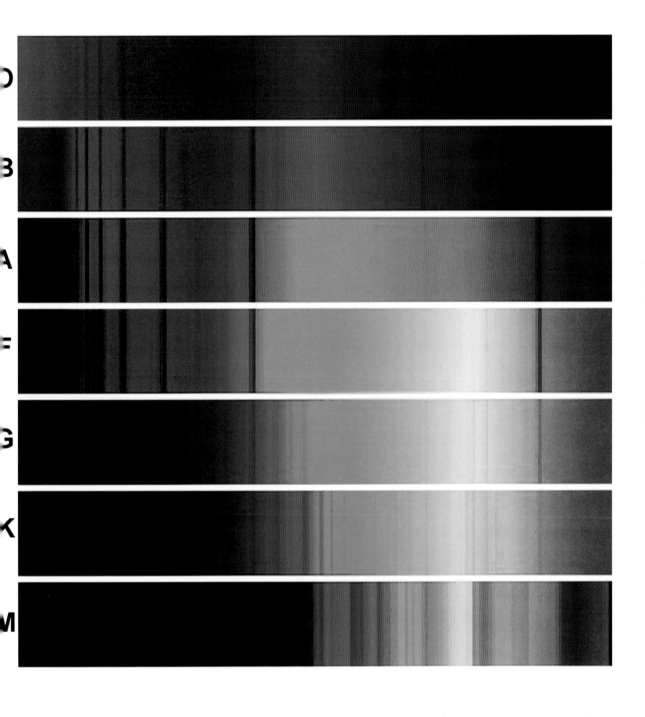

Himalia

Joseph-Louis Lagrange (1736–1813), **Édouard Roche** (1820–1883), **George William Hill** (1838–1914), **Charles Perrine** (1867–1951)

As seen by the discovery of **Phoebe** in 1898, astronomers were scanning ever larger volumes of space around the giant planets to search for new moons. Nineteenth-century astronomers such as Édouard Roche and George William Hill had expanded on the celestial mechanics work of Joseph-Louis Lagrange to come up with more accurate estimates of how far a moon could be from a planet and still stay in a stable orbit—a realm that is today known as a planet's Hill sphere.

In 1904 the Lick Observatory astronomer Charles Perrine discovered a faint, distant outer moon of Jupiter orbiting four times as far from the planet as **Callisto**, the outermost Galilean satellite. It was known simply as "Jupiter VI" until 1975, when it was finally named Himalia after the Greek nymph who bore three of Zeus's sons.

Himalia is too far from **Jupiter** for any resolved images to have been acquired from the **Voyager** or **Galileo** mission cameras, but the **Cassini** mission, en route to Saturn, took distant photos of Himalia that revealed its size to be around 93 miles (150 kilometers) across. It has since been discovered to be the largest and brightest member of a veritable swarm of more than 50 small, so-called irregular satellites now known to be in distant orbits around Jupiter. Jupiter is not alone in hosting irregular satellites, though: Saturn has 38; Uranus, 9; and Neptune, 6.

Many irregular moons orbit backward (retrograde) relative to the main moons of the giant planets, and many are in significantly tilted orbits relative to the equator of their primary planet. Unlike the main moons (and Earth's **Moon**), none seems to be tidally locked, with one face always pointed toward the primary planet. These characteristics lead astronomers to believe that outer irregular satellites such as Himalia are captured bodies. Perhaps they formed nearby but eventually wandered close enough to become gravitationally trapped in a giant planet's Hill sphere. Or perhaps they were main belt asteroids or **Kuiper Belt Objects** that got gravitationally diverted and then captured. To find out for sure, dedicated space missions may need to target these tiny worlds.

SEE ALSO Lagrange Points (1772), Phoebe (1899), Kuiper Belt Objects (1992).

A snapshot looking down on the swarm of small, irregular moons now known to orbit Jupiter, from the University of Maryland's online Solar System Visualizer. Himalia is just above the large moon Callisto in this view.

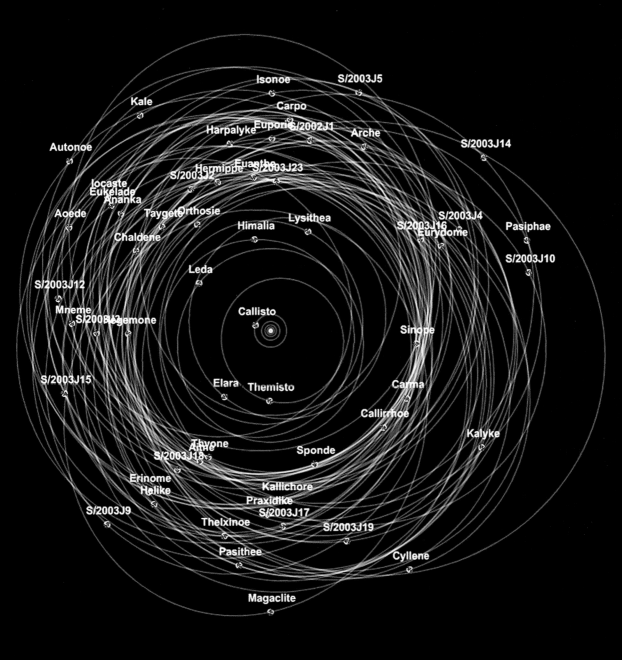

Einstein's "Miracle Year"

Albert Einstein (1879–1955)

Imagine knowing something brand new about the very nature of the universe, of space and time, that no one else knows. And then imagine if no one would believe you, or even understand you, when you talked about it. Would the immense frustration of being so far ahead of your time overpower the sheer unbridled joy of having made one of the most important discoveries in all of science? This was the quandary faced by the physicist and visionary Albert Einstein.

Einstein was born in Germany to middle-class parents and showed an early aptitude in school for math and physics, as well as for unconventional thought and clashes with authority. After graduating from a Swiss college with a degree in physics, he was unsuccessful in finding a teaching job, so in 1902 he began working in the patent office in Bern, specializing in electromagnetics. Still he continued to dabble in physics.

And boy, did he dabble! In one magnificent year—1905—Einstein worked out the details of the particle nature of light and the so-called photoelectric effect that is the basis of all modern digital camera CCD (Charge-Coupled Device) Detectors; he explained the tiny random movements of molecules called Brownian motion; he introduced the theory of special relativity, which set the speed of light as immutable— meaning that both space and time change in bizarre ways as one approaches light speed; and, perhaps most notably, he showed that energy and mass are fundamentally related through the famous equation $E = mc^2$, a powerful new concept that would eventually lead to the nuclear era that we live in today. This prolific outpouring of new ideas and explanations earned him a PhD from the University of Zurich and shortly thereafter led to a Nobel Prize in recognition of his genius.

Throughout his life, Einstein continued to develop new ideas and to extend his previous ones. Among his most famous later theories was one called general relativity, which explained gravity as a change in the shape of normal, four-dimensional space and time—what physicists call the space-time continuum. Astronomers and physicists have worked for more than a century testing and extending Einstein's relativity theories and other concepts, almost all of which have proven to be correct.

SEE ALSO Speed of Light (1676), Quantum Mechanics (1900), Astronomy Goes Digital (1969), Gravitational Lensing (1979).

Albert Einstein during a lecture in Vienna in 1921.

Jupiter's Trojan Asteroids

Max Wolf (1863–1932), **Johann Palisa** (1848–1925), **Joseph-Louis Lagrange** (1736–1813)

The advent of photography in astronomy enabled both the discovery of faint objects—because of the increased sensitivity of photographic plates compared to the human eye—as well as fast objects such as asteroids, which move through the sky at a different rate from the stars. Among the leaders in late-nineteenth and early-twentieth-century astrophotography was the German astronomer Max Wolf.

Over the course of his career, Wolf discovered almost 250 asteroids using his photographic methods. One of his most important discoveries was made on February 22, 1906, when he identified the 588th known "minor planet." Unlike most of the others, however, this one was far beyond the **Main Asteroid Belt**, with an average distance from the Sun of around 5.2 astronomical units, almost the same as Jupiter. The Austrian astronomer Johann Palisa, also a prolific discoverer of asteroids, found that Wolf's discovered object was orbiting the Sun in approximately the same orbit as Jupiter, but about 60 degrees ahead. After more asteroids were found in the same area, as well as more in a corresponding area some 60 degrees behind Jupiter, it became clear that the idea of gravitationally balanced stability regions, or **Lagrange Points**, proposed by the French mathematician Joseph-Louis Lagrange in 1772, had been verified. Wolf had discovered the first of what were later named by Palisa the Trojan asteroids, in honor of the heroes of the Trojan War. Wolf's asteroid number 588, which orbits near the L4 Lagrange point of the Jupiter-Sun system, was eventually given the name Achilles after the Greek hero from Homer's *Iliad*. Other asteroids in the L4 region were named after other warriors from the Greek camp; those in the L5 region were given names of heroes from the Trojan camp.

More than four thousand Trojans have now been discovered between the two camps, and astronomers estimate that there may be upward of a million asteroids greater than 0.6 mile (1 kilometer) in size trapped in what astronomers refer to as the L4 and L5 clouds of Jupiter Trojans. Many are dark and reddish and potentially like inactive comets, but little else is known about this enormous population of small bodies.

SEE ALSO Main Asteroid Belt (c. 4.5 Billion BCE), Lagrange Points (1772), Milky Way Dark Lanes (1895), Himalia (1904).

A graphical representation of the positions of 4,079 Jupiter Trojan asteroids (yellow dots) with well-known orbits as of 2011. The leading (L4) cloud of "Greeks" is at top; the trailing (L5) "Trojans" are at lower right. Jupiter is at center right. (The plot was made using the Celestia planetarium program.)

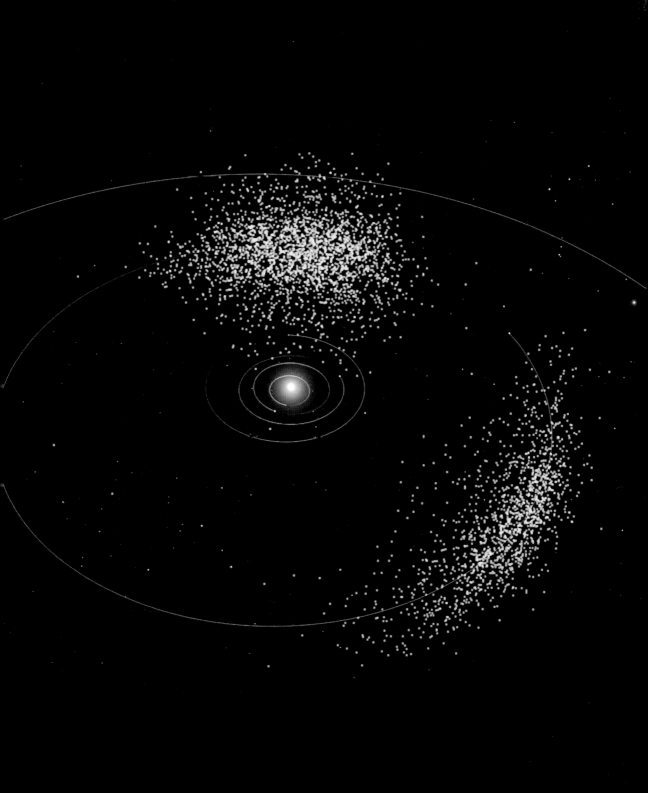

Mars and Its Canals

Giovanni Schiaparelli (1835–1910), Percival Lowell (1855–1916)

The Martian moons **Phobos** and **Deimos** were discovered during the opposition of 1877 partially because it was an excellent and relatively rare closer-than-normal alignment between Earth and Mars. Others took advantage of the opportunity to focus on Mars itself, including the Italian astronomer Giovanni Schiaparelli, who created numerous maps in which he gave the planet's "seas" (darker areas) and "continents" (brighter areas) historical Latin and Mediterranean names. He also observed fine, dark, linear features on the planet that he thought might be channels of some kind. The Italian word for "channels," *canali*, was widely mistranslated into English as "canals."

The Massachusetts industrialist, author, and astronomy enthusiast Percival Lowell became smitten with Schiaparelli's and others' descriptions of such linear features on Mars, and was convinced that higher-quality observations could reveal the features

in more detail. In 1894 he used part of his personal family fortune to found a high-altitude observatory in Flagstaff, Arizona, and to equip it with a 24-inch (61-centimeter) Alvan Clark and Sons refracting telescope. Lowell Observatory quickly became one of the world's premier centers for astronomical research, and Lowell himself focused much of his own observing time on detailed observations and drawings of Mars.

To Lowell's eyes, Mars was crisscrossed by intricate networks of dark linear markings that could only be the work of an intelligent species. In his popular 1906 book, *Mars and Its Canals*, he imagined that these were alien waterways designed to bring meltwater from the polar caps down to the great equatorial cities of the Red Planet. The idea that Mars was inhabited by a race of highly skilled engineers was widely popularized.

Modern photography and space missions have showed that Schiaparelli's and Lowell's canals were likely just an optical illusion. But public (and scientific) fascination with Mars as a potential abode of life has remained just as strong.

SEE ALSO Mars (c. 4.5 Billion BCE), Phobos (1877), Deimos (1877), First Mars Orbiters (1971), *Vikings on Mars* (1976), First Rover on Mars (1997), Mars Global Surveyor (1997), *Spirit* and *Opportunity* on Mars (2004), First Humans on Mars? (~2035–2050).

LEFT: *A photograph of Percival Lowell, from around 1895.* RIGHT. *One of Percival Lowell's sketches of Mars from around 1900, showing examples of the fine, linear markings that he and some others interpreted as evidence of a large-scale network of irrigation channels on Mars. Through his work, Lowell did much to popularize the idea of life on the Red Planet.*

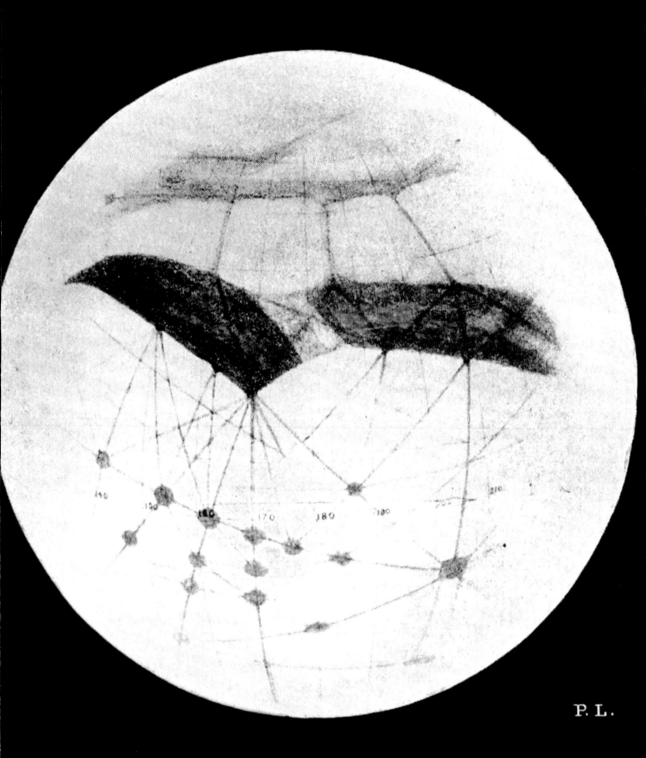

P. L.

Tunguska Explosion

Leonid Kulik (1883–1942)

Many of the residents of what is now the Krasnoyarsk Krai region of remote central Siberia, near the Tunguska River, were startled awake on the morning of June 30, 1908, by a spectacular event. Around 7.15 a.m. the skies above Siberia erupted in what eyewitnesses reported as a blinding flash of light followed by thunderous explosions. The ground shook with the force of a magnitude 5.0 earthquake; a fierce hot wind and a rain of fire stripped and felled 80 million trees over more than 811 square miles (2,100 square kilometers)—an area half the size of Rhode Island. Seismic shock signals from the event were recorded across Asia and Europe, and night skies around the world glowed with an eerie light for several days afterward.

Scientists suspected that the residents of Tunguska had experienced a meteoroid impact. The first scientific group to study the remote, uninhabited region didn't arrive until 1927, however, when the Russian mineralogist Leonid Kulik searched in vain for the resulting impact crater and potentially valuable iron-nickel meteorite deposits. Apparently, the event was primarily an airburst explosion, with surface damage caused by shock waves, heat, and fire—but with no associated crater formed.

Planetary scientists have debated the nature of the impactor for more than a century, with many claiming that it must have been an icy comet fragment that disintegrated catastrophically from atmospheric entry, and others claiming that the object must have been a small, rocky asteroid, perhaps a rubble-pile object that also was too weak to survive to the surface. Whatever its origin, an object only about 33 feet (10 meters) across traveling at around 6 miles (10 kilometers) per second exploded around 6 miles (10 kilometers) above the surface with an energy of about 10 megatons of TNT— or about 1,000 times the explosive yield of a World War II atomic bomb.

Amazingly, no one was known to have been killed by the Tunguska explosion. Tunguska was a wake-up call for understanding impact events, especially the catastrophic effects that even small objects traveling at extreme velocities can have on our environment when they occasionally slam into our planet.

SEE ALSO Cambrian Explosion (550 Million BCE), Dinosaur-Killing Impact (65 Million BCE), Arizona Impact (c. 50,000 BCE).

LEFT: *1927 photo from the Kulik expedition.* RIGHT: *Artist and planetary scientist William K. Hartmann's impression of the Tunguska forest one minute after the airburst explosion. The painting was made at Mount St. Helens, where the 1980 blast from that volcano produced a Tunguska-like scene.*

Cepheid Variables and Standard Candles

Henrietta Swan Leavitt (1868–1921), **Edward Charles Pickering** (1846–1919), **Ejnar Hertzsprung** (1873–1967)

Astronomers had known for some time that it should be possible to measure the distances to the nearest stars using their **Parallax** Shift from Earth's motion around the Sun over the course of a year. But how could the distances be measured to farther-away stars that don't show any detectable parallax, even when viewed through the world's largest telescopes?

The answer came from the work of Henrietta Swan Leavitt, one of Harvard College Observatory director Edward **Pickering's "Harvard Computers"**—female staffers who were tasked with the menial work of analyzing millions of stars on thousands of photographic plates. Leavitt's analysis was focused on the brightness changes in periodic, or pulsating, variable stars. She examined thousands of variables, and discovered an

interesting pattern among a certain class of them called Cepheid variables (after the prototype, Delta Cephei). In 1908 she published her initial finding that the brighter or more luminous the Cepheid, the longer was its period. This period-luminosity relationship meant that Cepheids with the same period that vary in brightness must simply be at varying distances from us. Cepheids could be used as distance yardsticks, or so-called standard candles, if the distance to any one of them could be independently determined.

In 1913 the Danish astronomer Ejnar Hertzsprung independently determined the distances to several Cepheid variables using sensitive parallax measurements; these provided the key to estimating the distances to any other Cepheids. Leavitt's unfortunately largely unheralded discovery, and the subsequent analysis of Cepheids in the Andromeda galaxy and elsewhere, led to the proof that Andromeda and other similar nebulae were actually independent galaxies of their own, at vast distances of millions of light-years beyond the Milky Way.

SEE ALSO Stellar Magnitude (c. 150 BCE), Stellar Parallax (1839), Mira Variables (1596), Pickering's "Harvard Computers" (1901), Main Sequence (1910), Hubble's Law (1929).

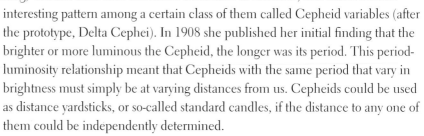

LEFT: *Portrait of Henrietta Swan Leavitt.* RIGHT: *May 1994 Hubble Space Telescope photos of a Cepheid variable star (insets) changing brightness in the spiral galaxy Messier 100 (M100). By treating the Cepheid as a so-called standard candle, the distance to M100 was estimated to be 56±6 million light-years.*

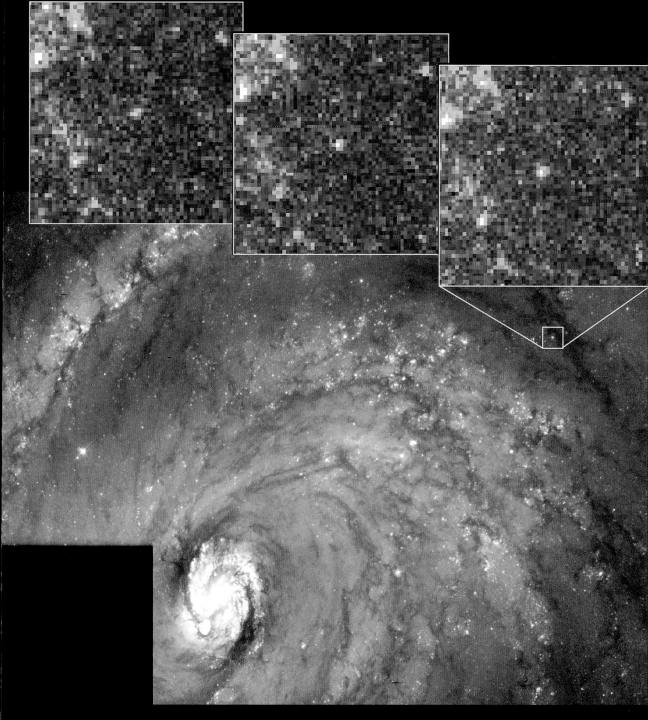

Main Sequence

Ejnar Hertzsprung (1873–1967), Henry Norris Russell (1877–1957)

In the early part of the twentieth century, astronomers worldwide were characterizing and classifying enormous numbers of stars in terms of their colors and spectroscopic lines, expanding on the methods pioneered by Edward Pickering's Group at Harvard. Among the most important advances was the observation, noticed independently by the Danish astronomer Ejnar Hertzsprung and the American astronomer Henry Norris Russell, that when the spectral classes or temperatures of the stars were plotted against their actual brightness (that is, after their apparent brightness in the sky was corrected for their distance from us), most of the stars cluster in a broad sequence from upper left to lower right. Hertzsprung coined the term "main sequence" to describe this prominent trend among the stars. Such plots began being used around 1910 and are called Hertzsprung-Russell (H-R) diagrams.

Over the next few decades astronomers began to understand that the main sequence was more than just a random clustering—it represents an evolutionary pathway for tracking the age and eventual fate of the stars. Most stars are born when their central pressures and temperatures are high enough for the **Nuclear Fusion** of hydrogen atoms into helium. During this hydrogen-fusing phase of its lifetime, a normal star will plot on the main sequence at a position that depends on its mass, with luminous stars a few to ten times the mass of the Sun (blue giants) on the upper left end of the plot and dim stars from about one-tenth to one-half the Sun's mass (red dwarfs) on the lower right. As stars age and run out of hydrogen fuel, they diverge off the main sequence and eventually "die" in characteristic (and often spectacular) ways that again depend on their mass.

As the details of stellar interiors later became understood by astrophysicists such as Arthur Eddington and Hans Bethe, it became possible to predict how stars of specific masses would live and die. Our **Sun** turns out to be an average mass, middle-aged, main sequence star that appears destined, in about 5 billion more years, to bloat up into a red giant, expel its outer layers into a **Planetary Nebula**, and then fade away as a **White Dwarf**.

SEE ALSO Stellar Magnitude (c. 150 BCE), "Daytime Star" Observed (1054), Planetary Nebulae (1764), Mira Variables (1596), Star Color = Star Temperature (1893), White Dwarfs (1862), Pickering's "Harvard Computers" (1901), Cepheid Variables (1908), Eddington's Mass-Luminosity Relation (1924), Nuclear Fusion (1939), End of the Sun (5–7 Billion).

Plots of the intrinsic luminosity of stars (on the y axis, normalized so the Sun's luminosity = 1) versus their color—or equivalently, their temperature (on the x axis)—reveal a prominent diagonal band of stars known as the main sequence, bracketed by brighter blue and red giants and dimmer white dwarfs.

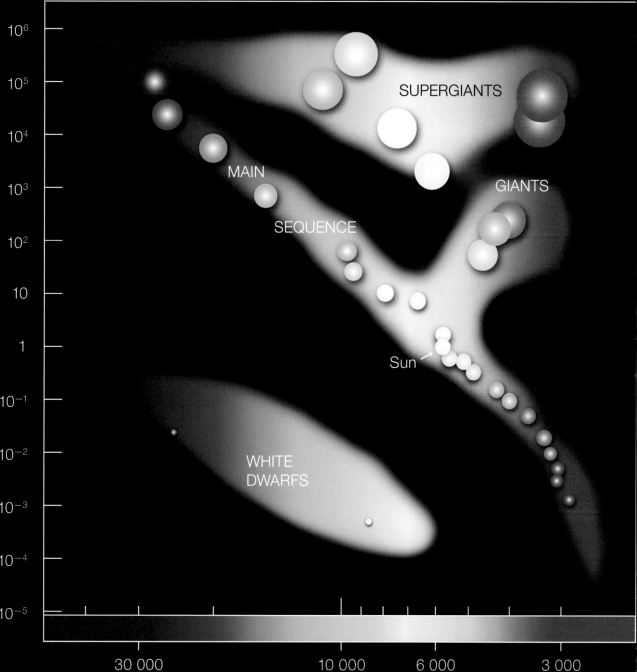

10^6
10^5
10^4
10^3
10^2
10
1
10^{-1}
10^{-2}
10^{-3}
10^{-4}
10^{-5}

SUPERGIANTS

MAIN

GIANTS

SEQUENCE

Sun

WHITE
DWARFS

30 000   10 000   6 000   3 000

Size of the Milky Way

Harlow Shapley (1885–1972), Edwin Hubble (1889–1953)

During the decade since **Cepheid Variable** stars were found in 1908 to be useful as measuring sticks for determining cosmic distances, a number of astronomers worked to determine distances to spiral nebulae, Globular Clusters, and other enigmatic objects in order to get a handle on whether they were inside or outside of the **Milky Way**. Indeed, the size of the Milky Way itself was the subject of intense debate, with many astronomers believing that it essentially was the universe, while many others believed it to be just one of many separate "island universes," as the distant nebulae had been dubbed by the eighteenth-century philosopher Immanuel Kant.

The first astronomer to make an experimental estimate of the size of our galaxy was the American astronomer Harlow Shapley, who studied the distribution of globular clusters in the sky. Cepheid variable stars had been used to determine the distance to a nearby globular cluster, so Shapley assumed that they were all the same size and used the changing apparent diameter of other clusters to estimate their distances. By 1918 he had determined that the globular clusters form a sort of halo around the platelike disk of our galaxy, enabling him to estimate that the Milky Way is about 300,000 light-years across, with the Sun not centered but offset by about 50,000 light-years (so much for heliocentrism). This was astonishingly larger than many had thought the galaxy to be, and it convinced Shapley that there were no island universes: the globular clusters and spiral nebulae must all be bordering or within the Milky Way.

It turned out that Shapley's estimate of the size of the Milky Way was about three times too large, mostly because his assumption that all globular clusters are the same size is not really valid. The disk of our galaxy is actually about 100,000 light-years across and about 1,000 light-years thick (with a somewhat thicker central bulge), and the Sun is offset from the center by about 27,000 light-years. Shapley was right about the globular clusters being in and near the Milky Way in a diffuse halo, but he was wrong about the spiral nebulae. As shown by **Edwin Hubble** and others in subsequent decades, spirals and many other forms of "nebulae" are actually separate galaxies, some like our own, some not, but all are millions to billions of light-years away.

SEE ALSO Milky Way (c. 13.3 Billion BCE), Andromeda Sighted (c. 964), Globular Clusters (1664), Cepheid Variables (1908), Hubble's Law (1929).

A Hubble Space Telescope photo of the face-on spiral galaxy NGC 1309. With bright bluish regions of new star formation in its spiral arms, and a central region of older, yellowish stars, this is probably close to what our Milky Way galaxy would look like if we could view it from far above.

"Centaur" Asteroids

Walter Baade (1893–1960)

Even as the total number of discovered asteroids began to approach 1,000 in the early twentieth century, astronomers were still occasionally surprised by the discovery of new, enigmatic objects. A case in point was the discovery in 1920 of asteroid 944 by the German astronomer Walter Baade. The asteroid, which was eventually named Hidalgo after the Mexican priest and independence leader Miguel Hidalgo, turned out to have an extremely comet-like, tilted, and eccentric orbit (eccentricity of 0.66) that takes it from the inner edge of the **Main Asteroid Belt** (1.95 astronomical units, or AU) out a to a distance almost as far from the Sun as Saturn (9.5 AU).

In 1977, another asteroid—2060 Chiron—was discovered in a similarly comet-like orbit, moving between the distances of Saturn and Uranus. Since then, several hundred more eccentric asteroids with orbits in the zone roughly between Jupiter and Neptune have been discovered. Collectively, members of this part-asteroid, part-comet population are now called Centaurs, after the mythical part-horse, part-human creatures.

Centaurs have a range of colors, suggesting a range of compositions; indeed, telescopic spectra of some of them reveal water ice, methanol ice, and tholins, or organic residues from the solar-UV irradiation of methane or ethane ices. Many of these components are also found on comets. Three Centaurs, including Chiron, have been found to have a weak coma (a fuzzy, comet-like head), a sign of comet-like activity.

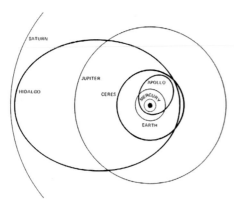

No Centaurs have been encountered by spacecraft, but many astronomers think that Saturn's irregular moon **Phoebe** might be a captured Centaur and thus might be an excellent example of what one looks like up close. We might have to act fast to study more: because they cross the orbits of the giant planets, they have lifetimes of only a few million years before they are flung into new orbits.

SEE ALSO Main Asteroid Belt (c. 4.5 Billion BCE), Phoebe (1899), Jupiter's Trojan Asteroids (1906), Kuiper Belt Objects (1992).

LEFT: *Orbit of 944 Hidalgo compared to Jupiter's.* RIGHT: *Painting by artist and planetary scientist William K. Hartmann imagining a close flyby of the Centaur asteroid 5145 Pholus. Like other Centaurs, Pholus has physical and orbital properties that make it seem like part asteroid and part comet.*

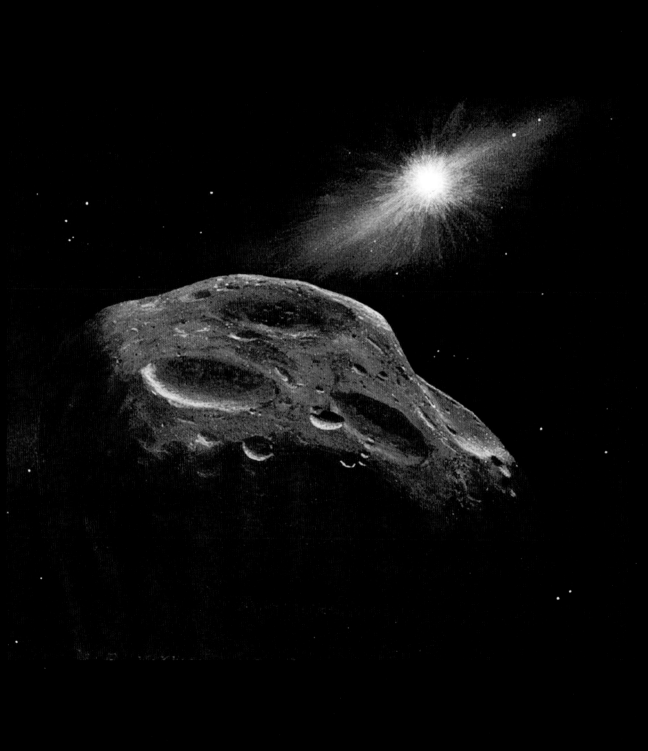

Eddington's Mass-Luminosity Relation

Arthur Stanley Eddington (1882–1944)

Even though astronomers were classifying stars based on properties such as color, temperature, and intrinsic brightness, it took much longer to figure out how stars actually work. Why do they shine? Where to do they get their energy? What's happening inside a star? One scientist whose work was important in answering such questions was the British astrophysicist Arthur Stanley Eddington.

Eddington was interested in understanding the process of gravitational collapse of nebular gas and dust to form a star. Astronomers knew that gravity was responsible for pulling the gas into a sphere and causing it to compress. But what stopped gravity from collapsing stars down to smaller and smaller sizes? In 1924 Eddington published a detailed model of the way radiation pressure—an outward force created by the super-high temperatures and pressures inside a star—balances gravity and allows the stars to reach their equilibrium sizes.

Eddington's stellar interior models allowed him also to determine that there is a simple relationship between a **Main Sequence** star's luminosity and its mass: a star that is twice as bright as another will be more than ten times as massive (specifically, LOC $M^{3.5}$). Thus, based on his work, it was possible to estimate the mass of a star just by measuring its apparent brightness and its distance, and therefore astronomers could show that the main sequence is the mass-dependent "lifeline," an evolutionary track taken by about 90 percent of all stars.

Neither Eddington nor anyone else at the time knew why this so-called mass-luminosity relationship worked, or how and why radiation pressure was generated inside a star. A leading theory was that gravitational contraction provided the energy. Eddington speculated that perhaps **Nuclear Fusion** could generate the required energy, but the idea was met with skepticism until the late 1930s.

SEE ALSO Birth of the Sun (c. 4.6 Billion BCE), Star Color = Star Temperature (1893), Main Sequence (1910), Nuclear Fusion (1939).

LEFT: *Photo of Arthur Stanley Eddington.* RIGHT: *The NASA Solar Dynamics Observatory obtained this spectacular ultraviolet view in August 2010 of a dark coronal hole in the Sun's atmosphere, a region where solar wind particles stream out into space, generating auroral displays in our atmosphere.*

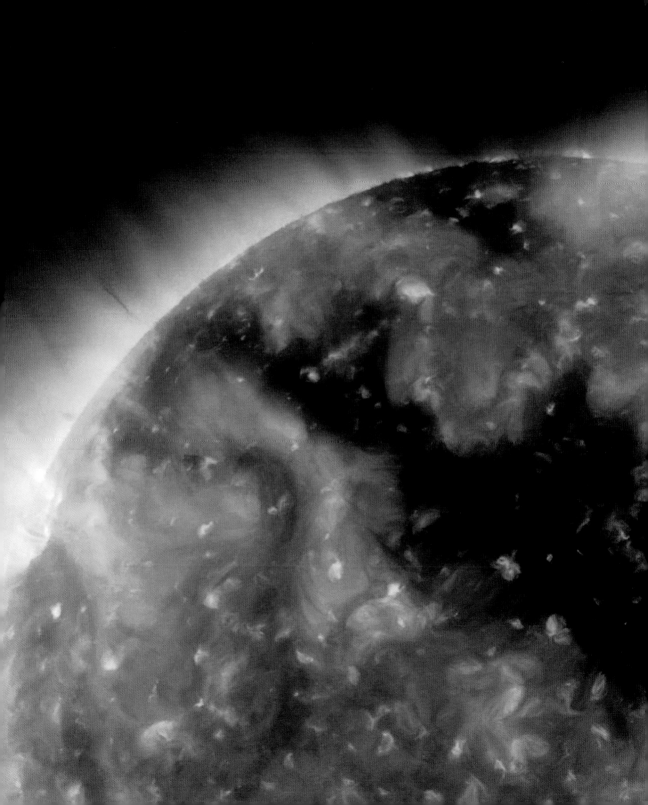

Liquid-Fueled Rocketry

Konstantin Tsiolkovsky (1857–1935), **Robert Goddard** (1882–1945), **Wernher von Braun** (1912–1977)

Propelled by the burning of gunpowder, rockets have been around for more than a thousand years. The Chinese are credited with first using rockets in battle as well as for entertainment (fireworks). But in 1903 the Russian mathematician Konstantin Tsiolkovsky wrote the first scholarly work that envisioned them as more than weapons; he also saw them as a potential means of space travel. He worked out much of the theory of rocketry and was among the first to propose using liquid fuels instead of gunpowder to maximize the combustion efficiency as well as the rocket's thrust-to-weight ratio. Tsiolkovsky is widely regarded as the father of modern rocketry in Russia and the Soviet Union.

However, it was the American rocket scientist and Clark University physics professor Robert Goddard who was first able to test Tsiolkovsky's—and his own—theories and show that liquid-fueled rockets were feasible and could provide the thrust needed to lift significant mass to high altitudes. He developed and patented key designs for rockets powered by gasoline and liquid nitrous oxide, as well as designs for the concept of multistage rockets, which he claimed could eventually be used to reach "extreme altitudes." Even though the flights of his own rockets were modest by today's standards, Goddard's methods were sound, and others—including a group of postwar space-race engineers led by rocketry pioneer Wernher von Braun—were able to expand on his designs to enable longer, higher, and eventually, orbital (and beyond) flights.

Like many inventors, Goddard was a visionary, often working alone and seeing possibilities that others overlooked. He was an early advocate of rocketry for atmospheric science experiments and, like Tsiolkovsky, for eventual travel into space. It is perhaps ironic that it was war that was the impetus for the eventual development of the rockets that would posthumously achieve Goddard's dream of space travel.

SEE ALSO Newton's Laws of Gravity and Motion (1687), *Sputnik 1* (1957), First on the Moon (1969), Space Shuttle (1981), First Humans on Mars? (~2035–2050).

Robert Goddard poses with his first liquid-fueled rocket, which was launched from Auburn, Massachusetts, on March 16, 1926. Unlike conventional rockets today, this model's combustion chamber and nozzle were at the top, and the fuel tank below. It flew for 2.5 seconds and rose 41 feet (12.5 meters).

The Milky Way Rotates

Bertil Lindblad (1895–1965), Jan Oort (1900–1992)

In 1918 the American astronomer Harlow Shapley had obtained the first quantitative estimate of the **Size of the Milky Way** galaxy by measuring the distances and directions to the halo of **Globular Clusters** that surround the galaxy's disk. Shapley's study also allowed him to estimate the position of the approximate center of the galaxy, which he placed in the brightest part of the Milky Way's vivid band of stars, in the direction of the constellation Sagittarius.

As it became more clear that we were embedded inside a **Spiral Galaxy** much like others that were now being studied in detail with **Astrophotography** and **Spectroscopy**, it dawned on some astronomers that, presumably like the other spiral galaxies, individual stars in the Milky Way could be spinning around a common galactic gravitational center. In the 1920s the Swedish astronomer Bertil Lindblad was the first to work out this hypothesis in detail.

In 1927 the Dutch astronomer Jan Oort provided the first observational proof of Lindblad's hypothesis by carefully measuring the movements of many hundreds of individual stars. He confirmed that the Milky Way rotates, and further, that the rotation is differential—that is, stars at different distances from the rotation axis orbit the center at different speeds, with farther-away stars lagging behind the closer-in ones. The Sun, about halfway out from the galactic center, takes about 250 million years to orbit the galaxy's center.

Oort's and Lindblad's work helped to refine the exact rotational center of the galaxy based on Shapley's earlier estimate. It was difficult for astronomers at the time to learn much more from visual observations, however, because much of the galactic center is obscured by the **Dark Lanes** of dusty nebulosity that E. E. Barnard and Max Wolf had studied in the late nineteenth century. Later astronomers would use X-ray, infrared, and radio telescopes to study this region intensely, eventually learning that an enormous energy source—called Sagittarius A* (pronounced "Sagittarius A-star")—probably powered by a 4-million-solar-mass **Black Hole**, lurks at the center of our galaxy.

SEE ALSO Milky Way (c. 13.3 Billion BCE), Globular Clusters (1665), Milky Way Dark Lanes (1895), Size of the Milky Way (1918), Dark Matter (1933), Spiral Galaxies (1959), Black Holes (1965).

A cluster of stars, gas, and dust surrounding the center of our Milky Way galaxy. This photo is a composite of infrared images from the ground-based Two Micron All Sky Survey (2MASS) program; infrared images like this allow astronomers to peer deeper into this dusty region.

Hubble's Law

Edwin Hubble (1889–1953), Vesto Slipher (1875–1969)

The 1848 discovery of the **Doppler Shift of Light** provided a tool that astronomers could use to determine the velocity of an astronomical object relative to Earth. All that was required was the ability to detect and measure the shift of a suitable absorption line or lines using **Spectroscopy**. In 1912 the Lowell Observatory astronomer Vesto Slipher obtained the first spectra of spiral nebulae and other objects that were later recognized as other galaxies. Slipher discovered that most spiral nebulae have spectral lines Doppler-shifted toward longer (redder) wavelengths—they are red-shifted and receding from us.

The American astronomer Edwin Hubble was also interested in studying the spiral nebulae, and beginning in 1919 he had the advantage of access to the brand-new 100-inch (254-centimeter) Hooker reflecting telescope at the Mount Wilson Observatory in Southern California—then the largest and most sensitive telescope in the world. Hubble studied Slipher's galaxy red-shift data and spent a decade painstakingly collecting more data of his own.

In 1929 Hubble published a landmark paper describing his initial results. He had found that, remarkably, the red shift of the galaxies increases the farther away a galaxy is from the Earth. Everything seemed to be moving away from us, and the farthest stuff the fastest. The implication of this observation, which has become known as Hubble's law, is that the volume of the observable universe is expanding. This amazing result was consistent with earlier theoretical predictions about the expansion of space-time by the Russian cosmologist Alexander Friedmann, based on **Albert Einstein**'s theory of general relativity.

Hubble's law means that space was smaller in the past. Today, cosmologists interpret the data to indicate that all of space and time—the universe as we know it—began in an enormous explosion called the **Big Bang**, some 13.7 billion years ago. Hubble had profoundly changed our understanding of the cosmos.

SEE ALSO Big Bang (c. 13.7 Billion BCE), Birth of Spectroscopy (1814), Doppler Shift of Light (1848), Einstein's "Miracle Year" (1905), Age of the Universe (2001).

LEFT: *Edwin Hubble.* RIGHT: *The Hubble Space Telescope, named in honor of the discoverer of the expanding universe, took this unprecedentedly sensitive photograph during more than 11 days of exposure on the same tiny patch of the sky in 2004. Almost every dot and smudge in this photo is a galaxy!*

Discovery of Pluto

Clyde Tombaugh (1906–1997), **Percival Lowell** (1855–1916), **William Henry Pickering** (1858–1938)

The 1846 **Discovery of Neptune** was enabled by French mathematician Urbain Le Verrier's calculations of the likely position of a planetary-mass object that was causing variations in the orbit of Uranus. Subsequent observations of Uranus and Neptune led some astronomers to speculate that Neptune was not responsible for all the discrepancies in Uranus's orbit—there could still be an Earth-size planet lurking way out there.

One of the astronomers who believed in this "Planet X" was Percival Lowell, the New England businessman who had founded Lowell Observatory in Flagstaff, Arizona, in 1894. Both he and the American astronomer William Henry Pickering generated predictions for where Planet X might be found. An unsuccessful search was carried out from Flagstaff from 1909 until Lowell's death in 1916, when the search was put on hold during a battle over Lowell's estate. Pickering's own 1919 search also failed.

The search in Flagstaff resumed in 1929 and was placed in the hands of 23-year-old Clyde Tombaugh, a new staff member who had impressed Lowell Observatory director Vesto Slipher by sharing with Slipher the observations and sketches he made while growing up in Kansas. Tombaugh used Lowell's 13-inch (33-centimeter) astrograph (a telescope designed to work with large photographic plates) to survey the expected regions of the sky for objects moving at the right trans-Neptunian rate. After almost a year's effort, on February 18, 1930, he found a small, new world in the right place. A British girl won a subsequent naming contest by choosing the moniker Pluto, after the Roman god of the underworld. Ironically, modern reanalysis of the orbit of Uranus showed that there was no discrepancy beyond that which Neptune could explain; Tombaugh's discovery of Planet X was pure skill and coincidence.

Nonetheless, for more than 75 years Pluto was the solar system's ninth planet, orbiting out near 40 astronomical units, and now known to have five moons (one, **Charon**, relatively large). In the 1990s, however, it became clear that Pluto is simply one of the largest of the small icy worlds in the **Kuiper Belt**. In 2006, the ninth planet was controversially and unceremoniously demoted to "dwarf planet" status.

SEE ALSO Pluto and the Kuiper Belt (c. 4.5 Billion BCE), Discovery of Neptune (1846), Triton (1846), Charon (1978), Kuiper Belt Objects (1992), Demotion of Pluto (2006), Pluto Revealed! (2015).

Clyde Tombaugh's original Pluto discovery plates from his photos taken at Lowell Observatory on January 23 (top) and January 29 (bottom), 1930. Pluto is indicated by the white arrow in each plate.

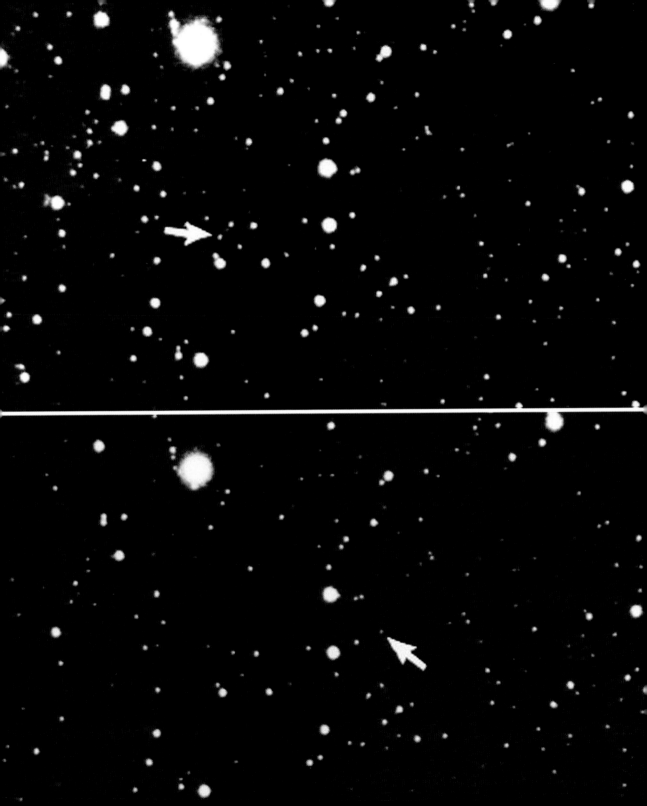

Radio Astronomy

Karl Guthe Jansky (1905–1950)

Young Karl Guthe Jansky grew up in Oklahoma surrounded by physics and radio science: his father was a professor of electrical engineering and dean of the engineering college at the university in Norman, and his older brother was a radio engineer. It was no wonder that he went to college, majored in physics, and in 1928 took a job with the fledgling Bell Telephone Laboratories, the relatively new research arm of Alexander Graham Bell's original American Telephone & Telegraph Company (AT&T).

At Bell Labs, Jansky studied the problem of how noise and static might interfere with possible long-distance transatlantic radio telephone service. He needed a way to monitor the intensity and direction of noise sources, so he built a radio telescope, 100 feet (30 meters) long, steerable on a set of four Ford Model T tires. The telescope could detect radio signals with a wavelength half its length, or a frequency around 20.5 megahertz (MHz).

In the summer of 1931 Jansky began "observing" with his radio telescope. He was successful in finding sources of background static, detecting radio signals from nearby and distant thunderstorms as well as a faint and relatively steady hiss that he couldn't initially identify. Over time he found that the strength of the signal varied with a period of 23 hours and 56 minutes—the exact duration of the sidereal day (the time it takes the Earth to spin on its axis relative to the fixed stars). He found that the hiss was most intense when looking toward the constellation Sagittarius, and specifically toward the region that astronomers had identified as the center of the **Milky Way** galaxy.

Karl Jansky had basically just invented radio astronomy. He had discovered what 40 years later was identified as the intense radio (along with X-ray and infrared) emission from Sagittarius A* (pronounced "A star"), a region thought to contain a 40-million-solar-mass **Black Hole** at the center of our galaxy. Jansky was not an astronomer, but his initiative, skill, and creativity with the world's first radio telescope inspired a new kind of astronomy, and a new way of seeing and studying the cosmos.

SEE ALSO "Daytime Star" Observed (1054), Size of the Milky Way (1918), Black Holes (1965).

The large radio antenna built in New Jersey in 1931 by Bell Labs radio engineer Karl Jansky to search for radio frequency noise. The antenna was 100 feet (30 meters) long and 20 feet (6 meters) high, and could rotate a full 360 degrees, earning it the nickname "Jansky's merry-go-round."

The Öpik-Oort Cloud

Ernst Öpik (1893–1985), **Jan Oort** (1900–1992)

By the early twentieth century, hundreds of years of careful observations had enabled accurate orbits to be calculated for dozens of bright comets. They appeared to come in two varieties: periodic short-period comets, with orbits that bring them back into the inner solar system roughly every 20 to 200 years; and periodic long-period comets, with long, highly eccentric orbits that could take hundreds to thousands of years or more to complete (or not to repeat at all, for the related nonperiodic single-apparition comets).

The trajectories of the longest-period comets take them extremely far from the Sun, with some traveling as far out as 50,000 to 100,000 astronomical units (almost a third of the way to the nearest star) when near their aphelion, or farthest point in their orbit. Several researchers independently noted that the cluster of aphelion distances at those extreme ranges probably meant that there was a supply or reservoir of comets that originated at those extreme distances. The fact that long-period comets come from all directions in the sky (not just along the plane of the ecliptic, like the planets and most short-period comets) also meant that this reservoir was likely spherical, like a huge cloud surrounding the solar system.

In 1932 the Estonian astrophysicist Ernst Öpik was the first to postulate the existence of this vast reservoir of comets in a paper describing the role that passing stars might have in gently nudging comets from this distant cloud into new orbits that would take them in toward the Sun. Independently, in 1950, the Dutch astronomer Jan Oort also came up with a similar idea, but expanded to include the role of **Jupiter** and the other giant planets in flinging inner solar system comets out into this enormous distant cloud.

Subsequent studies of new long-period comets (about one new one is discovered every year) confirm the idea: even though it's never been directly seen, a vast cloud of distant comets appears to surround the Sun. Astronomers now call this the Oort cloud (or Öpik-Oort cloud). By some estimates there could be a few Earth masses in a trillion or more kilometer-size comet nuclei out there, some formed closer in to the Sun but ejected into a perennial deep freeze, others formed at the edge of the Sun's gravity and waiting for their first gentle stellar nudge into the warmth.

SEE ALSO Halley's Comet (1682), Kuiper Belt Objects (1992), "Great Comet" Hale-Bopp (1997).

Artist's representation of the Oort Cloud (or Öpik-Oort Cloud) of comets surrounding the Sun at distances of 5,000 to more than 50,000 astronomical units. The cloud seems to have an inner (flatter) and outer (more spherical) population. The much smaller Kuiper belt (inset) is indicated for comparison.

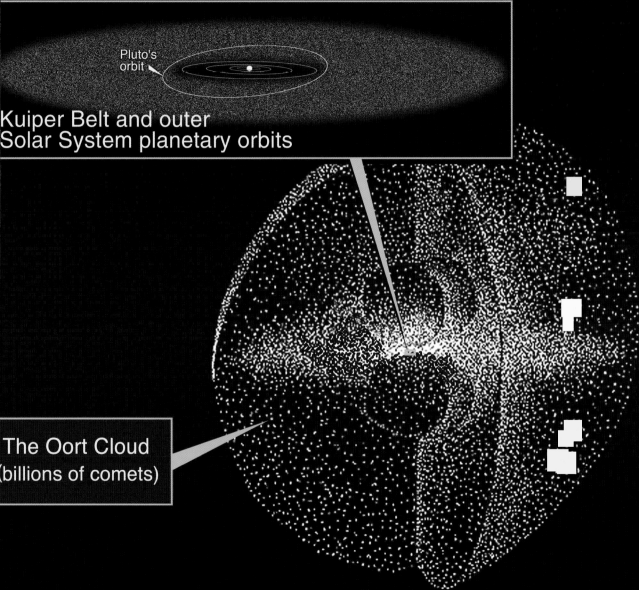

Kuiper Belt and outer
Solar System planetary orbits

Pluto's
orbit

The Oort Cloud
(billions of comets)

Neutron Stars

James Chadwick (1891–1974), **Walter Baade** (1893–1960), **Fritz Zwicky** (1898–1974)

Advances in astronomy on the grandest cosmic scales were paralleled in the early twentieth century by advances in physics and chemistry at the molecular, atomic, and subatomic levels. Indeed, atomic physics helped astronomers to make theories and predictions about processes that were difficult to observe directly.

A wonderful example of this synergy between the macroscopic and microscopic came about shortly after the English physicist James Chadwick discovered the neutron in 1932. Neutrons are subatomic particles with about the same mass as protons, but unlike protons and electrons, neutrons do not carry any electrical charge. Without the strong nuclear binding force contributed to an atomic nucleus by neutrons, positively charged protons would repel each other and atoms would be unstable and fly apart.

The discovery of the neutron had a profound effect on astronomy. For example, in 1933 the astronomers Walter Baade and Fritz Zwicky began thinking in detail about the processes that could lead to the gravitational collapse and explosion of massive stars — explosions that Zwicky dubbed supernovae. They speculated that the enormous central pressures and temperatures of such explosions could unbind atomic nuclei and leave compact, remnant objects behind composed mostly of bare neutrons. They called these hypothetical objects neutron stars.

By Baade and Zwicky's calculations, neutron stars should be rapidly spinning, extremely dense objects, packing one to two times the mass of the Sun into a sphere only about 6 to 7.5 miles (10 to 12 kilometers) across, and having a surface gravity more than 100 billion times Earth's gravity! Their theory was vindicated when a tiny, massive stellar remnant spinning at 30 revolutions per second was found in 1968 at the heart of the Crab Nebula, created in the **Supernova of 1054**. Thousands of other hot, spinning neutron stars — **Pulsars** — have since been discovered, providing astronomers with very precise "cosmic clocks" with which to study the extreme astrophysics of compact objects.

SEE ALSO "Daytime Star" Observed (1054), White Dwarfs (1862), Eddington's Mass-Luminosity Relation (1924), Pulsars (1967).

Hubble Space Telescope visible light image of a dim, lone neutron star, which had been first identified in X-ray telescope data as a highly energetic source. To explain both the visible and X-ray data requires the star to be extremely hot and extremely small — the perfect fit for a neutron star.

Dark Matter

Fritz Zwicky (1898–1974)

We deal with unseen forces all the time—the wind blowing through our hair; gravity pulling us downward. We can make observations or run experiments that tell us these forces are there, however, and that can ultimately reveal their source. In 1933 the Swiss-American astronomer Fritz Zwicky encountered some new evidence for the action of unseen forces in the cosmos, but ran into a major, paradigm-changing roadblock because there was no apparent, or perhaps even measurable, explanation for what he saw.

Zwicky studied clusters of galaxies—some of the largest structures known in the universe. In one study, using **Spectroscopy,** he was able to measure the red shifts and relative velocities of the members of the Coma cluster, a group of about one thousand co-moving galaxies about 320 million light-years from Earth. He discovered that the galaxies were moving relative to each other in a way that was inconsistent with their inferred masses. Even when Zwicky accounted for all the mass that he could see in visible light photographs, there still seemed to be an enormous amount—about four hundred times more than what he could see—of "missing mass" that was needed to explain the gravitational movements of the individual galaxies. This missing-mass problem led Zwicky to postulate that there must be some form of unseen matter, undetectable with then-modern methods, that was causing observed motions.

Even as new methods of radio, infrared, X-ray, and gamma-ray astronomy were developed, the missing mass in galaxy clusters—including apparently, within our own galaxy, based on the motions of nearby **Globular Clusters**—remained unseen. Astronomers now call this ubiquitous unseen material dark matter.

Many studies now require the existence of apparently undetectable material that has mass and exerts a gravitational influence on "normal" matter. Cosmologists believe it accounts for about 80 percent of all matter, making the mystery extremely profound and humbling. We appear to be a minor piece of a universe made of stuff that we simply don't yet understand.

SEE ALSO Globular Clusters (1665), Newton's Laws of Gravity and Motion (1687), Birth of Spectroscopy (1814), Hubble's Law (1929), Spiral Galaxies (1959).

Images of the "bullet" galaxy cluster from the ground-based Magellan telescope and the Hubble Space Telescope (orange stars) and the Chandra X-ray observatory (pink gas) have been augmented with blue regions depicting the areas where computer calculations show that most of the cluster mass appears to be concentrated. However, the mass associated with the theoretical "blue" regions here remains unseen, or "dark."

Elliptical Galaxies

Edwin Hubble (1889–1953)

The work of the astronomers Harlow Shapley, Vesto Slipher, Edwin Hubble, and others on determining the scale of the galaxy and collecting spectroscopic data for large numbers of spiral nebulae during the first few decades of the twentieth century eventually led to the realization that they are other galaxies—other Milky Ways—each harboring hundreds of billions of stars of their own. As more galaxies were identified and studied, and as it became clear that they were not all the same, astronomers naturally sought to classify them into distinct categories, as they had done for the stars.

As a leading observer of galaxies and with access to some of the best telescopic facilities in the world, Hubble was in a particularly unique position to take the lead on galaxy classification. And lead he did. In a series of papers and lectures, eventually compiled into a landmark 1936 book called *The Realm of the Nebulae*, Hubble outlined a scheme for the morphologic (shape, size, brightness) classification of extragalactic nebulae, now called the Hubble sequence.

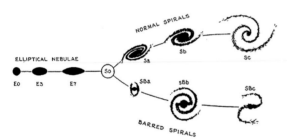

On one end of the Hubble sequence were the elliptical nebulae, now known as elliptical galaxies. Ellipticals are one of three main classes, the others being **Spiral Galaxies**, like our own Milky Way, and lenticular (lens-shaped) galaxies, intermediate in form between ellipticals and spirals.

Elliptical galaxies, as the name implies, are ellipsoidal to spherical in shape, and vary smoothly in brightness from a bright central core to diffuse outer edges. Modern surveys reveal that roughly 10–15 percent of galaxies in the local neighborhood consist of elliptical galaxies, but that fewer existed in the early universe. Ellipticals consist mostly of older, lower-mass stars, and are mostly devoid of the gas and dust needed for new star formation. The origin of elliptical galaxies is controversial, but some astronomers hypothesize that elliptical galaxies might be the end result of ancient mergers and collisions between former spiral galaxies.

SEE ALSO Cepheid Variables (1908), Size of the Milky Way (1918), Hubble's Law (1929), Spiral Galaxies (1959).

LEFT: *Edwin Hubble's original "tuning fork" galaxy classification scheme diagram, from* The Realm of the Nebulae. RIGHT: *Hubble Space Telescope image of the massive elliptical galaxy M87—home to trillions of stars, 15,000 globular star clusters, and a massive central black hole.*

Nuclear Fusion

Hans Bethe (1906–2005), Carl Friedrich von Weizsäcker (1912–2007)

While some of the major characteristics of stellar interiors—including their extremely high pressures and temperatures—had been worked out empirically by astrophysicists such as **Arthur Stanley Eddington** in the 1920s, substantial uncertainties remained about just how stars generated their energy. Eddington had thought about the possibility that nuclear fusion (the fusing of lighter elements into heavier ones, with the release of energy in the process) might power stars like the Sun. But this was only speculation, based partly on early nuclear transmutation experiments (the conversion of one element to another) by Ernest Rutherford and others.

Physicists soon had the means to work out and test theories for energy generation in stars in more detail, however. Among the pioneers in this field were the German physicist Carl Friedrich von Weizsäcker and the German-American nuclear physicist Hans Bethe. Between about 1937 and 1939, Bethe (in the United States) and Weizsäcker (in Germany) worked out the details of the ways that hydrogen atoms could

be fused into helium at the extreme conditions near the centers of stars. Bethe's 1939 paper, "Energy Production in Stars," described the specific nuclear chain reactions that were likely at work in the interiors of average-mass stars like the Sun, as well as in high-mass stars.

As exciting as these discoveries were scientifically, Weizsäcker, Bethe, and their colleagues knew that the nuclear genie was out of the bottle. Some nuclear synthesis chain reactions based on the same physics as reactions inside the stars could also be created artificially, releasing enormous amounts of energy. As World War II broke out, both the American and German governments set their physicists at work on the weaponization of nuclear fusion. The US effort was dubbed the Manhattan Project and involved Bethe as a leading theorist; it led to the development and detonation of the atomic bombs that ended the war, as well as the hydrogen bombs that began the long Cold War that followed.

SEE ALSO Radioactivity (1896), Eddington's Mass-Luminosity Relation (1924), Neutrino Astronomy (1956).

The direct result of figuring out the details of how energy is generated by nuclear fusion in stars like the Sun (inset, in an extreme ultraviolet color composite from the NASA and ESA Solar and Heliospheric Observatory, or SOHO, satellite) was the development of nuclear fusion weapons. This eerie nuclear fireball photograph was taken about one millisecond after the detonation of a nuclear bomb test in the Nevada desert in 1952.

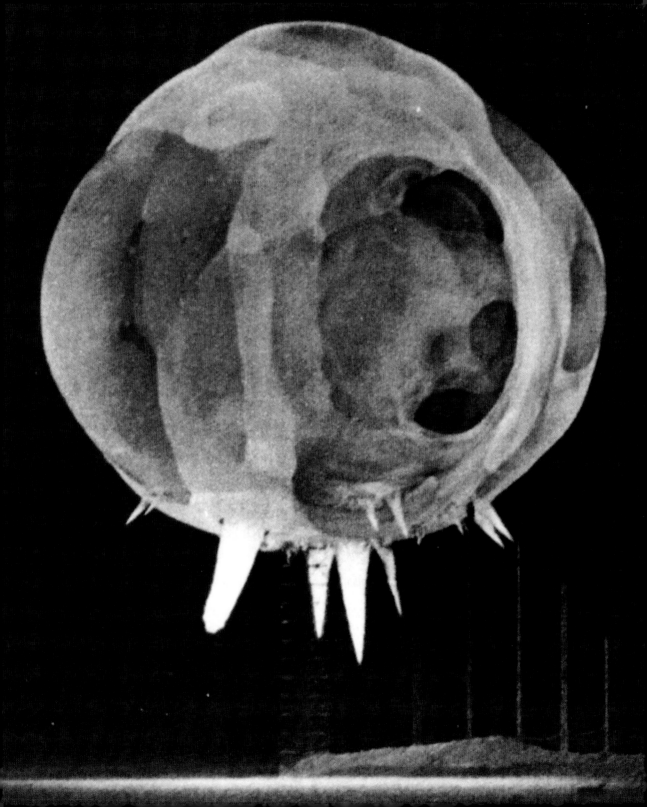

Geosynchronous Satellites

Hermann Oberth (1894–1989), **Herman Potočnik** (1892–1929), **Arthur C. Clarke** (1917–2008)

Newton's Laws of Gravity and Motion and Kepler's **Laws of Planetary Motion** in particular apply to artificial satellites just as well as to planets in orbit around a star or moons in orbit around a planet. Rocket technology and astronautics—the study of navigating through space—advanced quickly after the first **Liquid-Fueled Rockets** capable of reaching high altitudes were developed in the 1920s by Robert Goddard.

Several of Goddard's contemporaries were already beginning to think about the mechanics and dynamics of orbital (and beyond) rocket flights. Two of those contemporaries were the Hungarian-German physicist Hermann Oberth and the Austro-Hungarian rocket engineer Herman Potočnik, who expanded on and worked out the details of concepts first described by the Russian mathematician Konstantin Tsiolkovsky. One of those concepts was the idea of a geostationary, or geosynchronous, orbit.

A satellite in a geosynchronous orbit will complete one orbital revolution in the same amount of time that it takes the Earth to spin once on its axis. From the vantage point of an observer on the Earth's surface, such a satellite would appear to be parked high in the sky, never moving. Given the Earth's mass and rotation rate, Newton's second law can be used to derive the orbital altitude of a geosynchronous satellite—it turns out to be about 22,000 miles (36,000 kilometers) above the surface.

The British science fiction author and futurist Arthur C. Clarke was one of the first to grasp what would ultimately be one of the most practical applications of such satellite orbits: global telecommunications, described in a 1945 magazine article called "Extra-Terrestrial Relays—Can Rocket Stations Give Worldwide Radio Coverage?" Clark's popularization of the idea helped it to gain wide attention and support. Starting in 1964, the actual use of geosynchronous satellites has now gone far beyond just radio relays. Today they also relay TV, Internet, and global positioning system (GPS) signals and help us to monitor Earth's weather and climate.

SEE ALSO Three Laws of Planetary Motion (1619), Newton's Laws of Gravity and Motion (1687), Liquid-Fueled Rocketry (1926).

LEFT: *Space shuttle* Discovery *deploying the AUSSAT-1 communications satellite in 1985.* RIGHT: *A snapshot of satellites currently being tracked by the NASA Orbital Debris Program Office; Earth's ring of geosynchronous satellites can be clearly seen.*

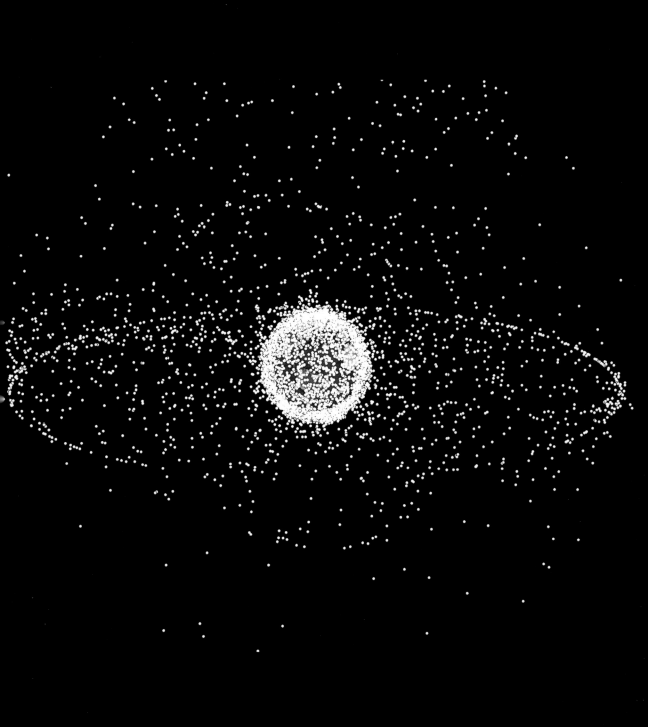

Miranda

Gerard P. Kuiper (1905–1973)

The rate of discovery of large (hundreds of miles in size) planetary moons dropped precipitously after the discovery of Saturn's moon **Phoebe** in 1898. After several decades with no major new moons discovered, solar system studies went out of fashion, probably because many early-twentieth-century astronomers probably thought that the census of the solar system's main bodies was relatively complete.

One scientist who continued to study and observe the planets, however, was the Dutch-American astronomer Gerard P. Kuiper. Starting in the late 1930s, Kuiper worked at the University of Chicago's Yerkes Observatory (home of the largest refracting telescope in the world) and the new McDonald Observatory in Texas, which had a telescope with the resolution and sensitivity that Kuiper needed to search for faint new planetary satellites. In 1948 he discovered the fifth, and innermost, satellite of **Uranus**. Following the established theme, he named the new moon Miranda, after a character from Shakespeare's *The Tempest*.

Little was known about Miranda besides its orbit, small size, and likely icy composition until the ***Voyager 2* Flyby** of the Uranus system in 1986. Miranda is indeed small, with a relatively spherical shape about 292 miles (470 kilometers) in diameter. But its surface is surprisingly diverse, consisting of a patchwork of regions covered by bright and dark ridges and cliffs interspersed with bland, heavily cratered, icy terrain. It looks like Miranda was taken apart and then put back together, clumsily. Some astronomers think an ancient impact may have done just that.

In 1949 Kuiper also discovered a second moon of Neptune, which he named Nereid. He was a founding father of modern planetary science, pioneering the use of **Spectroscopy** to study planets and moons. Kuiper discovered carbon dioxide in the atmosphere of Mars, a methane atmosphere around Saturn's moon **Titan**, and helped pick the Apollo lunar landing sites in the 1960s.

SEE ALSO Discovery of Uranus (1781), Titania (1787), Oberon (1787), Birth of Spectroscopy (1814), Ariel (1851), Umbriel (1851), *Voyager 2* at Uranus (1986).

LEFT: *Whole-disk* Voyager 2 *view of the jumbled surface of Miranda.* RIGHT: *High-resolution mosaic of part of the icy surface of Miranda, photographed during the* Voyager 2 *flyby of Uranus in 1986. The incredibly sheer cliff at the lower right is possibly more than 66,000 feet (20 kilometers) high.*

Jupiter's Magnetic Field

Just as in a standard wire-wound electric motor, rotating electrically conductive (metallic) interiors and/or cores of planets and satellites are prime locations for the generation of magnetic fields. The magnetic fields of Earth and **Mercury** are thought to be generated by electrical currents within their spinning, partially molten, iron-rich cores. High electrical conductivity in a metallic core deep within **Ganymede** may explain that moon's magnetic field as well.

Our solar system's gas- and ice-giant planets have been found to have even stronger magnetic fields. The first such discovery was in 1955, when radio astronomers from the Carnegie Institution in Washington, D.C., noticed strong radio frequency emissions coming from **Jupiter**. Starting with Karl Jansky's pioneering discovery of the strong radio source at the center of the Milky Way galaxy in 1931, radio astronomers had been scanning the skies looking for other natural sources of extraterrestrial radio waves. The Crab Nebula had been identified as another strong radio source. In fact, Jupiter's radio emission, interpreted to come from a strong magnetic field, was accidentally discovered in radio telescope observations intended to study the Crab Nebula.

Significantly more details about Jupiter's field came from flying spacecraft through it, first with the *Pioneers* in the 1970s, then the *Voyagers* in the 1980s, and then the *Galileo* orbiter and *Cassini* Jupiter flyby in the 1990s and early 2000s. Sensitive magnetometers on these missions were able to discover that Jupiter's magnetic field is about 10 times stronger than Earth's, and generates about 100 terawatts of power, or more than a million times Earth's magnetic field radio power. The field is generated by electrical currents flowing through metallic hydrogen in Jupiter's outer core. It interacts with the solar wind (as does Earth's field), but also interacts with the moons—especially **Io**, where volcanic eruptions inject sulfur dioxide into a doughnut-shaped "plasma torus" that encircles Io and is ionized by Jupiter's magnetic field.

And it is enormous: the volume of space encompassed by Jupiter's magnetic field—its magnetosphere—is the largest continuous structure in the solar system (not counting the Sun's own magnetic envelope). If we could see Jupiter's magnetosphere with our eyes, it would be five times larger than the full Moon!

SEE ALSO Violent Proto-Sun (c. 4.6 Billion BCE), Mercury (c. 4.5 Billion BCE), Jupiter (c. 4.5 Billion BCE), "Daytime Star" Observed (1054), Io (1610), Ganymede (1610), Solar Flares (1859), Radio Astronomy (1931), *Pioneer 10* at Jupiter (1973), *Galileo* Orbits Jupiter (1995).

A schematic cartoon illustration of Jupiter's magnetic field. Strong magnetic fields generated in Jupiter's deep interior extend outward and interact with the satellites and rings. The field is connected to Io and its plasma torus, for example, via a stream of high-energy particles called a flux tube.

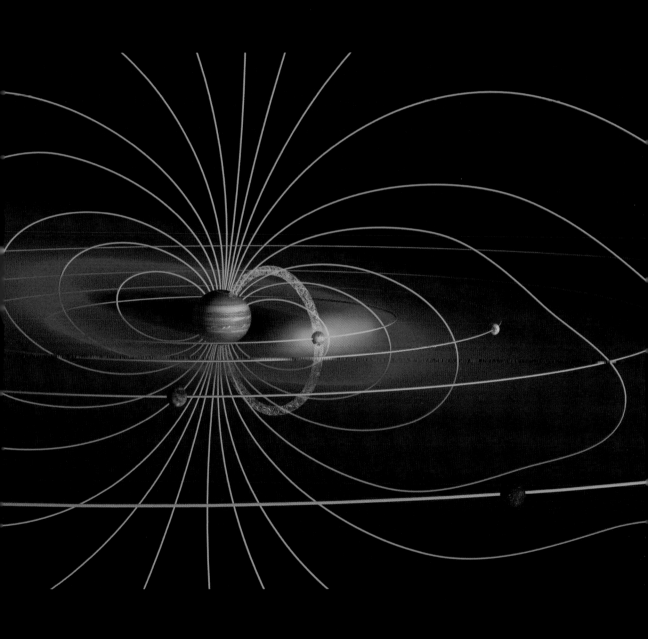

Neutrino Astronomy

Wolfgang Pauli (1900–1958)

Early-twentieth-century physicists, such as Wolfgang Pauli, who were trying to understand the nature of the radioactive decay of certain elements knew that they had big holes in their understanding of the atom. Energy released during radioactive decay, for example, couldn't be explained just by the protons and electrons created in the process. The discovery of the neutron in 1933 didn't help—it was too massive. An additional low-mass, electrically neutral elementary particle also had to be involved. Pauli had theorized the existence of a new subatomic particle that would later be called the neutrino ("little neutral one").

Actual evidence for the existence of the neutrino came in 1956, in high-energy particle accelerator collision experiments. Subsequent experiments in the 1960s showed that neutrinos come in different varieties, or "flavors," each associated with other kinds of elementary particles (including electrons), and that each flavor of neutrino also has an antiparticle. It was becoming evident that atoms are busy places.

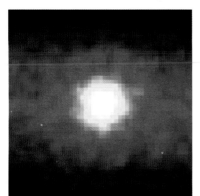

Discovery of the neutrino opened up the entirely new field of neutrino astronomy. Without a charge, and having almost negligible mass, neutrinos easily pass through even enormous quantities of ordinary matter at nearly the speed of light and with little attenuation. Neutrinos generated in the **Nuclear Fusion** reactions that occur deep inside the Sun and other stars thus pass almost effortlessly through the star and can be detected at Earth; by contrast, photons—sunlight—generated in the Sun's interior can take more than 40,000 years to bounce around and escape from that dense, opaque environment.

Today, special neutrino detectors enable us to understand more details of nuclear reactions occurring deep within the Sun, like supernovae, **Black Holes**, and even the **Big Bang**, as well as other inaccessible environments.

SEE ALSO Big Bang (c. 13.7 Billion BCE), Radioactivity (1896), Eddington's Mass-Luminosity Relation (1924), Neutron Stars (1933), Nuclear Fusion (1939), Black Holes (1965).

LEFT: *An image of the Sun taken by collecting 500 days of neutrino measurements using the detectors in the Japanese Super Kamiokande observatory.* RIGHT: *The interior of the Super Kamiokande chamber uses 11,200 photomultiplier tubes immersed in 50,000 tons of pure water to detect and measure neutrinos from the Sun and other cosmic sources.*

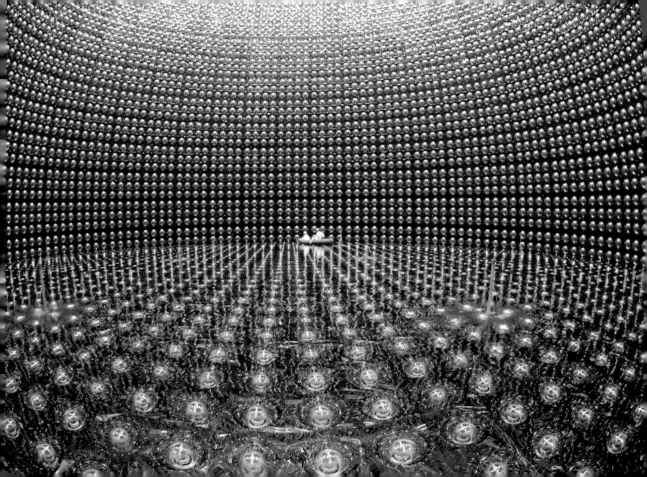

Sputnik 1

Sergei Korolev (1907–1966)

Americans tend to vividly recall where they were and what they were doing during key events that have come to define certain generations or eras. Examples include the bombing of Pearl Harbor, the assassination of John F. Kennedy, the explosion of the space shuttle *Challenger*, and of course the traumatic terrorist events of 9/11. For one generation of Americans, the defining event came in the fall of 1957.

On October 4, the Soviet Union became the first nation to launch an artificial satellite into space. Leading Soviet rocket engineer Sergei Korolev headed the team that had created the USSR's first intercontinental ballistic missile (ICBM), and he lobbied the government to allow him and his team to modify the R-7 rocket so that it could launch a small scientific payload into Earth's orbit. The Soviet government approved Korolev's plan in the hopes that they could beat the Americans into space. The payload was called *Sputnik*, Russian for "satellite." The Space Age had officially begun.

Sputnik 1 circled Earth every 96 minutes for three months, emitting a telltale beep-beep-beep on its single-watt radio, which could easily be picked up by ham radio operators around the world. The satellite created a sort of mild hysteria in the United States, where the public was acutely aware of the Soviet Union's ability to launch ICBMs—outfitted with nuclear warheads—to any target on the planet. The US government stepped up its own space efforts, and America's first satellite, *Explorer 1*, was successfully launched about two weeks after *Sputnik* burned up.

Sputnik also launched an unprecedented mini revolution in science and technology funding and education in the United States, the effects of which are still experienced today. The Americans most influenced by *Sputnik*—often referred to as the Apollo generation—went on to see America win the space race, watching 12 people walk on the Moon between 1969 and 1972, followed by decades of other stunning achievements.

SEE ALSO Liquid-Fueled Rocketry (1926), Geosynchronous Satellites (1945), Earth's Radiation Belts (1958), First Humans in Space (1961).

A replica of Sputnik 1, *the world's first artificial space satellite, housed in the National Air and Space Museum of the Smithsonian Institution in Washington, D.C. The metallic sphere is about 23 inches (58 centimeters) in diameter and the antennae (only partially shown here) extend out 112 inches (285 centimeters).*

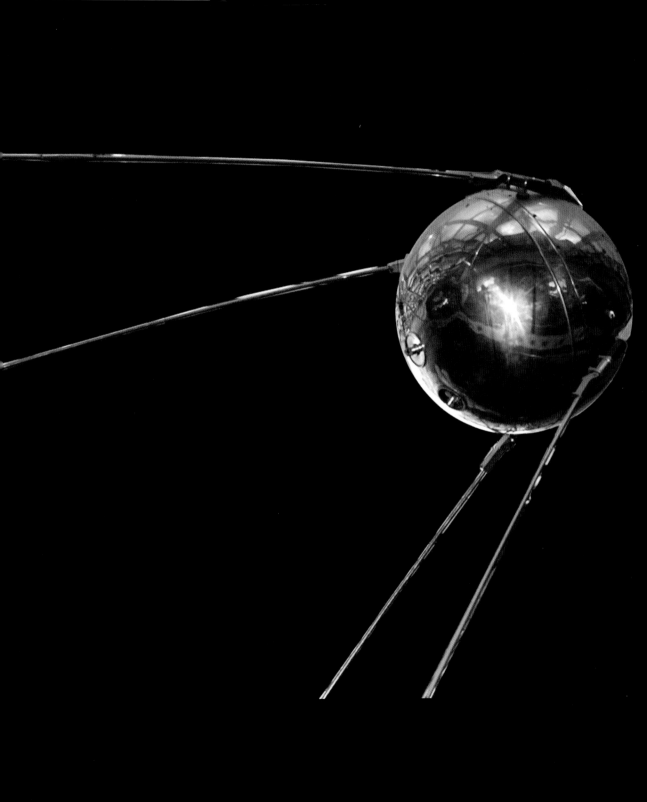

Earth's Radiation Belts

James Van Allen (1914–2006)

After the Soviet Union surprised the world with the successful launch and operation of *Sputnik 1* in the fall of 1957, the US government scrambled to catch up. The goal of launching and operating a small satellite with a minimal, simple science payload was passed along to a joint team composed of the Army Ballistic Missile Agency, responsible for the launch of the satellite on a modified Jupiter-Redstone intermediate-range ballistic missile, and an Army-Caltech (California Institute of Technology) facility near Pasadena called the Jet Propulsion Laboratory, which was responsible for the satellite—*Explorer 1*—and its scientific experiments.

Explorer 1's scientific payload—the brainchild of the American space scientist James Van Allen—consisted of a cosmic ray counter, a micrometeoroid impact detector, and some temperature sensors. The experiments were more complex than *Sputnik*'s simple radio transmitters, but low enough in mass, power, and volume to be deployed into orbit by the Jupiter missile.

America's first artificial satellite (and the world's third, after *Sputnik 2* in November 1957) was successfully launched from the Cape Canaveral Missile Annex in Florida on January 31, 1958. *Explorer 1* settled into a 115-minute elliptical orbit and operated its science instruments for more than three and a half months before its batteries died. During the mission the science instruments radioed back streams of data in real time to the JPL science team.

The data from *Explorer 1* were at first puzzling, because they appeared to be caused by dramatic increases in the number of energetic particles at certain altitudes and in certain locations around the Earth. Van Allen and his team interpreted the data to reveal the presence of a zone or belt of high-energy particles or plasma being confined by Earth's magnetic field. The results were confirmed a few months later, by *Explorer 3*. This was the first major space science discovery made by a satellite, and in honor of the science team's leader, the region of enhanced energetic particles in near-Earth space is now called the Van Allen Radiation Belt. *Explorer 1* was the first of 84 successful missions to date in the Explorer series.

SEE ALSO Violent Proto-Sun (c. 4.6 Billion BCE), Solar Flares (1859), Jupiter's Magnetic Field (1955), *Sputnik 1* (1957).

The aurora borealis, or northern lights, gleaming brightly over Bear Lake, Alaska, in January 2005. Auroral displays like this are the result of high-energy solar-wind particles interacting with Earth's magnetic field and with particles trapped in the Van Allen radiation belts.

NASA and the Deep Space Network

The success of **Sputnik 1** caused some soul-searching within the US government, which had been embarrassed by the Soviet launch. Part of the problem was that numerous federal agencies and branches of the military were splitting (or duplicating) efforts to try to make rapid progress in space exploration. As a result, in 1958 the US Congress and president Dwight Eisenhower established a new federal agency—the National Aeronautics and Space Administration (NASA)—to oversee the nation's civilian space and aeronautics programs. At the same time a military parallel, the Advanced Research Projects Agency (now DARPA, with *Defense* added to the name) was formed to oversee space-related technology for the armed forces.

Part of the consolidation of the civilian space effort also went into developing the critical communications infrastructure that would be needed to maintain contact and control of future space probes. For the early *Explorer* missions, the Army's Jet Propulsion Laboratory (JPL) at the California Institute of Technology (Caltech) had deployed portable radio tracking stations in California, Singapore, and Nigeria to maintain constant communication with the missions as they circled the globe. When control of JPL was transferred to NASA in late 1958, and JPL was assigned the leading role in coordinating an ambitious program of future robotic space missions, it became clear that a more permanent communications solution would be needed.

The solution was to establish the Deep Space Network (DSN), a set of small, medium, and large radio telescopes spaced roughly equally around the world so that they could remain in constant contact with NASA's space missions. DSN stations were established in Goldstone, California; near Madrid, Spain; and in Canberra, Australia, and each was outfitted with one large (230 feet [70 meters]) and several smaller (112 feet [34 meters]) radio telescopes and their transmitters and other equipment.

The DSN is the Earth's switchboard, the solar system's hub of interplanetary radio communications. Twenty-four hours a day, seven days a week, the stations and their dedicated and tireless staff collect data from, command, and sometimes even rescue civilian spacecraft that are part of what is today an incredible armada of more than 60 (soon increasing to more than 90) active Earth and planetary missions being run by NASA and other international space agencies.

SEE ALSO *Sputnik 1* (1957), Voyager Saturn Encounters (1980, 1981), *Voyager 2 at Uranus* (1986), *Voyager 2 at Neptune* (1989), Pluto Revealed! (2015).

The 230-foot (70-meter) diameter radio telescope of the NASA/JPL Deep Space Network (DSN) at Goldstone, California. Similar DSN telescopes are located in Spain and Australia to maintain a constant vigil over humanity's fleet of interplanetary exploration spacecraft.

Far Side of the Moon

The Moon orbits Earth with what is called synchronous rotation; that is, the Moon spins on its axis exactly once for every orbit about the planet. But from our perspective on Earth, it doesn't look like the Moon is spinning at all, because in synchronous rotation the same hemisphere (or "face") of a satellite is always pointed toward its primary planet. Our familiar view of the full Moon is thus a view of what astronomers call the near side, the side always facing us.

Until the space age, no one had ever seen the side of the Moon that always faces away from us—the far side. In fact, the only way to see it would be to send a spacecraft out beyond the Moon and turn it around to take a picture looking back. And that's exactly what the Soviet Union did in 1959 with the *Luna* 3 mission—another first for the Soviet space program.

Luna 3 was launched toward the Moon on October 4, 1959 (just two years after **Sputnik 1**). Three days later, after passing by the Moon's south pole, it became the first successful three-axis stabilized spacecraft when ground controllers directed it to take photos as it looped around the far side. Twenty-nine film pictures were taken on board and then later scanned, digitized, and radioed back to Earth.

The *Luna* 3 photos are fairly low quality compared to modern space photography, but they were good enough to enable Soviet space scientists to map and name features and to discover that the Moon's far side is dramatically different from the near side. The far side is more uniformly bright, with fewer dark mare (lava-filled impact basins; pronounced MAH-ray) than the near side. Much better follow-up photos of the far side from the Soviet *Zond* 3 satellite in 1965 showed the bright areas to be heavily cratered, rugged highlands regions.

Synchronous rotation turns out to be the usual situation among large planetary moons—all of the major satellites are synchronous rotators. Astronomers believe that this is because of the effects over time of tidal forces between a planet and its moon(s). Tides dissipate energy, slowing down a moon's spin until it reaches a stable, so-called tidally locked orbit, with a near side always facing the planet and a far side always facing away.

SEE ALSO Io (1610), Europa (1610), Ganymede (1610), Iapetus (1671), Origin of Tides (1686), Enceladus (1789).

LEFT: *A modern digital astrophotograph of the more familiar near side of the Moon.* RIGHT: *One of the first images of the far side of the Moon (the side always facing away from the Earth), taken by the Soviet* Luna 3 *satellite in October 1959.*

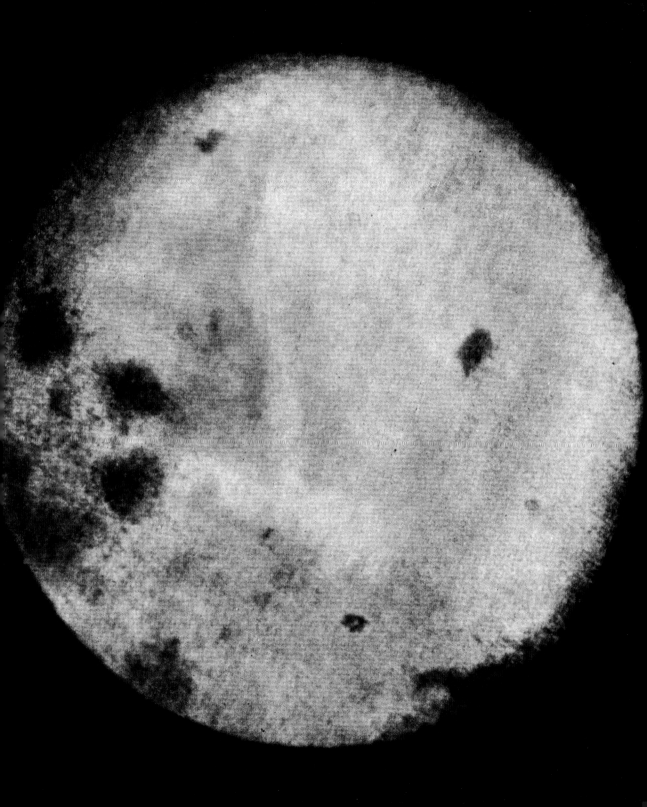

Spiral Galaxies

Harlow Shapley (1885–1972), **Edwin Hubble** (1889–1953), **Fritz Zwicky** (1898–1974), **Vera Rubin** (b. 1928)

In the galaxy classification scheme defined by Edwin Hubble in the 1920s and 1930s, spiral galaxies—immense numbers of stars bound together and sculpted by gravity, with two or more arms slowly spinning around a common center of mass—represent an extreme shape (the others being elliptical and lenticular galaxy forms). Some spiral galaxies are seen face-on, and others are seen edge-on. Edge-on galaxies reveal that the arms are confined to a wide, flat disk with a central bulge, all of which is surrounded by a halo of distant stars and **Globular Clusters**. As worked out by Harlow Shapley and others, our own galaxy—the **Milky Way**—is more than 100,000 light-years across.

Around 1959, radio astronomers developed techniques to use observations of the strong (21-centimeter) emission line of hydrogen detected by **Spectroscopy** to map out the rotational velocities of the arms of face-on spiral galaxies. The expectation was that stars farther from the galactic center would rotate more slowly, following **Kepler's Laws**, like planets orbiting a central star. But the observations revealed that the velocities of stars remain roughly constant with distance from the galactic center. This "galaxy rotation problem" was confirmed by more and more spiral galaxy observations.

In the mid-1970s, the American astronomer Vera Rubin proposed a solution: if most of the mass in spiral galaxies consisted not of the kind of matter that we could see through telescopes but instead of the same kind of unseen **Dark Matter** that Fritz Zwicky had proposed in 1933, then the observed non-Keplerian motions of spiral galaxies could make sense. Most astrophysicists today accept the existence of dark matter based on evidence presented by Zwicky, Rubin, and others.

Spiral galaxies are beautiful and ancient structures that are believed to have formed in the early universe and that continue to evolve today. Enshrouded in halos of mysterious dark matter and hosting million-solar-mass **Black Holes** in their cores, they are like enormous pinwheels in the cosmos, spinning along for billions of years on the virtual winds of the **Big Bang**.

SEE ALSO Andromeda Sighted (c. 964), Doppler Shift of Light (1848), The Milky Way Rotates (1927), Dark Matter (1933), Elliptical Galaxies (1936), Black Holes (1965).

M51, the Whirlpool galaxy, photographed by the Hubble Space Telescope, shows luminous hydrogen emission (red) in the strongest regions of new star formation.

SETI

Giuseppe Cocconi (1914–2008), **Philip Morrison** (1915–2005), **Frank Drake** (b. 1930)

Are we alone—in the solar system, the galaxy, or the universe? People have asked that question throughout (and before) recorded history, but only in the last century or so have we started to develop the means to find out. The 1931 discovery of natural sources of extraterrestrial radio waves (from the galactic center, and, later, from **Neutron Stars** and other high-energy objects) provided a route to address the question, because it became clear to astronomers that radio signals could travel vast distances across interstellar and intergalactic space.

The concept of radio communication across vast galactic distances became a serious topic of scientific study in 1959, when the physicists Giuseppe Cocconi and Philip Morrison published a paper in *Nature* entitled "Searching for Interstellar Communications." In that work they described ways that astronomers might search for radio signals transmitted by other civilizations. The challenge was almost immediately taken up by the radio astronomer Frank Drake, who in 1960 used the 85-foot (26-meter) radio telescope at the National Radio Astronomy Observatory in Green Bank, West Virginia, to conduct the first dedicated search for nonnatural radio transmissions from nearby, Sun-like stars. Drake's search was not successful, but it spurred what has now been a more than 50-year international observing campaign called the Search for Extraterrestrial Intelligence (SETI). Drake also developed a crude estimate of the potential number of civilizations in our galaxy (N_C): number of stars (N_*) × fraction with planets (f_P) × number of habitable planets (n_{HZ}) × fraction with life (f_L) × fraction with intelligent life (f_I) × fraction with technological civilizations (f_C) × the lifetime of those civilizations (L). Known as the Drake equation, this yields estimates ranging from 1 (Earth) to potentially millions of intelligent civilizations in the Milky Way.

Given recent discoveries of **Extremophile** life forms on Earth and of habitable **Extrasolar Planets** around other nearby stars, many SETI participants (now including the general public, thanks to the Internet) remain optimistic about the long-term potential to make contact. Of course, if we don't listen, we may never know. . . .

SEE ALSO *Mars and Its Canals* (1906), Radio Astronomy (1931), Study of Extremophiles (1967), First Extrasolar Planets (1992), Habitable Super Earths? (2007).

This Hubble Space Telescope view of star formation within giant molecular clouds in the Eagle Nebula forms a fitting backdrop for the famous Drake equation.

$$N_C = N_* \times f_P \times n_{HZ} \times f_L \times f_I \times f_C \times L$$

First Humans in Space

Yuri Gagarin (1934–1968), Alan Shepard (1923–1998)

The Soviet Union's successful launch of **Sputnik 1** in 1957 marked the beginning of the Space Age, as well as the beginning of an epic geopolitical race for technological, military, and moral superiority with the United States. The Russians had launched the first animal into space—a dog named Laika onboard *Sputnik 2*—and the US was launching monkeys and chimpanzees, but both governments knew that the next big victory in the space race could only be claimed by launching a person into space.

The Soviet human spaceflight program was called Vostok, and, like the original Sputnik effort, it was based on adapting existing intercontinental ballistic missile rockets to accommodate a small passenger capsule. About 20 Soviet Air Force pilots were secretly screened for the privilege of becoming the first cosmonauts ("space sailors" in Russian); the man chosen to be first was Senior Lieutenant Yuri Gagarin. At the same time, the US human spaceflight program, called Project Mercury, was on a parallel track, modifying the Redstone missile to accommodate its small single-passenger capsule. Seven test pilots, from the air force, navy, and marines, were ultimately selected and became instant celebrities, even before their flights. Navy test pilot Alan Shepard was chosen to fly the first *Mercury* mission.

Both Vostok and Mercury had early (unmanned) launch failures; both teams had to demonstrate that their rockets would work with an empty capsule before government leaders would authorize a human-piloted flight. Both teams were neck and neck in the race to launch a person first in early 1961, and once again the Soviets scored an enormous international victory by successfully sending Gagarin into space first, for one orbit of Earth in *Vostok 1* on April 12, 1961. Three weeks later, Shepard became the second person—and first American—launched into space with his successful suborbital flight in the *Freedom 7* capsule.

The Russians had again taken the lead. But America upped the ante shortly after Shepard's flight, when president John F. Kennedy, in an address to Congress, called for NASA to land a **Man on the Moon** before the decade was out.

SEE ALSO Liquid-Fueled Rocketry (1926), *Sputnik 1* (1957), Earth's Radiation Belts (1958), First on the Moon (1969).

Cosmonaut Yuri Gagarin preparing to board his Vostok 1 *spacecraft on the morning of April 12, 1961. Seated behind him was his backup, cosmonaut German Titov, who eventually piloted* Vostok 2 *in August 1961, becoming the second person to orbit the Earth.*

Arecibo Radio Telescope

Going back to the time of the first astronomical telescopes used by Galileo and others in the early seventeenth century, astronomers have known that the way to increase a telescope's sensitivity and resolution is to simply make it larger. But physical limitations on the strengths of materials and the ability to grind and polish large glass lenses or silvered mirrors make it technically impossible to build useful single-lens telescopes larger than about 3 feet (1 meter) in diameter, or single-mirror telescopes larger than about 16 feet (5 meters) in diameter.

In **Radio Astronomy**, however, "mirrors" that reflect or transmit radio waves can be made out of metal—just as antennas are. Thus they can be made much larger. A group of radio astronomers and atmospheric scientists at Cornell University in the late 1950s and early 1960s realized that it would be possible to build a giant stationary radio telescope out of wire mesh inside natural, bowl-shaped depressions. A suitably "radio quiet" location was identified in the mountains near the town of Arecibo on the island of Puerto Rico, also conveniently located near the equator, which enables it to have a view of most of the sky. With the help of the Advanced Research Projects Agency, the US military's parallel agency to NASA, construction of the enormous semispherical dish and support structures for steerable radio transmitters and receivers hanging above the dish began in 1960. The Arecibo radio telescope became operational in the fall of 1963 and remains the largest single-dish telescope in the world.

Many important astronomical discoveries have been made at Arecibo. The rotation rate of Mercury and the topography on the surface of cloud-covered Venus were both discovered using radar observations early in the telescope's operation. One of the first known **Pulsars**—rapidly spinning neutron stars—was discovered by Arecibo astronomers in the heart of the Crab Nebula, and the first millisecond pulsar (rotating around 500–1,000 times per second) was discovered at Arecibo as well. Arecibo astronomers bounced the first radar waves off near-earth asteroids to determine their size and shape, and the telescope is still the premier observatory on Earth for determining the orbits of potentially hazardous asteroids that could represent an impact threat to our planet.

SEE ALSO Radio Astronomy (1931), SETI (1960), Pulsars (1967).

The semispherical radio dish of the Arecibo Observatory, 1,000 feet (305 meters) in diameter, nestled into the mountains of northwestern Puerto Rico. The platform hanging above the dish is a 900-ton structure that contains radio and radar transmitters, receivers, and support structures.

1963

Quasars

Maarten Schmidt (b. 1929)

Like astronomers working in visible-light wavelengths (so-called optical astronomers), early radio astronomers were keen to survey the skies using their newly developed radio telescopes to identify the most interesting natural sources of radio waves. In addition to strong radio sources with well-identified optical counterparts—like the center of the Milky Way galaxy, or the Crab Nebula supernova remnant of 1054—astronomers in the 1950s began discovering hundreds of strong radio sources with no corresponding optical counterparts. Many appeared very small in the sky, almost starlike, but clearly not stars. Astronomers began calling them quasi-stellar objects, or quasars for short.

In 1962, the brightest quasar known, called 3C 273, passed behind the path of the Moon's limb on multiple occasions, allowing radio astronomers to determine its position with very high accuracy. The Dutch-American astronomer Maarten Schmidt used this high-precision location data to search for and identify the quasar optically using the 200-inch (5-meter) Hale Telescope at Mount Palomar, which had become the world's largest reflecting telescope since coming online in 1948. Schmidt took spectra of 3C 273 and in 1963 discovered that it was indeed not starlike at all, but rather showed Doppler-shifted emission lines from hydrogen—which suggested that it was both a great distance from us (following **Hubble's Law**) and full of extremely high-speed (about 16 percent the speed of light!) ionized gas.

Quasars have since been discovered to be the brightest objects in the observable universe. The quasar 3C 273 turns out to be one of the closest to us, but it is still 2.4 billion light-years away, telling us that quasars are ancient features that were more common in the early history of the universe. Astronomers now believe that quasars are the violent inner regions of the centers of ancient active galactic nuclei, where enormous amounts of gravitational energy are being released as matter falls into **Black Holes** at the centers of quasar host galaxies. Energy from a spiraling-in disk of material is sometimes released in intense jets of radiation perpendicular to the disk; the brightest quasars appear to be sources where these luminous energy jets are pointed almost directly toward us.

SEE ALSO Neutron Stars (1933), Black Holes (1965), Pulsars (1967).

Space artist Don Dixon painted this dramatic representation of two spiral galaxies colliding, feeding material into their central black holes to power intense, jet-like emission of radiation—quasars—in the central active galactic nucleus region surrounding each black hole.

Cosmic Microwave Background

Arno Penzias (b. 1933), Robert Wilson (b. 1936)

Most cosmologists used Edwin Hubble's 1929 discovery of the expansion of space as evidence that the universe was smaller in the past, and that if you could go far enough in the past (about 13.7 billion years ago, by the latest estimates), you would discover that the universe had sprung into being with a **Big Bang** from an extremely hot, dense, and small point.

Not all astronomers, however, initially embraced the big bang theory. For example, an alternate model was proposed in 1948 that hypothesized that even though space was expanding, new matter (mostly hydrogen) was constantly being created to keep the density (amount of matter per unit of volume) constant over time. Called the steady state theory, this model described a universe with no beginning and no end, and, as strange as it seemed, it was consistent with the available astronomical data of the time.

Cosmologists came up with ways to test between the **Big Bang** and steady state models of the universe. For example, in the big bang model, the universe today should still exhibit a faint residual glow, with a characteristic pattern from the early **Recombination Era,** when electrons were deionized and space became transparent to photons. Predictions were that the temperature of that glow would be around 3–5 kelvins, just above absolute zero. The steady state model was not consistent with background radiation of this predicted amount or pattern.

Radio astronomers knew that the radiation signal should be most detectable in the microwave part of the spectrum (around wavelengths of 1–2 millimeters), and the race was on to detect it. The race was won in 1964, when the astronomers Arno Penzias and Robert Wilson detected unexplained and near-uniform background radiation corresponding to a temperature near 3.5 kelvins. The discovery—made at Bell Labs, where **Radio Astronomy** had been invented in 1931 by Karl Jansky—earned them the 1978 Nobel Prize in Physics.

Subsequent space satellite measurements have revealed the cosmic microwave background to have a temperature of 2.725 kelvins, with very tiny fluctuations in matter and space that appear to be the "seeds" that eventually grew into stars and galaxies.

SEE ALSO Big Bang (c. 13.7 Billion BCE), Recombination Era (c. 13.7 Billion BCE), First Stars (c. 13.5 Billion BCE), Radio Astronomy (1931).

A photo of the historic horn-shaped radio antenna at Bell Labs in Holmdel, New Jersey, that was used by Arno Penzias and Robert Wilson to discover the faint glow of the cosmic microwave background emission that had been predicted by the big bang model for the origin of the universe.

HORN ANTENNA

HAS BEEN DESIGNATED A

NATIONAL HISTORIC LANDMARK

THIS SITE POSSESSES NATIONAL SIGNIFICANCE
IN COMMEMORATING THE HISTORY OF THE UNITED
STATES OF AMERICA. SOUTH OF THE ANTENNA,
AND BOB WILSON WITH THIS HORN THEY AS ARNO PENZIAS
ANTENNA WERE GIVING THE FIRST EVIDENCE
OF THE ORIGIN OF THE UNIVERSE AND MADE
OF THE SUPPORTING OF THE BIG BANG THEORY
WHICH IS ONE OF THE UNIVERSE.

Black Holes

Roger Penrose (b. 1931)

Even though they are among the strangest, most exotic, and perhaps most misunderstood objects in the universe, black holes can really be thought of simply as collapsed stars. Part of their mysterious appeal comes from their fundamentally unobservable nature and the fact that we can only learn about them by observing the strange and beautiful ways that they modify and interact with their surroundings.

Once a big enough star, maybe 5 to 10 times the mass of our Sun, converts all its hydrogen to helium and other heavier elements, nuclear fusion reactions can no longer counterbalance gravitational forces, and the star collapses. The collapse eventually causes an enormous supernova explosion, which ejects much of the star's material into space. Some of that explosive energy goes into further compressing the dense remaining core of the star, however, which continues to contract, to grow more dense, and to radiate more energy. If the collapsed star's mass continues to grow (perhaps by stealing material from a companion star), at some point not even light may be able to escape from the object's gravity. The region around the core would then appear, from the outside, to look like a black hole. Physicists know of no force in nature that can stop the collapse; in 1965 the British astrophysicist Roger Penrose proved mathematically that black holes can be formed from collapsing stars and that massive stellar should shrink to an infinitely small point known as a singularity. Weird stuff, indeed.

But since gravity falls off as one moves away from such an object, beyond some distance (called the event horizon) any light or other radiation related to the black hole can escape and be observed. Much of this observed radiation is a result of gas or dust being accelerated to enormous velocities by the black hole's huge gravitational and magnetic fields. **Quasars** are thought to form in such areas.

Einstein's theory of relativity predicts that many strange things happen near a black hole's event horizon, including time itself appearing to stand still as viewed by an observer from the outside. Unfortunately, since information can never escape from a black hole, it may never be possible to truly know what they are like up close.

SEE ALSO "Daytime Star" Observed (1054), Neutron Stars (1933), Quasars (1963), Gravitational Lensing (1979).

An artist's impression of a supermassive black hole "stealing" gas from a companion in a binary star system. The gas forms an accretion disk around the black hole that can radiate enormous amounts of energy into strong polar jets as the gas falls into the black hole's gravity well.

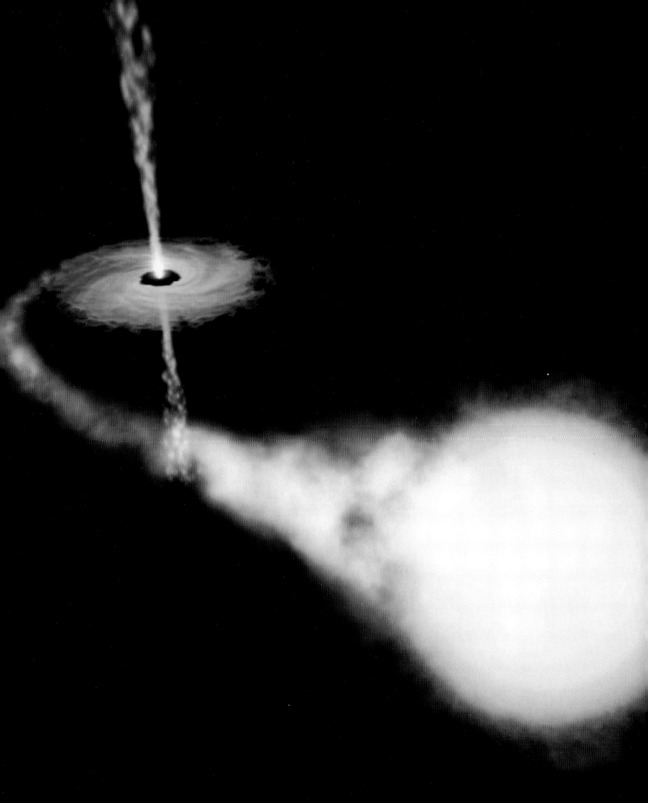

Hawking's "Extreme Physics"

Stephen W. Hawking (b. 1942)

Research by early- to mid-twentieth-century astronomers led not only to a detailed understanding of how the stars work, but also to the realization that they have finite lifetimes. Stars are born, they follow a fairly common lifestyle, and then they die, with all the details depending mostly on their mass. The most massive stars end up exploding as supernovae, leaving behind highly dense stellar cores that can become **Neutron Stars**, **Pulsars**, or—if massive enough—**Black Holes**.

Many astrophysicists are interested in understanding these kinds of compact, high-energy objects because they and their environs are places to study some of the most extreme physics in the universe. Among the most influential of these researchers is the British cosmologist Stephen W. Hawking, who began publishing papers about the physics of black holes while a Cambridge graduate student in 1965. He has also done important work refining theories of quantum gravity, wormholes, and the **Big Bang**.

Hawking is particularly interested in studying singularities, the infinitely small and dense remains of the runaway collapse of massive stellar cores. He and his Cambridge colleague Roger Penrose each made critical discoveries about these strange objects, which because of their super-high gravitational and magnetic fields turn out to be excellent places to also study Einstein's theory of general relativity and extreme examples of quantum mechanics. Hawking's theoretical studies have led to the realization that singularities are not only possible, but they are potentially abundant in the cosmos, from quasar host galaxies to individual black hole stellar remnants. Indeed, Hawking pointed out that the big bang itself began as a singularity, and so studying the origin and behavior of singularities is providing insight into the very origin of our universe.

When he was 21, Hawking was afflicted with a motor neuron disease related to amyotrophic lateral sclerosis (ALS), which left him paralyzed and relying on a computer to communicate. Notwithstanding an initially bleak prognosis, he has persevered and become a leading theoretical physicist, inspirational best-selling author, and staunch advocate for scientific education and literacy among the general public.

SEE ALSO Big Bang (c. 13.7 Billion BCE), Einstein's "Miracle Year" (1905), Eddington's Mass-Luminosity Relation (1924), Neutron Stars (1933), Black Holes (1965), Pulsars (1967).

Astrophysicist and cosmologist Stephen Hawking at Cambridge University in 2001. Mostly paralyzed by a motor neuron disease, Hawking uses facial muscles to control a computer and a voice synthesizer that enables him to write and to deliver lectures and speeches.

Microwave Astronomy

The search for evidence of the **Cosmic Microwave Background** radiation predicted by cosmologists spurred the development of sensitive new radio telescopes and receiver systems in the early 1960s. In particular, because the background radiation was predicted to be so cold (around 3 kelvins), instruments were built to make sensitive measurements at radio frequencies corresponding to wavelengths of about a millimeter to a meter—in the so-called microwave range of the electromagnetic spectrum.

In the process of scanning the sky to detect and isolate the faint cosmic microwave background radiation, microwave radio astronomers detected a number of much stronger and just as scientifically interesting signals, many of which were quite puzzling. For example, in 1965 microwave radio astronomers discovered strong sources of unknown radio emission at a wavelength near 18 centimeters (a frequency of 1665 MHz), soon identified to be isolated primarily to compact, high-energy radio sources (such as neutron stars and pulsars) embedded in dense interstellar clouds like the **Orion Nebula**. As the radio waves from those more distant sources passed through the intervening molecular clouds, strong spectroscopic absorption and emission lines appeared that provided diagnostic information about the compositions of those clouds. Because the radio waves at certain frequencies were intensified by passing through intervening materials, these kinds of objects became known as masers, an acronym analogous to "laser." Maser stands for "microwave amplification by stimulation of emission of radiation."

Maser spectroscopy allowed microwave astronomers to identify molecules in interstellar clouds for the first time. Initially, the detections included hydroxyl (OH) and water (H_2O) molecules, but later organic molecules like methanol (CH_3OH) and formaldehyde (H_2CO) as well as silicates like silicon monoxide (SiO) were identified. Masers were subsequently identified around some individual stars as well as entire galaxies, and even in the comas of some comets within our own solar system. Microwave astronomy now routinely enables water and other molecules to be widely mapped among a wide range of astrophysical objects and environments.

SEE ALSO Orion Nebula "Discovered" (1610), Radio Astronomy (1931), Cosmic Microwave Background (1964).

Part of an all-sky microwave emission map created by the European Space Agency's Planck *satellite in 2009.* Planck *measured the cosmic microwave background emission from the early universe as well as natural sources of radio energy, such as masers and the galactic disk (the white band in the image).*

Venera 3 Reaches Venus

Most of the media attention on the 1960s space race between the United States and the Soviet Union focused on the human spaceflight program and the quest to land the first astronauts on the Moon. But the race extended farther across the solar system as well. After the successes of the *Luna 3* and *Zond 3* lunar orbiters and their far-side photos and the *Ranger* 7, 8, and 9 lunar impact probes between 1959 and 1965, both the Soviets and the Americans began to focus more of their robotic missions on studying the nearby planets **Venus** and **Mars**.

Venus, Earth, and Mars form a triad of generally similar terrestrial planets that, when studied in detail, reveal important differences about planetary surface and climate evolution. Based on telescopic observations, the surface of Venus was known to be shrouded continuously in clouds; only radar observations from the **Arecibo Radio Telescope** in the early 1960s were able to reveal some details about the surface below. *Mariner 2* became the first robotic probe to fly past Venus in 1962, but much about Venus remained unknown, including a basic consensus about the planet's surface temperature and atmospheric pressure.

The Russians initiated the Venera robotic probe program to study Venus in more detail, using a series of flybys, orbiters, atmospheric probes, and landers. The *Venera 1* (1961) and 2 (1965) Venus flyby missions failed before reaching the planet, but the *Venera 3* mission did reach the planet and enter the atmosphere before losing radio contact with the Earth. While no scientific data were returned from the mission, *Venera 3* nonetheless became the first human artifact sent into another planet, crashing into Venus on March 1, 1966.

Persistence paid off for the Soviet space program, however; the follow-on missions of *Venera 4*, 5, and 6 in 1967–1969 were successful, with *Venera 4* providing the first direct measurements of the chemistry, temperature, and pressure of the atmosphere of another planet, and *Venera 5* and 6 providing more measurements of wind speeds, temperature, and pressure. These missions, and the successful 1967 US *Mariner 5* mission flyby, revealed Venus to be a hellish world, with a surface pressure more than 90 times that of the Earth and surface temperatures above 840°F (450°C).

SEE ALSO Venus (c. 4.5 Billion BCE), *Sputnik 1* (1957), Far Side of the Moon (1959), First on the Moon (1969), Second on the Moon (1969), Fra Mauro Formation (1971), Roving on the Moon (1971), Lunar Highlands (1972), Last on the Moon (1972), Venus Mapped by *Magellan* (1990).

A 1966 Soviet postage stamp commemorating the mission of Venera 3 *to the surface of Venus.*

ВЫМПЕЛ и МЕДАЛЬ НА ВЕНЕРЕ
1. III. 1966

ЗЕМЛЯ

ВЕНЕРА

ПОЧТА СССР

6 к

Pulsars

Antony Hewish (b. 1924), **Samuel Okoye** (1939–2009), **Jocelyn Bell** (b. 1943)

The astrophysicists Walter Baade and Fritz Zwicky proposed the concept of **Neutron Stars**—highly dense, compact stellar remnants from supernova explosions—back in 1933. However, it wasn't until 1965 that radio astronomers Antony Hewish and Samuel Okoye discovered the first observational evidence for a neutron star—a powerful but very small source of intense radio energy coming from the center of the Crab Nebula, the explosive remains of the famous **"Daytime Star"** supernova of 1054.

Hewish and colleagues at the University of Cambridge continued searching for new neutron stars and other radio sources. Just two years later, using the new, more sensitive four-acre radio telescope west of Cambridge, Hewish's student Jocelyn Bell discovered the first rapidly pulsating radio star ("pulsar") in the constellation Vulpecula, with a constant pulse rate of every 1.3373 seconds.

Bell and Hewish considered the possibility that the pulsar's eerily regular radio signal might be a sign of extraterrestrial intelligence (they had jokingly named the source LGM-1, for "Little Green Men-1"). However, by 1968 they and other astronomers had come up with a more plausible explanation, partly because the neutron star in the center of the Crab Nebula had also been discovered to be a radio pulsar, with a pulse rate of every 33 milliseconds. Pulsars were found to be rapidly spinning neutron stars with strong magnetic fields that "beam" some of their energy in specific directions (usually along or close to their rotation axis). If the beamed electromagnetic radiation from the spinning pulsar is aligned so that it sweeps past the Earth, it can "light up" radio telescopes like the spinning beacon of a lighthouse.

Several thousand pulsars have since been discovered, including several hundred-millisecond pulsars like the one in the Crab Nebula. Amazingly, variations in the timing of signals from the pulsar named PSR B1257+12 were interpreted in 1992 to be caused by the presence of planets orbiting the pulsar—the first examples of **Extrasolar Planets**.

SEE ALSO "Daytime Star" Observed (1054), Neutron Stars (1933), SETI (1960), Arecibo Radio Telescope (1963), First Extrasolar Planets (1992).

A high-resolution composite Hubble Space Telescope (red) and Chandra X-ray Observatory (blue) image of the central region of the Crab Nebula (Messier 1), a remnant from a supernova explosion in 1054. The central energy source is a pulsar—a rapidly rotating neutron star—with a 33-millisecond rotation period.

Study of Extremophiles

Thomas Brock (b. 1926)

Astrobiology is the study of the origin, evolution, and distribution of life and habitable environments in the universe. It is perhaps a unique discipline in that it has only one data point with which to conclusively justify its existence. So far, we know of only one example of life in the universe—that is, **Life on Earth**—all of which is fundamentally similar, based on similar RNA, DNA, and other carbon-based organic molecules.

The search for life elsewhere is more than just the search for complex life forms like us, however. It is a search for other planetary environments that could be suitable for the most dominant form of life on our own planet—bacteria and other "simple" life forms. The best place to start a search for those conditions is right here on our own planet, where significant advances in our understanding of habitability have been made in the past 50 years.

In 1967, the American microbiologist Thomas Brock wrote a landmark paper describing heat-tolerant bacteria (hyperthermophiles) that flourished within hot springs at Yellowstone National Park. He challenged the prevailing wisdom that the chemistry of life requires moderate temperatures to operate. Brock's work helped to spur the study of extremophiles   life forms that survive and even thrive in harsh environments.

Hyperthermophilic bacteria have since been identified in very hot water near deep sea hydrothermal vents as well; on the opposite extreme, psychrophiles have been found that live and thrive in near- or below-freezing temperatures. Life forms have also been found that exist over extremes of salinity (halophiles), acidity (acidophiles and alkaliphiles), high pressure (piezophiles), low humidity (xerophiles), and even high levels of UV or nuclear radiation (radioresistant).

The message for astrobiologists from the history of life on our planet is clear: life can thrive in an enormous range of environments. Thus, searching for evidence of past or present extremophiles or their habitable environments in extreme places such as Mars, the deep oceans of **Europa** and **Ganymede**, or the frigid, organic-rich surface of **Titan** is not as crazy an idea as it used to be.

SEE ALSO SETI (1960), An Ocean on Europa? (1979), Life on Mars? (1996), An Ocean on Ganymede? (2000), *Huygens* Lands on Titan (2005).

Morning Glory Pool, a hot spring in Yellowstone National Park, Wyoming. The colors along the spring's outer edges are from a variety of hyperthermophilic bacteria that can survive and thrive even at the high temperatures of the spring (above 176°F [80°C]).

First on the Moon

Neil A. Armstrong (1930–2012), **Edwin G. "Buzz" Aldrin** (b. 1930), **Michael A. Collins** (b. 1930)

After Yuri Gagarin became the **First Human in Space**, the race between the United States and the Soviet Union quickly became focused on the next big milestone: landing astronauts on the Moon and bringing them safely back to the Earth. The Soviet Vostok program was reoriented toward the larger rockets and landing systems needed for a lunar landing and return. In America, the challenge was to beat the Russians and to meet slain president John F. Kennedy's 1961 goal of doing it "before this decade is out."

A series of incrementally more advanced US missions were conducted between 1961 and 1969, starting with the Mercury single-astronaut flights, continuing with the Gemini two-person Earth orbital docking and rendezvous flights, and culminating in the three-person Apollo missions to the Moon. *Apollo* 8 achieved an important first in 1968 by sending the first humans to orbit the Moon and view the whole Earth and the lunar far side firsthand; the feat was repeated in early 1969 with the flight of *Apollo 10*,

a full dress rehearsal of a lunar landing that sent astronauts to within 10 miles (16 kilometers) of the lunar surface before returning home. Meanwhile, the Russians continued to make progress in their own secret lunar cosmonaut program. Several catastrophic unmanned launch failures in 1969 set their program back significantly, however, opening the door for an American victory.

That victory came on July 20, 1969, with the entire world watching as astronauts Neil A. Armstrong and Edwin G. "Buzz" Aldrin became the first humans to land, walk, and work on the moon. Armstrong and Aldrin landed on the ancient volcanic lava flows of the Mare Tranquillitatis (Sea of Tranquillity) impact basin (age dating of the samples showed them to be 3.6–3.9 billion years old), and spent about two and a half hours collecting samples and exploring the terrain. After less than a day, they took off and rejoined command module pilot Michael A. Collins back in lunar orbit for the three-day trip home—as world heroes.

SEE ALSO Birth of the Moon (c. 4.5 Billion BCE), Liquid-Fueled Rocketry (1926), First Humans in Space (1961).

LEFT: *Buzz Aldrin's bootprint in the fine, powdery lunar soil.* RIGHT: Apollo 11 *astronaut Aldrin unloads scientific equipment from the lunar module* Eagle *at the landing site in Mare Tranquillitatis (photo taken by Neil Armstrong).*

Second on the Moon

Charles "Pete" Conrad (1930–1999), **Alan Bean** (b. 1932), **Richard Gordon** (b. 1929)

Just four months after the successful voyage of *Apollo 11*, NASA astronauts were once again walking on the Moon. On November 19, 1969, astronauts Alan Bean and Charles "Pete" Conrad targeted the lunar module *Intrepid* toward a precision landing near NASA's 1967 *Surveyor 3* robotic lander, while command module pilot Richard Gordon circled high above. Conrad and Bean were able to successfully demonstrate pinpoint landing accuracy by setting their spacecraft down within 600 feet (180 meters) of *Surveyor 3*. Such accurate landings would be critical to future Apollo missions.

Conrad and Bean spent about 32 hours on the Moon, nearly a quarter of which was spent outside, taking samples and setting up experiments in the flat lava plains of Oceanus Procellarum (Ocean of Storms). Their longest excursion consisted of a walk to the *Surveyor 3* probe, where they took off some pieces and instruments to return to the Earth. *Surveyor 3* had been sitting on the lunar surface for more than three years; the returned parts provided detailed information on the long-term effects of vacuum, intense sunlight, and micrometeorite impacts on lunar surface equipment. Amazingly, there were even some dormant bacteria on some of the surfaces that had somehow made the trip to the Moon (and back) and that some tests showed still remained viable, even after three years in the Moon's harsh vacuum and ultraviolet (UV) environment.

The longer duration on the surface let the *Apollo 12* astronauts collect about 75 pounds (34 kilograms) of lunar samples (compared to 48 pounds [22 kilograms] from *Apollo 11*), including soils, small rocks, pieces of boulders, and impact crater deposits. Scientists analyzing those samples discovered that the Procellarum basin's dark volcanic rocks are much younger (3.1–3.3 billion years old) than those returned by the *Apollo 11* crew from the Tranquillitatis basin (3.6–3.9 billion years old), indicating that the Moon was volcanically active for at least the first 1.3–1.5 billion years of its history. The *Apollo 12* samples also contain different rock chemistries from those seen in Tranquillitatis, including dark glassy rocks, and the first samples of a new class of lunar rock made out of jumbled, impact-melted, and welded-together rocky fragments known as a breccia.

SEE ALSO Birth of the Moon (c. 4.5 Billion BCE), Arizona Impact (50,000 BCE), First on the Moon (1969).

Astronaut Alan Bean setting up part of a scientific experiment package on the plains of Oceanus Procellarum (Ocean of Storms) during the Apollo 12 mission in November 1969. The photo was taken by Bean's fellow astronaut Pete Conrad, whose shadow is in the foreground.

Astronomy Goes Digital

Willard Boyle (b. 1924), George Smith (b. 1930)

For millennia, astronomy was a visual craft, with practitioners relying on their keen eyesight and excellent night vision to find and characterize celestial objects. Even two hundred years after the invention of the Telescope, the only detector available to astronomers was still the human eye. The introduction of Astrophotography in 1839 provided astronomers with more-sensitive light-detection devices—first silvered glass plates and then sensitized photographic films. Although it was a huge advance in data archiving and observational repeatability, photography still provided only a modest improvement in the ability to record light from fainter astronomical sources.

A gigantic advance in astronomical detector sensitivity was made possible by the development of radar and navigation electronics for aircraft and weapons during World War II. These advances led to the invention of the first electronic switches—diodes—around 1939, and the first electronic amplifiers—transistors—in 1947. These devices relied on the properties of certain elements, such as silicon or germanium, that are not quite conductors (as metals are) and not quite insulators. Such semiconductors are materials that generally don't conduct electricity but that can be coerced, by applying the right kind of voltage or, in some cases, just by shining light on them, to conduct electricity.

The critical semiconductor advance for astronomers (and, ultimately, for digital camera and cell phone users) came in 1969, when the American physicists Willard Boyle and George Smith at AT&T Bell Labs created an array of semiconductors that converted incoming light—photons—into analog voltage signals that could then be stored, amplified, and converted into digital numbers. The invention was called a charge-coupled device (CCD) because the incident light on the array is coupled to the electrical charge generated.

Astronomers instantly fell in love with CCDs as they started becoming available in the 1970s and 1980s, partly because their output signal is linearly proportional to the brightness of objects shining on them, and partly because they are one hundred times more sensitive than photographic film. CCD cameras are now standard equipment for astronomical telescopes as well as space missions.

SEE ALSO First Astronomical Telescopes (1608), First Astrophotographs (1839), Einstein's "Miracle Year" (1905).

Example of a modern charge-coupled device (CCD) semiconductor detector used for astronomical as well as consumer electronics imaging.

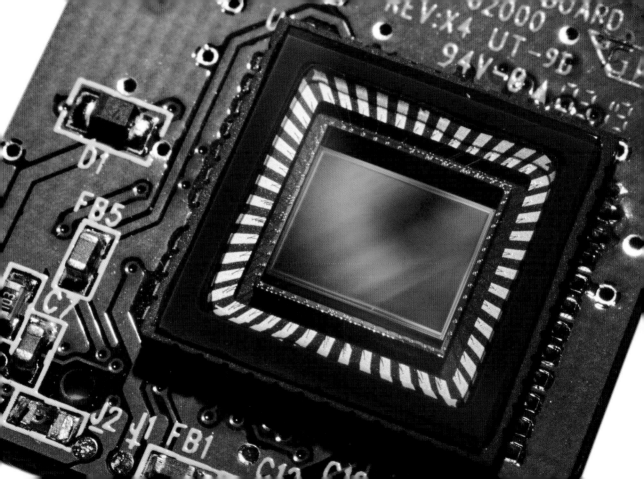

Organic Molecules in Murchison Meteorite

One of the motivators of space exploration is the search for life beyond our home planet. But how do we conduct such a search? One way is to search for the chemical elements that occur in life on our own planet—elements like carbon, hydrogen, nitrogen, oxygen, phosphorus, and sulfur. But these elements occur throughout the cosmos, in many places and environments (such as the insides of stars) that are unlikely to be conducive to life. A more effective strategy might be to search not for specific elements but for certain arrangements of them—molecules—that could reveal evidence for life's basic chemistry.

Life on Earth is based on organic molecules. Some organic molecules are simple, like methane (CH_4), methanol (CH_3OH), or formaldehyde (H_2CO), and others are much more complex, like proteins, amino acids, ribonucleic acid (RNA), and deoxyribonucleic acid (DNA). Over the last half century, astronomers have identified many simple organic molecules in dense interstellar clouds, comet tails, icy moons and rings of the outer solar system, and the atmospheres of Titan and the giant planets.

On September 28, 1969, a meteor and fireball streaked across the daytime sky and crashed to the ground near the town of Murchison in Victoria, Australia. More than 220 pounds (100 kilograms) of meteorite samples were found in the area. After detailed analysis of the samples, scientists announced in 1970 that the meteorite—from the most ancient and primitive class of meteorites, known as carbonaceous chondrites— contained some common amino acids. Later studies found that the Murchison meteorite contained more than 70 kinds of amino acids, plus many other simple and complex organic molecules.

Life as we know it requires liquid water, sources of energy like heat or sunlight, and abundant, complex organic molecules. The discovery of amino acids in Murchison and other meteorites supports the idea that molecules that are critical to life can form nonbiologically in environments such as a solar nebular disk, comet, or planetesimal. Life may or may not be abundant in the cosmos, but the stuff of life appears to be everywhere.

SEE ALSO Solar Nebula (c. 5 Billion BCE), Life on Earth (c. 3.8 Billion BCE), Saturn Has Rings (1659), Iapetus (1671), Halley's Comet (1682), Enceladus (1789).

X-ray image of magnesium (red), calcium (green), and aluminum (blue) in the Murchison meteorite, a more than 4.55-billion-year-old carbonaceous chondrite. This ancient rock contains primitive minerals condensed from the solar nebula, water, and complex organic molecules, including more than 70 kinds of amino acids.

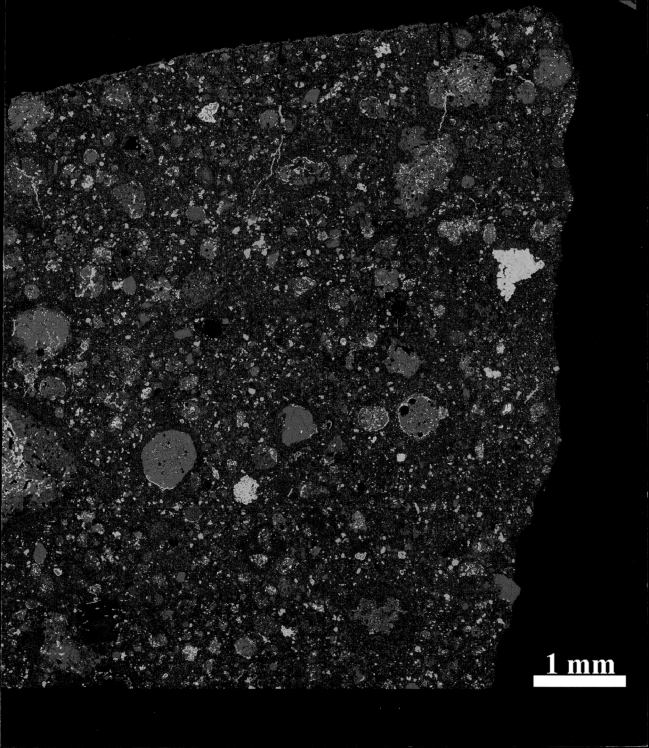

1 mm

Venera 7 Lands on Venus

Despite the failure of their lunar cosmonaut program in 1969, the Soviet Union continued to achieve remarkable successes with their robotic probes to the planets. The *Venera* 3–6 missions to Venus, undertaken from 1966 through 1969, did not survive to the surface, but still they provided enough new data on the planet's atmosphere to enable mission controllers to design subsequent missions to withstand the harsh conditions. Success came with the flight of *Venera* 7, which on December 15, 1970, became the first human-made object to land and return data from the surface of another planet.

Venera 7 floated down through the atmosphere on a parachute for 35 minutes before landing, after which it continued to return data for another 23 minutes. During that time the lander relayed temperature data showing the surface to be at about 870°F (465°C), with a surface pressure inferred to be about 90 times Earth's surface pressure. Under such scorching and crushing conditions, it's a wonder the lander survived at all.

The success of *Venera* 7 set the stage for an incredible string of successful follow-on Soviet Venera landers, orbiters, and atmospheric probes between 1972 and 1985 — the most ambitious long-term program of Venus robotic exploration ever conducted.

Highlights included geochemical measurements of the planet's surface composition (discovered to be similar at the landing sites to basaltic volcanic rock compositions found in places like Hawaii and Iceland); the first images of the rocky surface, dimly but adequately lit by sunlight filtering through the thick cloud layers; and the first large-scale maps of mountains, ridges, plains, and other tectonic and volcanic features, obtained using radar imaging from the *Venera 15* and *16* orbiters.

Besides the hellish surface conditions, the *Venera* probes also helped planetary scientists discover that the wind speeds in Venus's middle and upper atmosphere exceed 220 miles (100 meters) per hour, which is much faster than the planet itself rotates (one Venus day is about 243 Earth days). The cause of this super rotation of the Venus atmosphere is unknown, and it is one focus of ongoing and planned future missions to study Earth's so-called twin, Venus.

SEE ALSO Venus (c. 4.5 Billion BCE), *Venera* 3 Reaches Venus (1966), Venus Mapped by *Magellan* (1990).

LEFT: *An engineering test model of the Soviet* Venera 7 *landing capsule.* RIGHT: *A reprocessed view of part of the Venus surface panorama obtained by the Soviet* Venera 13 *lander on March 1, 1982. A lander leg and piece of a jettisoned camera cover are in the foreground.*

Don P. Mitchell

Lunar Robotic Sample Return

The Soviet Union's robotic space exploration program in the 1960s and 1970s achieved a string of scientifically important "firsts" in missions to the Moon, Venus, and Mars. Among the most important and most technologically impressive of these were *Luna 16*, *Luna 20*, and *Luna 24*, the world's first robotic sample return missions. These were small "automatic stations" that were launched from Earth on a five-day cruise to the Moon. They autonomously performed a soft landing on the lunar surface, drilled shallow holes into the surface to collect lunar samples, and then launched small sample return capsules back to Earth, where they were collected after a three-day return trip and parachute landing.

Luna 16, launched in September 1970, was the first of these autonomous sample-collection-and-return missions, returning 3.5 ounces (100 grams) of lunar soil and rock fragments from the dark lava plains of the Mare Fecunditatis impact basin. *Luna 20* repeated the feat in 1972, collecting 2 ounces (55 grams) of samples from a bright highlands region also near Mare Fecunditatis, and *Luna 24* did it again in 1976, returning 6 ounces (170 grams) of samples from Mare Crisium, a lava-filled impact basin near the Moon's eastern limb. The Luna specimens complement the much larger Apollo-returned sample collection, because they include some different and unique lunar surface compositions and mineralogies. Combined, the Apollo and Luna samples provide the fundamental information that drives current theories for the origin and evolution of the Moon—including providing the key pieces of supporting evidence for the giant impact model for the **Formation of the Moon**.

The Soviet Luna sample return missions were the most complex robotic missions yet attempted at the time. Other sample return missions have since been performed to bring back pieces of a comet tail (the Stardust mission), particles of the solar wind (the Genesis mission), and tiny fragments of a near-Earth asteroid (the Hayabusa mission). Still, the Luna sample return missions remain among the most complex planetary exploration missions ever conducted, an achievement made even more remarkable by the fact that they relied on 1960s technology. Future robotic sample return missions from other regions of the Moon, and from Mars, Venus, and near-Earth asteroids, continue to be proposed and planned.

SEE ALSO Birth of the Moon (c. 4.5 Billion BCE), Far Side of the Moon (1959), First on the Moon (1969), *Genesis* Catches Solar Wind (2001), *Stardust* Encounters Wild-2 (2004), *Hayabusa* at Itokawa (2005).

A model of the Soviet Luna sample return lander, used for three completely robotic lunar sample return missions in 1970, 1972, and 1976.

Fra Mauro Formation

Alan Shepard (1923–1998), **Edgar Mitchell** (b. 1930), **Stuart Roosa** (1933–1994)

Following the successes of *Apollo 11* and *Apollo 12* in 1969, NASA's plans for *Apollo 13* in 1970 were to send the next astronaut crew for a two-day stay to explore part of a geologic terrain on the Moon known as the Fra Mauro formation (near the crater Fra Mauro, named after a fifteenth-century Italian mapmaker). But an explosion inside the *Apollo 13* command module during the three-day cruise to the Moon caused the cancellation of the lunar landing and placed the lives of the crew in peril. Heroic efforts by the astronauts and their supporting teams on Earth helped get them home safely.

After a short delay to understand and fix the cause of the mishap, NASA resumed the landings by sending *Apollo 14* to *Apollo 13*'s original destination. The Fra Mauro formation is a widespread, bright, hilly terrain located between several of the large, dark-floored impact basins on the near side of the Moon. Lunar geologists speculated that the materials there might be impact ejecta—debris dug up and redeposited on the surface by the Moon's largest impact craters and basins. By sampling Fra Mauro, then, it might be possible to obtain samples from different impact events and different depths in the lunar subsurface without having to drill or traverse very far.

Alan Shepard, one of the original Project Mercury astronauts and the first American to travel in space, and his crewmate Edgar Mitchell landed the lunar module *Antares* on Fra Mauro on February 5, 1971. While colleague Stuart Roosa orbited above in the command module *Kitty Hawk*, Shepard and Mitchell spent nearly 10 hours walking on the Moon and collecting 93 pounds (42 kilograms) of samples. Shepard also made sports history by hitting the first golf balls on the Moon.

Lunar geologists and geochemists were delighted by the samples brought back by the crew of *Apollo 14*. Like some *Apollo 12* samples, many of the Fra Mauro rocks are impact breccias, samples from large and small asteroid impact events (including the gigantic 4-billion-year-old Mare Imbrium basin) spanning more than 500 million years. As such, the samples are a treasure trove of geologic history of both the Moon and our own planet, which was similarly bombarded early on.

SEE ALSO First on the Moon (1969), Second on the Moon (1969).

The Apollo 14 lunar module Antares, *photographed by astronauts Alan Shepard and Edgar Mitchell on the lunar plains near Fra Mauro crater. Tracks from the astronauts' two-wheeled handcart gleam in the sunlight because the powdery soil was compressed and flattened by the cart's wheels.*

First Mars Orbiters

Robotic planetary exploration for the past 50 years has followed a progression of ever more daring and technologically challenging space missions. Initially the goals were simply to learn how to navigate remote probes in space and make them fly by (or fly into) the Moon and other planets, sending back data in the form of pictures and other measurements. The next logical step is to try to establish orbiting satellites around other worlds—not only so that we can spend time learning about an alien environment but also so that we can map an extraterrestrial surface or atmosphere.

The first satellite to take this leap to the next level of planetary exploration and orbit another planet was NASA's *Mariner 9* space probe. *Mariner 9* arrived at **Mars** in November 1971, during the throes of what astronomers had noticed was a planetwide dust storm. While the spacecraft's spectrometers were taking data on dust properties and atmospheric temperatures, the probe's television cameras were seeing primarily a bland, dusty cue ball of a world, with a few dark spots poking out of the dust.

After almost an Earth year in orbit, however, the dust had cleared enough for the mission to begin to map the planet's surface in unprecedented detail. The Mars that was revealed in the *Mariner 9* images was a geologic wonderland of giant towering

volcanoes (the dark spots seen during the dust storm), enormous tectonically formed canyon systems, ancient river channels, and countless impact craters. It was a far cry from the impact-crater-dominated glimpse of Mars that had come from the previous Mariner flyby missions in 1965 and 1969.

The Soviet Union also took advantage of the opportunity to launch two Mars orbiters of their own in 1971, called *Mars 2* and *Mars 3*. Both went into orbit a few weeks after *Mariner 9*, and both returned useful scientific information about the atmosphere and surface (once the dust had cleared). Both also deployed small landers and mini rovers to explore the surface. Though neither lander was successful, they became the first human artifacts to impact the surface of Mars.

SEE ALSO Mars (c. 4.5 Billion BCE), *Vikings* on Mars (1976), Mars Global Surveyor (1997).

Part of a Mariner 9 photograph of a maze of ridges, troughs, mesas, and impact craters called Noctis Labyrinthus. This region is about 190 miles (300 kilometers) across and is near the largest canyon system on Mars, Valles Marineris (Mariner Valleys), named after the Mariner 9 probe (left).

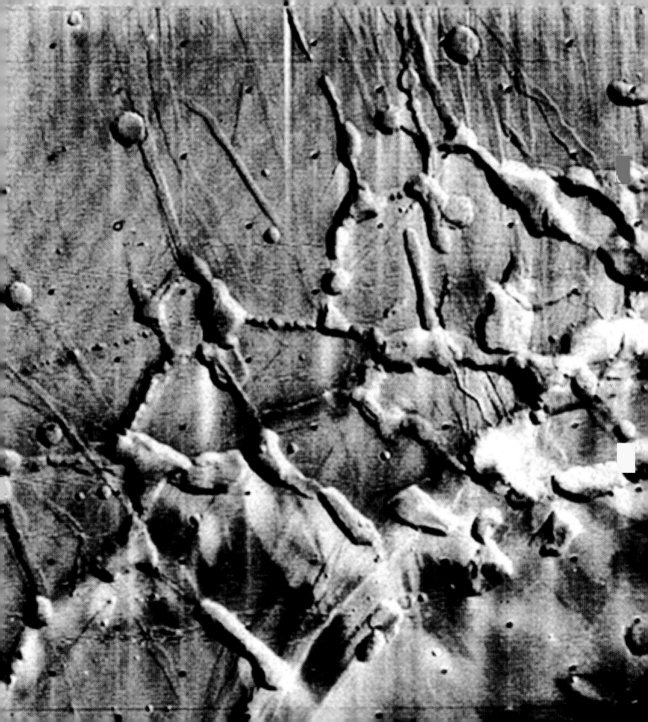

Roving on the Moon

James B. Irwin (1930–1991), **David R. Scott** (b. 1932), **Alfred M. Worden** (b. 1932)

NASA's first three Apollo Moon-landing missions were designed to be short, "plant the flag" kinds of visits, in which the main goal was to land accurately and safely, and then return home. The astronauts performed a limited amount of scientific activities, but they had strict limits on their mobility and time spent on the surface.

That changed for the last three missions in the Apollo lunar-landing program, however. For *Apollo 15, 16,* and *17,* NASA configured the gigantic *Saturn V* launch rocket so that it could carry nearly twice the mass to the Moon, enabling the astronauts to bring more supplies for a longer stay, to set up more experiments, and to substantially increase their mobility on the surface by bringing along a lunar roving vehicle. The last three Apollo missions were thus much more focused on lunar science; in many ways they were the first—and last—great human exploration voyages in space.

Apollo 15 was the first of these extended voyages, sent to explore a rugged region in the Apennine Mountains between the Mare Serenitatis and Mare Tranquillitatis impact basins. Lunar geologists wanted astronauts David R. Scott and James B. Irwin to use the lunar rover to explore a 60-mile-long (100 kilometers) ancient collapsed lava tube known as Hadley Rille that snaked along the flat floor of one of the Apennine valleys. On July 30, 1971, Scott and Irwin undocked the lunar module *Falcon* from the command module *Endeavor,* bidding a temporary farewell to pilot Alfred M. Worden, and executed a pinpoint landing, avoiding the steep mountain peaks, setting down only about 0.6 miles (1 kilometer) from the edge of Hadley Rille.

The *Apollo 15* mission was a great success. Scott and Irwin spent nearly three days on the Moon, almost 19 hours of which was spent driving the rover to different sample and vista sites and collecting nearly 170 pounds (77 kilograms) of precious lunar rocks and soils. The samples confirmed the volcanic origin of Hadley Rille and helped to reveal that the Moon was volcanically active—including with what must have been spectacular fire-fountaining volcanic eruptions—as "recently" as 3.3 billion years ago.

SEE ALSO Birth of the Moon (c. 4.5 Billion BCE), First on the Moon (1969), Second on the Moon (1969), Fra Mauro Formation (1971).

Astronaut Jim Irwin, photographed by fellow astronaut Dave Scott, loading up their lunar rover with equipment unpacked from the lunar module Falcon. *Scott and Irwin drove the rover nearly 17.5 miles (28 kilometers) in their exploration of the Hadley Rille lava tube.*

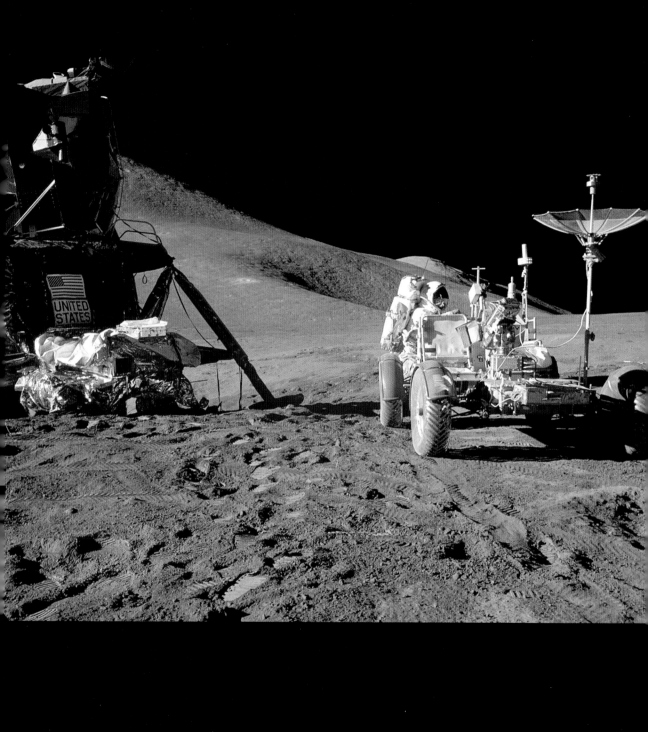

Lunar Highlands

John W. Young (b. 1930), **Charles M. Duke, Jr.** (b. 1935), **T. Kenneth Mattingly II** (b. 1936)

The four Apollo lunar-landing missions prior to 1972 had primarily targeted the Moon's flat, dark, volcanic plains or gently rolling hills close to major volcanic regions. The decision to choose those areas as landing sites was based mostly on landing safety considerations. Lunar scientists knew that the dark mare (flat, volcanic) regions cover less than 20 percent of the Moon's surface, however, and so more than 80 percent of the Moon's more typical bright, mountainous geology—the highlands—was going unexplored. The goal of the *Apollo 16* mission was to rectify that imbalance.

Astronauts John W. Young and Charles M. Duke, Jr., guided their lunar module *Orion* down into a rugged highlands region near the crater Descartes, in a smoother area containing a unit that geologists call the Cayley Plains. The hypothesis before landing was that the Cayley Plains, a relatively flat unit that fills in many craters and valleys among the highlands in many places across the Moon, might be a different kind of volcanic deposit that is unique to the lunar highlands. Because of its widespread nature, sampling this material was important to understanding the Moon overall.

Young and Duke spent more than 20 hours, over the course of four "extravehicular activity" excursions, walking on the Moon and driving their lunar rover more than 17 miles (27 kilometers) to collect a wide variety of highlands samples. After three days on the surface, they redocked *Orion* with the command module *Casper* and rejoined pilot T. Kenneth Mattingly II for the three-day return to Earth.

Surprisingly, the *Apollo 16* samples did not show evidence for widespread highlands volcanism. Instead, the highlands are dominated by lower-iron, lower-density silicate minerals than the mare regions, a discovery that led to the idea that the Moon once had a molten "magma ocean" crust, enabling the heavier elements to differentiate or sink into the mantle and core. The Cayley Plains turned out to be a widespread deposit of ejecta—smashed-up, mostly highlands material redistributed over the landscape by the bombardment of the lunar surface by asteroids and comets over the eons.

SEE ALSO Birth of the Moon (c. 4.5 Billion BCE), First on the Moon (1969), Second on the Moon (1969), Fra Mauro Formation (1971), Roving on the Moon (1971).

The Apollo 16 command and service module Casper, *piloted by Ken Mattingly and photographed in lunar orbit by astronauts John Young and Charlie Duke after they undocked the lunar module* Orion *and prepared to land on the Cayley Plains.*

Last on the Moon

Eugene A. Cernan (b. 1934), **Harrison H. "Jack" Schmitt** (b. 1935), **Ronald E. Evans, Jr.** (1933–1990)

The last great voyage of lunar discovery in the Apollo Moon-landing series was conducted in December 1972. Astronauts Harrison H. "Jack" Schmitt and Eugene A. Cernan landed the lunar module *Challenger* in a narrow valley south of Littrow crater in the Taurus Mountains, which ring the southeast rim of the Mare Serenitatis. The area was chosen because it is along a boundary between dark volcanic mare materials and bright highlands materials, and evidence from orbital photographs suggested that the region would provide a rich diversity of geologic information about the Moon.

Schmitt and Cernan did indeed have an amazing adventure during their stay on the Moon. They set Apollo program records by being out on the surface for nearly 22 hours, by driving their lunar rover more than 22 miles (35 kilometers), and by collecting about 242 pounds (110 kilograms) of rocks and soils during three long, busy traverses around the valley. By the time they rejoined pilot Ronald E. Evans, Jr., back in the command module *America* three days later, they were at the brink of physical exhaustion.

Jack Schmitt was the first and only scientist by training (a geologist) to go to the Moon, and his expert eye was critical in identifying a number of key samples, including a patchy and hard-to-spot orange-colored soil that he had noticed near a small impact crater called Shorty. Later analysis of these samples, and related black soils from the same region, showed them to contain tiny titanium-rich glass spheres that had been formed by explosive volcanic eruptions on the Moon. Some of these spheres have been found to contain tiny amounts of water, proving that the Moon's interior isn't completely dry.

Schmitt and Cernan were the last people to visit the Moon; they and Evans were also the last humans to travel beyond low Earth orbit—a journey that took place more than 40 years ago.

SEE ALSO First on the Moon (1969), Second on the Moon (1969), Fra Mauro Formation (1971), Roving on the Moon (1971), Lunar Highlands (1972).

Apollo 17 astronaut Gene Cernan bounces along to his next sampling spot in this photograph from a large panorama taken by fellow astronaut Jack Schmitt near a boulder field along the rim of Camelot crater in the Taurus-Littrow Valley.

Gamma-Ray Bursts

Physicists studying the spontaneous decay of radioactive elements around the turn of the twentieth century recognized three kinds of particles or radiation that were emitted by **Radioactive** elements, such as uranium and radium. Some elements emitted helium nuclei (alpha particles) when they decayed; others emitted higher-energy electrons or positrons (beta particles), and when a third, even higher energy form was discovered in still other radioactive decay events, physicists called those particles gamma rays. Like X-rays, gamma rays are a form of electromagnetic radiation of very high energy (very short wavelength). Later twentieth-century physicists discovered that X-rays are emitted by electrons surrounding an atom undergoing radioactive decay, and higher-energy gamma rays are emitted by the atomic nucleus itself.

Because gamma rays are generated in the nucleus, physicists predicted and observed the creation of gamma rays during **Nuclear Fusion** bomb tests. Indeed, the United States and the Soviet Union both deployed gamma-ray detectors in space in the 1960s to verify that the terms of the 1963 Partial Test Ban Treaty were being upheld.

The US satellites were part of the military's Vela series, which could detect gamma rays coming from anywhere in space. Surprisingly, beginning in 1967, the satellites began to occasionally—a few times a year—detect mysterious short gamma-ray bursts (GRBs), lasting from milliseconds to several minutes, that were later determined to be coming from random directions in deep space. The military declassified the data and alerted civilian scientists to the phenomena in 1973.

Astrophysicists were puzzled by GRBs for decades, because the energies were far higher than gamma ray events produced by radioactive decay or stellar nuclear fusion reactions. The detailed nature of GRBs would remain mysterious until space telescope observations from NASA's Compton Gamma Ray Observatory in the 1990s revealed their origin. GRBs appear to be created during supernova explosions when massive stars collapse, or during the mergers of pairs of colliding **Neutron Stars**. Like **Pulsars**, GRBs occur when jets of highly focused energy stream out of the exploding star and are pointed toward Earth. Gamma-ray bursts appear to be the most violent and energetic events in the cosmos.

SEE ALSO Radioactivity (1896), Neutron Stars (1933), Nuclear Fusion (1939), Black Holes (1965), Pulsars (1967), Gamma-Ray Astronomy (1991).

An artist's impression of the gamma-ray burst event GRB 080319B, detected on March 19, 2008. The burst of energy is thought to have come from jets of gas accelerated to a speed of 99.9995 percent of the speed of light during the supernova collapse of a massive star 7.5 billion light-years away.

Pioneer 10 at Jupiter

Up until the 1970s, the robotic exploration of the solar system was confined to the inner solar system, and, even more specifically, to the exploration of the Moon, Venus, and Mars. **Jupiter**, which makes up more than 70 percent of the mass of everything in the solar system besides the Sun, was an obvious next target.

Humanity's first virtual voyage of outer solar system exploration was the mission of the *Pioneer 10* space probe. Launched on March 2, 1972, the nuclear-powered probe was designed to study the properties of interplanetary space beyond Mars, including making an assessment of the nature of the **Main Asteroid Belt** and the safety of flying spacecraft through it on their way to the outer solar system. *Pioneer 10* and its twin spacecraft, *Pioneer 11*, which was launched April 4, 1973, were also designed to fly closely past Jupiter and study the high-radiation environment of Jupiter's magnetic field.

The probe carried 11 instruments to take photographs and collect data on temperature, magnetic fields, solar wind, cosmic rays, and micrometeorites. It found that dust and micrometeors in the asteroid belt would not pose much of a threat to future spacecraft. During the Jupiter flyby, *Pioneer 10* got to within 125,000 miles (200,000 kilometers) of the cloud tops and took photographs revealing great detail in the planet's atmosphere.

After passing Jupiter, *Pioneer 10* began its interstellar mission to study the far outer solar system. The last signals were received from the probe in 2003, but it is still traveling at more than 7.5 miles (12 kilometers) per second on an escape trajectory from the solar system, now more than 100 astronomical units from the Sun. *Pioneer 10* carries a golden plaque with drawings of a man and a woman and symbols designed to provide information about the probe's origin. Perhaps someone (or something) will read the message in a few million years, when *Pioneer 10* reaches the star Aldebaran.

SEE ALSO Jupiter (c. 4.5 Billion BCE), Io (1610), Europa (1610), Ganymede (1610), Callisto (1610), Great Red Spot (1665), *Voyager* Saturn Encounters (1980, 1981), *Galileo* Orbits Jupiter (1995).

LEFT: *An approximate true-color photograph of Jupiter and its Great Red Spot, taken on December 1, 1973, by the* Pioneer 10 *spacecraft from a range of 1.6 million miles (2.6 million kilometers).* RIGHT: *Artist rendition of the* Pioneer 10 *spacecraft approaching Jupiter. The truss pointing down in this view is about 10 feet (3 meters) long.*

Vikings on Mars

The short but successful mission of the *Mariner 9*, in 1971, formed the basis of a new NASA project to explore Mars in greater detail than ever before. The project was called Viking, and its goals were to launch two orbiters and two landers to **Mars** to significantly enhance our knowledge of the Red Planet's surface, atmosphere, and potential as a past or present abode of life.

Launched in August and September 1975, the *Viking 1* and *Viking 2* probes arrived at Mars in June and August 1976, respectively. During the first month in orbit, the *Viking 1* team took photos of the surface, searched for a safe place to deploy the nuclear-powered lander, and eventually chose a flat region known as Chryse Planitia (Plains of Gold). *Viking Lander 1* set down safely on July 20, 1976, becoming the first successful mission to the surface of Mars. *Viking Lander 2* successfully landed a few months later about 3,000 miles (4,800 kilometers) away, in flat and rocky Utopia Planitia.

With a cost of about $1 billion, Project Viking was the most complex and expensive Mars mission attempted up to that time, and it was a phenomenal success. The orbiters provided detailed, global maps of the surface, down to hundreds-of-meters-or-

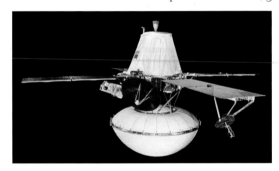

better resolution, revealing ancient water-carved valley networks, finely layered polar cap deposits, and details of the younger volcanic mountains and giant canyons discovered by *Mariner 9*. The idea that early Mars may have been warmer, wetter, and more Earthlike was based on Viking orbiter data.

The landers also developed a lasting paradigm for Mars studies. Meteorology experiments provided detailed information on the (hostile to humans) surface conditions and weather patterns. Most important, however, the landers' search for evidence of organic molecules—potential indicators of life, or at least of habitable environments—came up empty. The setback for astrobiologists was temporary, however, helping to refine new experiments for future missions.

SEE ALSO Mars (c. 4.5 Billion BCE), *Mars and Its Canals* (1906), First Mars Orbiters (1971), First Rover on Mars (1997), *Spirit* and *Opportunity* on Mars (2004).

LEFT: *A model of the* Viking *orbiter, including the entry capsule at bottom that housed the lander. The probe, including solar panels, was about 30 feet (9 meters) across.* RIGHT: Viking Lander 2 *image of the reddish, boulder-covered surface of Utopia Planitia, where it landed on September 3, 1976.*

Voyager "Grand Tour" Begins

As robotic missions to the outer solar system were being contemplated in the late 1960s and early 1970s, mission planners and celestial dynamics experts realized that a fortuitous alignment of **Jupiter**, **Saturn**, **Uranus**, and **Neptune** would make it possible for a single probe to use gravity assists to potentially fly past all four giant planets in the 1970s and 1980s. This historic "grand tour" concept generated a lot of excitement among NASA researchers, and the idea was eventually implemented in a new pair of missions called Voyager.

Voyager 2 was launched first, on August 20, 1977, on a trajectory that would get it to Jupiter in mid-1979, Saturn in mid-1981, Uranus in early 1986, and Neptune in mid-1989. *Voyager 1*'s launch was on September 5, 1977, on a faster trajectory that would get it to Jupiter in early 1979 and Saturn in late 1980. *Voyager 1*'s tour included just Jupiter and Saturn flybys because the desire to perform a close flyby of Saturn's large, thick-atmosphered moon **Titan** required a course correction that made it impossible to subsequently encounter Uranus and Neptune.

The missions of *Voyager 1* and *Voyager 2* are among the most exciting and successful adventures in the history of space exploration—or any other kind of exploration, for that matter. The probes enabled scientists to make new discoveries about

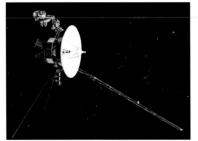

the giant planets' atmospheres, magnetic fields, ring systems, and their planet-size moons **Io**, **Europa**, **Ganymede**, **Callisto**, **Titan**, and **Triton** (plus many smaller moons). Voyager discoveries enabled even more discoveries at Jupiter and Saturn by the follow-on *Galileo* and *Cassini* orbiters, respectively, and provide the data needed for eventual orbital missions to Uranus and Neptune sometime in the future.

Like the *Pioneers*, each *Voyager* also has a message in a bottle of sorts—a golden record, containing encoded images, voices, and music—a cosmic time capsule bearing greetings from planet Earth to whoever finds it in the far future.

SEE ALSO Jupiter (c. 4.5 Billion BCE), Saturn (c. 4.5 Billion BCE), Uranus (c. 4.5 Billion BCE), Neptune (c. 4.5 Billion BCE), Io (1610), Europa (1610), Ganymede (1610), Callisto (1610), Titan (1655), Saturn Has Rings (1659), Triton (1846), *Pioneer 11* at Saturn (1979), *Voyager 2* at Uranus (1986), *Voyager 2* at Neptune (1989).

LEFT: *Artist's conception of the* Voyager *spacecraft. The long piece extending to the lower right is the 42-foot (13-meter) boom for the magnetic field instrument.* RIGHT: *Montage of a* Voyager *Grand Tour trajectory diagram, superimposed on greatest-hits* Voyager *images of the giant planets.*

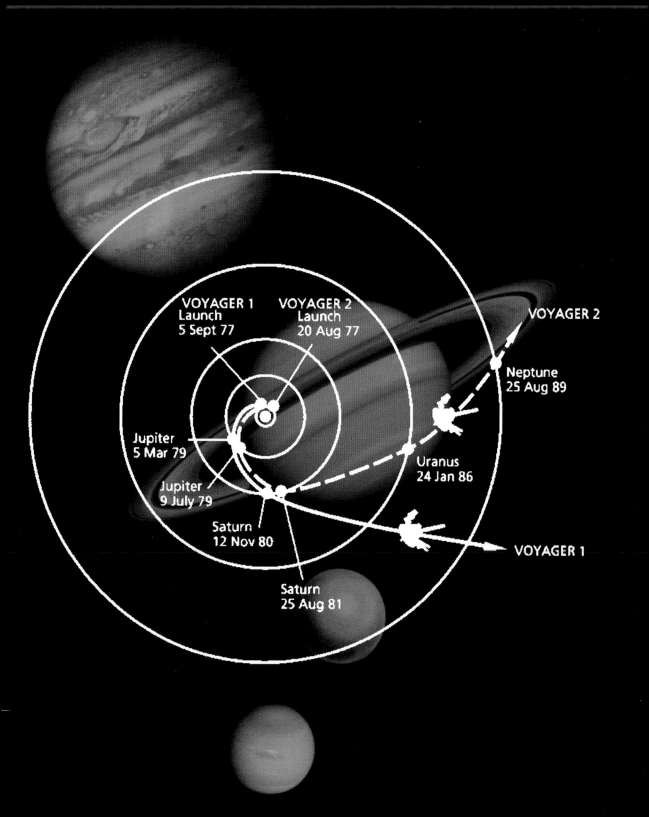

VOYAGER 1
Launch
5 Sept 77

VOYAGER 2
Launch
20 Aug 77

VOYAGER 2

Neptune
25 Aug 89

Jupiter
5 Mar 79

Jupiter
9 July 79

Uranus
24 Jan 86

Saturn
12 Nov 80

VOYAGER 1

Saturn
25 Aug 81

Uranian Rings Discovered

James L. Elliot (1943–2011), **Edward W. Dunham** (b. 1952), **Douglas J. Mink** (b. 1951)

In 1659, the Dutch astronomer Christiaan Huygens observed and explained the **Rings of Saturn** as a thin disk of material circling the sixth planet. Saturn appeared to be the only planet with a ring system until 1789, when, shortly after discovering **Uranus**, the English astronomer William Herschel thought he detected a faint ring around that planet as well. But subsequent visual and photographic observations of Uranus couldn't verify Herschel's claim.

Nearly two hundred years later, on March 10, 1977, the American team of planetary scientists James Elliot, Ted Dunham, and Douglas Mink was preparing to observe Uranus pass in front of a relatively bright star—an eclipse of sorts known as an occultation. To ensure that they were in the precise location to catch the rare occultation, the team observed the event from the Kuiper Airborne Observatory, a telescope mounted in a NASA C-141A jet that flew missions in the stratosphere, above most of our atmosphere's clouds and water vapor.

The occultation yielded an exciting surprise. Just before the star was eclipsed by Uranus, its brightness briefly dropped dramatically, five separate times. Then, just after the star reemerged from the eclipse, it happened five times again. Analysis by Elliot and his team revealed that the drops in starlight occurred at the same radial distance from Uranus on either side of the planet. They had discovered a faint set of narrow rings— meaning that the seventh planet had become the second known planet with rings.

Follow-up observations revealed four more narrow rings around Uranus, bringing the total to nine. Two more rings were discovered during *Voyager 2*'s 1986 flyby—which revealed that the rings are extremely dark and are likely made of centimeter- to meter-size icy blocks that have been darkened by interactions between Uranus's magnetic field and icy, organic molecules—and two more were discovered in the early twenty-first century by astronomers using the Hubble Space Telescope, bringing the total to 13. Faint rings were later also discovered around **Jupiter** and **Neptune** (also in *Voyager* data), meaning that all the giant planets in our solar system—not just Saturn—are ring worlds.

SEE ALSO Uranus (c. 4.5 Billion BCE), Saturn Has Rings (1659), Jovian Rings (1979), Rings around Neptune (1982).

Voyager 2 image of the rings of Uranus, acquired during the probe's flyby in January 1986. Background stars appear as streaks in this long-exposure photograph. The blinking on and off of blocked starlight passing through the rings is what earlier had led to their discovery in 1977.

Charon

James W. Christy (b. 1938), **Robert S. Harrington** (1942–1993)

At nearly 40 times farther away from the Sun than Earth, **Pluto** was a challenging object to observe in the decades after its discovery by Clyde Tombaugh in 1930. Nonetheless, astronomers continued to observe Pluto to learn more about its origin and properties. Many of these observations were performed from telescopes at the Lowell Observatory in Flagstaff, Arizona, where Pluto had been discovered.

While analyzing a series of such observations in June, 1978, Lowell astronomer James W. Christy noticed a bulge in some of the photographs of Pluto. Christy noted that the bulge appeared to move around the main image of Pluto in a period of about 6.4 days. After some further analysis, he and colleague Robert S. Harrington published their discovery of a satellite orbiting Pluto. Christy named the moon Charon, after the mythological Greek ferryman of the dead, but also in honor of his wife (Charlene, or "Char"), which is why the name is pronounced "Sharon," not "Karon."

Charon turns out to be quite large—750 miles (1207 kilometers) in diameter, about half of Pluto's diameter—making Pluto-Charon more like a double planet system.

Charon also turned out to have a surface dominated by water ice and possibly small amounts of hydrated ammonia—a combination quite different from Pluto's surface, which is dominated by nitrogen and methane ice.

Observations of Pluto and Charon by the Hubble Space Telescope in 2005 revealed another surprise—more moons! Two additional but much smaller satellites, named Nix and Hydra after other Greco-Roman mythological figures, were discovered to be orbiting the center of mass of the Pluto-Charon system; two more, called simply P4 and P5, were discovered in 2011 and 2012. Excitement is mounting to see what these distant worlds look like when the NASA New Horizons probe flies past them in July 2015.

SEE ALSO Pluto and the Kuiper Belt (c. 4.5 Billion BCE), Discovery of Pluto (1930), Kuiper Belt Objects (1992), Pluto Revealed! (2015).

LEFT: *Hubble Space Telescope photograph of Pluto and Charon (bright pair) and the fainter moons Nix (closer) and Hydra (farther).* RIGHT: *An artist's conception of Pluto (larger) and Charon (smaller), viewed from the surface of one of the system's smaller moons, discovered in 2005.*

Ultraviolet Astronomy

Astronomers often make observations of various colors in an effort to learn details about astronomical bodies, such as the temperatures of stars or the compositions of planets and moons. These colors often extend beyond the visible part of the electromagnetic spectrum, outside the blue, green, and red wavelengths familiar to human vision. Wavelengths longer than the red are called infrared colors and those shorter than the blue are called ultraviolet (UV). Infrared observations allow astronomers to detect cooler (lower-energy) objects, while ultraviolet observations are sensitive to hotter, higher-energy sources.

Observing in the UV is problematic, however, because gases in Earth's atmosphere, like oxygen, carbon dioxide, and water vapor, almost completely absorb UV photons, making such observations impossible from ground-based telescopes. The advent of high-altitude balloons and space satellites in the 1960s and 1970s gave astronomers the chance to open up the entirely new field of ultraviolet astronomy by launching UV telescopes into space.

The first successful UV space telescope mission was a joint project involving NASA, the European Space Agency, and the British Science Research Council; it was

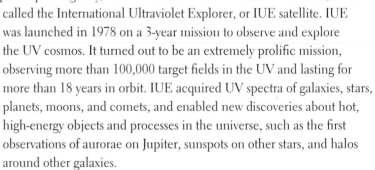

called the International Ultraviolet Explorer, or IUE satellite. IUE was launched in 1978 on a 3-year mission to observe and explore the UV cosmos. It turned out to be an extremely prolific mission, observing more than 100,000 target fields in the UV and lasting for more than 18 years in orbit. IUE acquired UV spectra of galaxies, stars, planets, moons, and comets, and enabled new discoveries about hot, high-energy objects and processes in the universe, such as the first observations of aurorae on Jupiter, sunspots on other stars, and halos around other galaxies.

Ultraviolet astronomy has continued to flourish since IUE. For example, a new NASA UV astronomy space telescope, called the Galaxy Evolution Explorer (GALEX), has been operating successfully since 2003, letting astronomers follow up and expand upon IUE's initial pioneering discoveries.

SEE ALSO First Astronomical Telescopes (1608), Radio Astronomy (1931), Hubble Space Telescope (1990), Gamma Ray Astronomy (1991), Chandra X-Ray Observatory (1999), Spitzer Space Telescope (2003).

The spiral galaxy M81 observed in the ultraviolet by the NASA GALEX observatory, the successor to the IUE observatory (left). Hot young stars forming in this galaxy's spiral arms show up as blue, and older main sequence stars show up as yellow in this false-color composite.

Active Volcanoes on Io

1979

When *Pioneer 10* flew by Jupiter in 1973, the probe took spectacular images of the giant planet but was not able to get very good views of Jupiter's large satellites **Io**, **Europa**, **Ganymede**, and **Callisto**, discovered by Galileo back in 1610. The *Voyager 1* probe that flew by Jupiter in March 1979 contained a much more sophisticated and higher-resolution camera system, however, and its flyby trajectory would take it closer to both the planet and its large moons.

The *Voyager 1* encounter with Jupiter revealed many surprises, perhaps the biggest of which was the discovery of active volcanic plumes and lava flows on Jupiter's innermost moon, Io. At first seen in silhouette in distant images, the surface of Io—as eventually revealed in high-resolution, close-approach views—is covered by more than four hundred active volcanoes. Io's volcanoes ooze out fresh, molten, rocky, lava flows tens to hundreds of miles long, often colored by the presence of molten sulfur at a range of temperatures, and they sometimes erupt explosively, hurling an umbrella-shaped cloud of ash and rocky debris hundreds of miles out into space. The images from *Voyager 1* and then *Voyager 2* revealed that Io is the most volcanically active world in the solar system.

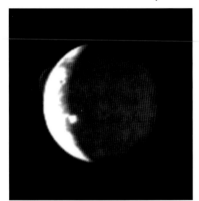

But why? Io, Europa, and Ganymede orbit Jupiter in a 4:2:1 pattern called an orbital Laplace resonance that causes the moons to occasionally tug on each other gravitationally and make them wobble a little in slightly eccentric orbits. Jupiter's gravity, on the other hand, tries to lock the moons into synchronous rotation, with the same side always facing the planet (like Earth's Moon). The result is that strong tidal forces constantly squeeze and stretch the moons, especially innermost Io, and the friction from all that squeezing heats their interiors. Io has been completely melted by tidal heating, boiling off all its water and other ices and causing global-scale volcanic eruptions that continually resurface the satellite and spew sulfur and dust into a giant doughnut-shaped disk around Jupiter. Io is a tortured little world that is literally being turned inside out by relentless tidal forces.

SEE ALSO Io (1610), Europa (1610), Ganymede (1610), *Pioneer 10* at Jupiter (1973), Jovian Rings (1979), An Ocean on Europa? (1979), *Voyager* Saturn Encounters (1980, 1981), *Voyager 2* at Uranus (1986), *Voyager 2* at Neptune (1989), *Galileo* Orbits Jupiter (1995).

LEFT: *March 8, 1979, discovery photo of volcanic plumes erupting on Io.* RIGHT: Voyager 1 *image of volcanic vents and lava flows on Jupiter's inner moon Io. These lava flows can be hundreds of miles long and are made of molten sulfur and silicate minerals at temperatures above 1800°F (1000°C).*

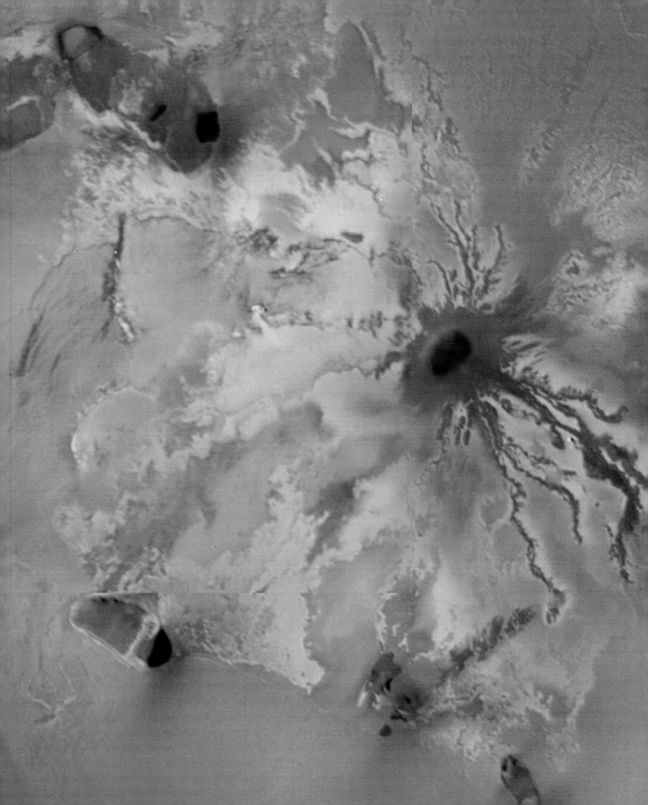

Jovian Rings

The first space probes to encounter **Jupiter**, *Pioneer 10* and 11 made detailed measurements of the planet's magnetic field and energetic particle environment as they sped past in 1973 and 1974. The data showed some unexplained variations in protons and electrons in Jupiter's equatorial plane near the orbit of **Amalthea** and the other inner moons, suggesting that something—perhaps a ring—was absorbing particles at those distances from the planet.

Alerted by this enigma, mission planners for the *Voyager 1* flyby of Jupiter in March 1979 designed special observations to take long-exposure photos of the region. To the team's delight, the photos captured the tip of a narrow ring system orbiting Jupiter, making it the third known giant planet with rings. The *Voyager 2* probe took more views of Jupiter's ring system during its flyby in July 1979.

Even more detailed photos from the subsequent NASA *Galileo* **Jupiter Orbiter** and the *Cassini* and *New Horizons* mission flybys of Jupiter showed that the ring system extends from about 1.4 to 3.8 Jupiter radii and has four main parts: a dusty, doughnut-shaped halo ring closest in to the planet; the bright and thin (19–186 miles [30–300 kilometers] in width) main ring; and two diffuse gossamer rings farthest away, associated with the small moons Amalthea and Thebe.

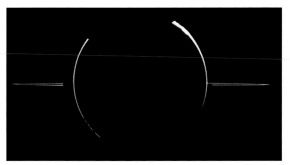

Unlike the rings around Saturn and Uranus, Jupiter's rings are dusty rather than icy. The main and halo rings appear to be made out of dust and small rocky particles blasted off the moons Metis and Adrastea by comet or asteroid impacts. The gossamer rings are made of dust-size grains blasted off Amalthea and Thebe. Bands and other structures indicate that small embedded moonlets or clumps of ring material also supply and shape the rings. Because dust should only survive about 1,000 years or so before being ejected by one of the inner moons, Jupiter's rings must be young and constantly replenished by new impacts.

SEE ALSO Jupiter (c. 4.5 Billion BCE), Saturn Has Rings (1659), Amalthea (1892), *Pioneer 10* at Jupiter (1973), Uranian Rings Discovered (1977), *Voyager* Saturn Encounters (1980, 1981), Rings around Neptune (1982), *Voyager 2* at Uranus (1986), *Voyager 2* at Neptune (1989), *Galileo* Orbits Jupiter (1995).

LEFT: *This Galileo Jupiter orbiter mosaic of Jupiter's ring system, viewed edge-on, looking toward the Sun, was created while the spacecraft was passing behind the planet, through Jupiter's shadow.* RIGHT: *A false-color view of the fine structure in Jupiter's rings, photographed by* Voyager 2.

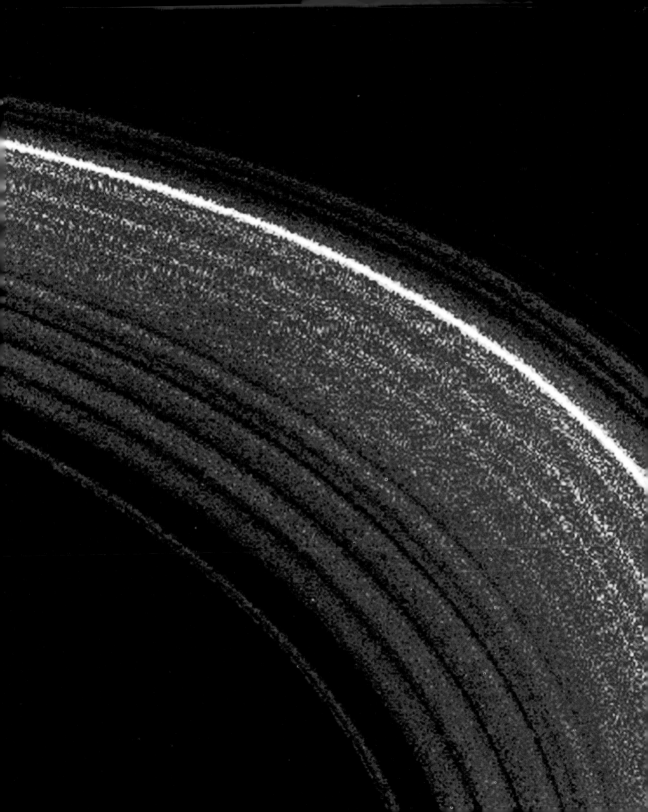

An Ocean on Europa?

Since its discovery in 1610 by Galileo, very little was known about **Jupiter's** second large satellite, **Europa**, except that it was locked into a 4:2:1 (**Io**:Europa:**Ganymede**) orbital resonance, and thus likely to experience interesting tidal forces, and that **Spectroscopy** from telescopes revealed it to have a bright, water-ice-covered surface. In terms of the detailed nature of Europa's surface, the slate was almost completely blank when the *Voyager 1* and 2 probes made the first detailed reconnaissance of Jupiter's system.

Voyager's discoveries at Europa didn't disappoint. The planet-size moon (only slightly smaller than **Mercury**) turned out to have one of the smoothest surfaces in the solar system. But that smooth surface is covered by a dense and intersecting network of cracks and low ridges that appear to divide the surface into icy plates that move relative to each other. This was one of the first clues that the surface is a relatively thin, tectonically active shell of ice that is floating on a thick layer of liquid water—an ocean—in the subsurface. Other clues that Europa is an ocean world come from the observation that there are very few impact craters on the surface, suggesting that it is geologically young and actively resurfacing itself. The NASA *Galileo* **Jupiter Orbiter** enabled much more detailed study of Europa during its mission from 1995 to 2003. More evidence for the presence of a deep ocean came from *Galileo* spectrometer observations of salty mineral deposits within some of the cracks and fractures—the kind of mineral deposits that would form from the evaporation of salty ocean water. *Galileo* magnetic field data also showed that the subsurface of Europa is electrically conductive, yet another indicator or the presence of salty ocean water beneath the icy crust.

The available data so far are consistent with Europa having a deep (60-mile [100-kilometer]) global ocean underneath a relatively thin (6- to 20-mile [10- to 30-kilometer]) icy shell, but direct proof of that ocean must await future missions that will orbit or land on (and possibly drill through) the Europan crust. In the meantime, astrobiologists have become excited about the prospects for Europa as a potential abode of life. With energy from tidal heating by Jupiter, organic molecules from a steady rain of comets and asteroids, and abundant liquid water, we may one day discover that Europa's ocean is a habitable—or perhaps even inhabited—environment.

SEE ALSO Io (1610), Europa (1610), Ganymede (1610), Jovian Rings (1979), *Voyager* Saturn Encounters (1980, 1981), *Voyager 2* at Uranus (1986), *Voyager 2* at Neptune (1989), *Galileo* Orbits Jupiter (1995), An Ocean on Ganymede? (2000).

Voyager 2 *photo of part of the flat, icy surface of Jupiter's second moon, Europa. The surface is crisscrossed with countless cracks that look like plates of sea ice floating on liquid water in many places. The surface has very few impact craters, meaning it is geologically very young.*

Gravitational Lensing

One of the fundamental features of physicist **Albert Einstein**'s early-twentieth-century theory of general relativity is that space and time are curved near extremely massive objects. The curvature of space-time led Einstein and others to predict that light from distant objects would be bent by the gravitational field of massive foreground objects. The prediction was verified in 1919 by the British astrophysicist Arthur Stanley Eddington, who noticed that stars observed near the Sun during a solar eclipse were slightly out of position. Einstein continued to study this effect in the 1930s, and he and others, including the Swiss-American astronomer Fritz Zwicky, speculated that more massive objects, such as galaxies and clusters of galaxies, could bend and amplify light from distant objects almost as a lens bends and magnifies normal light.

It took many decades for astronomers to find observational evidence of such gravitational lensing, however. The first example was discovered in 1979 by astronomers at the Kitt Peak National Observatory in Arizona, who found an example of what appeared to be twin **Quasars**—two active galactic nuclei very close to each other in the sky. The two quasars were shown to actually be a single object whose light was bent and split into two parts by the strong gravitational field of a foreground galaxy.

Since then, many more examples of gravitational lensing have been found, and the effect seems to occur in three ways: strong lensing is when distinct multiple or partial (usually arc-like) images are formed; weak lensing has been detected by observing small and subtle shifts in star or galaxy positions over large regions; and microlensing events have been detected when random distant stars (or even planets) have their brightness temporarily amplified by the gravitational lensing effect of a large foreground mass, such as another star or galaxy.

Gravitational lenses were initially discovered and studied as accidental, serendipitous events. Recently, however, a number of astronomical surveys have been conducted to intentionally search for gravitational lensing events, in order to obtain unique measurements of the properties of distant galaxies that would not be visible without the amplification from the lens, as well as the properties (such as mass) of the lensing galaxies and clusters themselves.

SEE ALSO Einstein's "Miracle Year" (1905), Dark Matter (1933), Quasars (1963), Black Holes (1965).

The thin arcs seen here are gravitationally lensed galaxies in the galactic cluster Abell 2218, photographed by the Hubble Space Telescope in 1999. These so-called Einstein rings are the smeared-out light from distant galaxies being bent by a massive foreground galaxy.

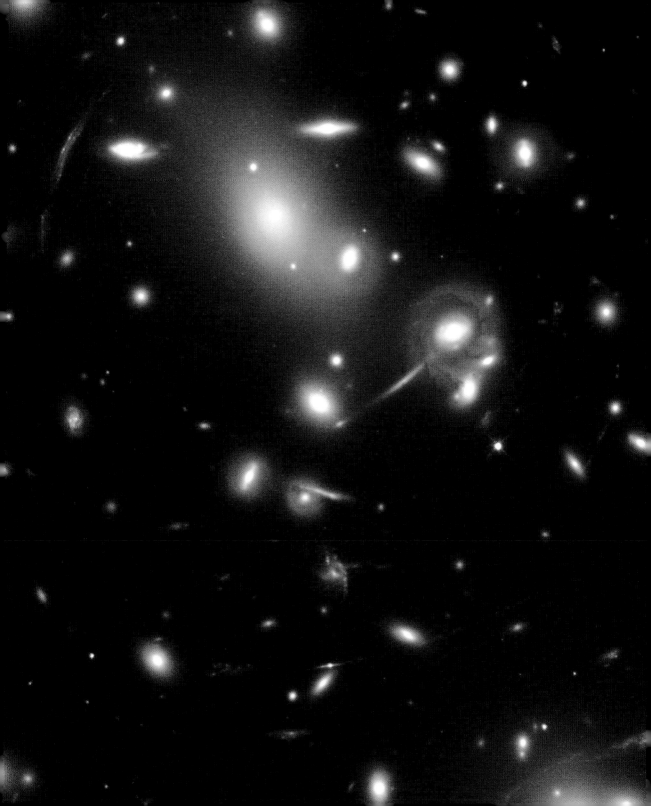

Pioneer 11 at Saturn

The *Pioneer 10* and *11* missions, launched in 1972 and 1973 respectively, were designed to provide the first reconnaissance of the outer solar system and beyond. *Pioneer 10* made the first encounter with Jupiter in 1973, setting the stage for *Pioneer 11*'s flyby of that giant planet in 1974–1975. Unlike *Pioneer 10*, however, *Pioneer 11* was on a trajectory that could be programmed to use Jupiter's gravity to slingshot it on to a first-ever encounter with **Saturn** in 1979.

The *Pioneer 11* encounter with Saturn was a great success, with the probe passing within about 13,000 miles (21,000 kilometers) of Saturn's cloud tops on September 1, 1979. The mission carried cameras, magnetic-field and charged-particle instruments, cosmic-dust-particle and radiation counters, and other scientific instruments that provided planetary scientists with their first views of the environment at and around the ringed planet.

In a way, the *Pioneer* missions were pathfinders for the more ambitious *Voyager* probes to the outer solar system. For example, during the Saturn flyby, *Pioneer 11* was directed to travel through Saturn's ring plane to determine if small dusty or icy ring particles might pose a threat to spacecraft. They did not, thus enabling mission planners to direct *Voyager 2* through the same region of the rings in order to get it on a trajectory that would allow subsequent encounters with Uranus and Neptune. *Pioneer 11* data yielded other discoveries at Saturn as well, like the observation of the large, atmosphere-bearing moon **Titan**'s extremely low temperature (90 kelvins; perhaps too cold for life), the discovery (and near collision with) a new small moon and an additional ring; and detailed mapping of Saturn's magnetic field, a large charged-particle structure similar in some ways to **Jupiter's Magnetic Field**.

Like *Pioneer 10*, *Pioneer 11* is on an escape trajectory from our solar system, now more than 83 astronomical units from the Sun and heading toward the center of the Milky Way galaxy. Contact was lost with the probe in late 1995, but the spacecraft carries a plaque, like *Pioneer 10*'s, that will, we hope, serve as an informative greeting for any galactic neighbors who discover the probe in the far future.

SEE ALSO Saturn (c. 4.5 Billion BCE), Titan (1655), Saturn Has Rings (1659), Jupiter's Magnetic Field (1955), *Pioneer 10* at Jupiter (1973), *Pioneer 10* beyond Neptune (1983), *Cassini* Explores Saturn (2004).

Pioneer 11 *false-color image of part of Saturn and its rings, taken on September 1, 1979, when the probe was about 250,000 miles (400,000 kilometers) from the planet. Saturn can be seen through the rings' Cassini Division, and the shadow of the rings can be seen silhouetted against the planet.*

Cosmos: A Personal Voyage

Carl Sagan (1934–1996)

Astronomy and space exploration are interesting and exciting topics, yet throughout most of recent history, scientists have not been compelled or encouraged to share their discoveries (or, pointedly, their failures) with the public. Publishing their results in books or academic journals, or presenting their findings at scientific conferences, was usually seen as enough. Many even appeared to have a certain arrogance about their work—the public just wouldn't understand, so why tell them?

Even in the 1960s and 1970s, when the huge public interest and international appeal of space exploration were heightened by the media coverage of the Apollo missions, it was still difficult for average folks to keep up with the latest observations and discoveries. In the United States, the three main broadcast TV networks aired mostly entertainment and news shows, and a fourth—the Public Broadcasting Service (PBS)—aired a few good general-interest science shows, but nothing focusing on space.

It was against that backdrop that a new television series, focusing specifically on astronomy and space exploration and hosted by a charismatic and thought-provoking American astronomer named Carl Sagan, became a smash hit in 1980. The show, called *Cosmos: A Personal Voyage*, was the most watched PBS series in the world, and was seen by an audience of more than 500 million people. Through *Cosmos*, Sagan had an enthusiastic and educational conversation with the public about the latest observations and theories concerning some of the biggest questions we all ponder: What is going on up there? Where did it all come from? Why are we here? Are we alone?

Sadly, Sagan met significant resistance from many of his scientific contemporaries for his tireless work popularizing the value of science and space exploration, and was reportedly denied membership in the National Academy of Sciences because of the petty jealousies of other scientists. But Sagan's ideals and legacy have since spread to a new generation of astronomers and planetary scientists (many who grew up watching *Cosmos*); his ideals have been promoted worldwide by members of a public space-advocacy organization called the Planetary Society, which he helped to found in 1980; and they have been embraced by a scientific community that now regards as essential the public communication and understanding of science in our modern world.

SEE ALSO *Mars and Its Canals* (1906), SETI (1960), *Vikings* on Mars (1976), Life on Mars? (1996).

Carl Sagan—astronomer, planetary scientist, author, science popularizer, and host of the acclaimed television series Cosmos—*next to a full-scale model of the* Viking *Mars lander in 1980.*

Voyager Saturn Encounters

The *Pioneer 11* robotic probe flyby of **Saturn** in 1979 was a dress rehearsal of sorts for an even more ambitious and detailed study of the planet's atmosphere, moons, rings, and magnetic field by the NASA *Voyager 1* and *Voyager 2* probes, which were launched in 1977. *Voyager 1* flew past Saturn in November 1980, and *Voyager 2* followed close behind, in August 1981.

The *Voyager* Saturn encounters dramatically increased our knowledge about all of the diverse bodies and processes in the Saturn system. High-resolution imaging revealed details about the impact cratering record, composition, and geologic histories of the large icy moons **Iapetus**, **Rhea**, **Tethys**, **Dione**, **Enceladus**, **Mimas**, and **Hyperion**. Seven small new moons of Saturn were discovered in the *Voyager* images, several of which are embedded in what turned out to be a much more spectacularly complex ring system than had been previously imagined. **Saturn's Rings** were discovered to consist of thousands of individual rings and ringlets separated by thin gaps and kept organized by the gravity of a number of **Shepherd Moons** that co-orbit in the ring system.

Voyager data confirmed the telescopic and *Pioneer 11* discovery that Saturn's largest moon, **Titan**, has a thick and hazy atmosphere. Titan's atmosphere was found to have more than 50 percent higher surface pressure than Earth's (making it the solar system's only moon with a thick atmosphere) and to be composed mostly of nitrogen. The orangish color of Titan's haze was found to be caused by small amounts of methane, ethane, propane, and other organic molecules, some of which were predicted to exist as surface liquids at Titan's very low temperature. *Voyager*'s cameras could not penetrate Titan's haze to view the surface, however. Those discoveries would come some 15 years later, during the *Cassini* **Saturn Orbiter** and *Huygens* **Titan Lander** missions.

Voyager 1's planetary exploration mission ended at Saturn. The probe's trajectory was taken off its **"Grand Tour"** path (including a possible late-1980s flyby of Pluto) in order to fly it close enough to Titan to study its early Earthlike atmosphere in great detail. *Voyager 1* is still operating some 120 astronomical units from the Sun and is the most distant object that humans have ever sent into the cosmos.

SEE ALSO Saturn (c. 4.5 Billion BCE), Titan (1655), Saturn Has Rings (1659), Iapetus (1671), Rhea (1672), Tethys (1684), Dione (1684), Enceladus (1789), Mimas (1789), Hyperion (1848), *Voyager* "Grand Tour" Begins (1977), *Pioneer 11* at Saturn (1979); *Cassini* Explores Saturn (2004), *Huygens* Lands on Titan (2005), Shepherd Moons (2005).

Mosaic of images obtained by Voyager 1 *during its flyby of the Saturn system in November 1980. Six of Saturn's largest moons are shown here: Dione (foreground), Tethys and Mimas (lower right), Rhea and Enceladus (upper left), and Titan (upper right).*

Space Shuttle

The dream of human space flight as first envisioned by rocketry pioneers Konstantin Tsiolkovsky and Robert Goddard has never been about one-way trips into space. Rather, the full impact of people voyaging beyond our planet is only truly realized when they return home and share their adventure stories (and scientific findings and samples) with the rest of us. But to get people (or payloads) to and from space requires the rocket to carry return capsules, parachutes, and supplies. In the Vostok, Mercury, Gemini, and Apollo astronaut programs, each return capsule and system was only used once.

For as long as engineers have thought about rocket designs, they've thought about reusable rocket designs, not only as ways to slash the cost of getting people and equipment into space but also as a means of making access to space a matter of routine, as commercial airline travel is today. This was the motivation for NASA's development of the National Space Transportation System, known as the space shuttle, in the 1970s.

Space shuttles used a reusable orbiter vehicle for the crew and equipment (with rocket engines for ascent and airplane-like wings for descent), two reusable solid-fuel rockets, and a large expendable tank to fuel the orbiter's engines during launch. Five orbiters were built and flown in space (*Columbia*, *Challenger*, *Discovery*, *Atlantis*, and *Endeavour*) between the first launch in 1981 and the 135th and last in 2011, carrying 355 astronauts (some on multiple flights) into low Earth orbit—about 250 miles (400 kilometers) above the surface. Fourteen astronauts lost their lives in the line of duty: seven in *Challenger*, which exploded on launch in 1986, and seven in *Columbia*, which broke up on reentry in 2003.

Despite never becoming routine, and never realizing the hope of more economical access to space, the missions of the space shuttle were still incredibly successful overall. Shuttles were critical for construction of the **International Space Station**, for the repair and servicing of the **Hubble Space Telescope**, for the launch of numerous Earth and planetary satellites, and for important space-related biological, astronomical, and Earth sciences research. Now that the space shuttle fleet is retired, NASA plans to build a new rocket system capable of carrying astronauts past the shuttle's limit of low Earth orbit—back to the Moon or on to new destinations, such as near-Earth asteroids and Mars and its moons.

SEE ALSO First on the Moon (1969), Second on the Moon (1969), Fra Mauro Formation (1971), Roving on the Moon (1971), Lunar Highlands (1972), Last on the Moon (1972), Hubble Space Telescope (1990), International Space Station (1998).

*The first NASA space shuttle (*Columbia*) launched from Cape Kennedy, Florida, on April 12, 1981. Astronauts John Young and Robert Crippen piloted the orbiter to a safe landing two days later.*

Rings around Neptune

With the discovery of the **Uranian Rings** in 1977 and then the **Jovian Rings** around Jupiter in 1979, three of the four giant planets were known to have ring systems. The British astronomer William Lassell, who had discovered Neptune's large moon Triton in 1846, also reported seeing a ring around the planet itself. But the observation was never confirmed. The hunt to find out if Neptune really had rings would go on for over 140 years.

Starting in the 1960s, a number of astronomers had been observing stellar occultations of Neptune—distant stars appearing to slip behind and be eclipsed by the planet. A stellar occultation of Uranus had led to the discovery of that planet's ring system, so, astronomers figured, perhaps the method would also reveal rings around Neptune. Most of the results were ambiguous and not repeatable, but in 1982 two separate groups, one in New Zealand and the other in Arizona, reported detecting potential rings or partial ring arcs around Neptune based on analysis of observations made in 1968 and 1981. Then, in 1984, two groups of astronomers from Arizona and France were able to observe the same occultation event, and for the first time they independently confirmed the detection of telltale dips in starlight near the planet—a strong indicator of the possibility of rings. It wasn't until the **Voyager 2** flyby of Neptune in August 1989, however, that solid proof of the existence of Neptune's rings was provided by the spacecraft's cameras.

Voyager images and additional stellar occultation data revealed Neptune to have five distinct and generally faint and dark rings, named after prominent astronomers who contributed to early Neptune science: Galle, LeVerrier, Lassell, Arago, and Adams. The Adams ring is farthest from Neptune, and it contains at least five partial segments or arcs that are brighter than the rest of the ring. These arcs appear to be what were detected and identified as possible rings by earlier ground-based astronomers.

Neptune's rings are dark and dusty, resembling the rings of Jupiter more than the rings of Saturn or Uranus. Like the rings of Uranus, however, Neptune's rings are currently thought to be relatively young remnants of a recent impact breakup of small inner moons or moonlets. Astronomers are still actively trying to figure out the cause of the puzzling arcs in the Adams ring.

SEE ALSO Saturn Has Rings (1659), Uranian Rings Discovered (1977), Jovian Rings (1979), *Voyager* 2 at Neptune (1989).

Voyager 2's confirmation of clumpy, tenuous rings around Neptune came from wide-angle photography of the rings taken during the spacecraft's August 1989 flyby. The three brighter arcs within the outer Adams ring were also detected by ground-based astronomers.

Pioneer 10 beyond Neptune

On June 13, 1983, the NASA space probe *Pioneer 10*, launched from Florida in 1972, became the first human-made object to pass beyond the orbit of Neptune. The spacecraft was then about 3.5 billion miles (5.6 billion kilometers) from Earth, or about 38 times the distance between Earth and the Sun, or 38 astronomical units (AU). Faint radio signals from the still-operating probe took more than four hours to get to Earth at the speed of light. NASA set up a special phone line for the occasion, and for 50 cents callers could listen to *Pioneer 10*'s radio signals from beyond Neptune, converted into audible beeps and screeches. The spacecraft was last heard from in 2003, when it was 80 AU from Earth; it is now more than 105 AU from Earth and is expected to reach the vicinity of the star Aldebaran in the constellation Taurus in a few million years.

Pioneer 10 is one of just five human artifacts that have been accelerated to escape velocity from our solar system. Three others, generally heading in the opposite direction, are contemporaries of *Pioneer 10*; they were also launched in the 1970s. *Pioneer 11*, last heard from in 1995, is currently about 85 AU from the Sun and heading toward the center of the Milky Way. *Voyager 2* is about 100 AU from the Sun and is still sending back occasional

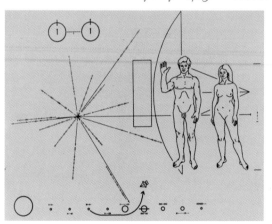

data from deep space. *Voyager 1* holds the record, though: it is more than 120 AU from the Sun and still transmitting scientific data as it speeds outward at 38,000 miles (61,155 kilometers) per hour. The fifth mission to achieve escape velocity from the solar system is the NASA New Horizons probe, launched in 2006 and on course to fly past Pluto and its moons in the summer of 2015, and then, it is hoped, to continue on into the Kuiper Belt to encounter other Pluto-like worlds there in the 2020s.

Voyager 1 will likely be the first of these robotic emissaries from Earth to actually leave the solar system, as it is predicted to cross the heliopause—the gentle, fuzzy boundary where the Sun's faint solar wind merges into the general background of interstellar space—sometime around 2014.

SEE ALSO Kuiper Belt Objects (1992), *Pioneer 10* at Jupiter (1973), *Pioneer 11* at Saturn (1979), *Voyager* Saturn Encounters (1980, 1981), *Voyager 2* at Uranus (1986), *Voyager 2* at Neptune (1989), Pluto Revealed! (2015).

LEFT: *A plaque carried by the* Pioneer *probes, with greetings and information from and about Earth.* RIGHT: *Space artist Donald E. Davis painted this lovely rendering of the* Pioneer *spacecraft heading to the stars. In 1983* Pioneer 10 *became the first human artifact to travel beyond the known planets.*

DON
DAVIS
5-9-83

Circumstellar Disks

Bernard Lyot (1897–1952)

The prevailing theory for the formation of our solar system is that an enormous cloud of gas and dust—perhaps the remnants from the supernova explosion of a previous-generation star—slowly began to gravitationally contract, spin, and flatten into a disk of condensing material. More than 99 percent of the mass in this **Solar Nebula** disk went into the Sun. Most of the rest formed Jupiter, and we all live on a tiny speck of the leftovers. If this model is correct, then it should also have occurred around other stars, especially around Sun-like stars, which are very common in the Milky Way galaxy. Astronomers looked for evidence of disks, rings, or halos of dust around stars, but the search was effectively impossible because direct starlight is a million to a billion times brighter than the light reflected from any disks or planets that might orbit them.

A critical breakthrough was the use of a special telescopic attachment called a coronagraph, invented by the French astronomer Bernard Lyot in 1930 to help block direct sunlight so that astronomers could observe the Sun's upper atmosphere, or corona. Using a smaller version of the device, astronomers could block out the direct light from a star and detect the faint light from objects close to the star.

In 1983 the joint NASA, Dutch, and British space telescope called the Infrared Astronomical Satellite (IRAS) conducted the first all-sky survey of infrared heat energy emitted by cosmic objects. IRAS data revealed an unusual excess of low-temperature infrared heat energy around the young star Beta Pictoris, speculated to be from dusty or rocky material in orbit around the star. The speculation was confirmed in 1984, when astronomers observing with a specially designed Lyot-style coronagraph from the 8.2-foot (2.5-meter) telescope at Las Campanas Observatory in Chile observed a spectacular dusty and rocky circumstellar disk extending some 400 astronomical units (AU) from the center of the star—evidence of a solar nebula around another star.

The Beta Pictoris disk is now one of many known circumstellar disks believed to be young solar systems caught in the act of formation. In 2008 astronomers imaged even closer to Beta Pic and discovered a giant planet about 8 times the mass of Jupiter orbiting just 8 AU from the star—one of the first directly imaged **Extrasolar Planets**.

SEE ALSO Solar Nebula (c. 5 Billion BCE), First Extrasolar Planets (1992), Spitzer Space Telescope (2003).

A composite near-infrared image of the dusty disk around Beta Pictoris, obtained from the 12-foot (3.6-meter) telescope at the European Southern Observatory (ESO), superimposed on an ESO Very Large Telescope view that reveals a large planetary companion (blue dot) orbiting close in, near the (blocked-out) star.

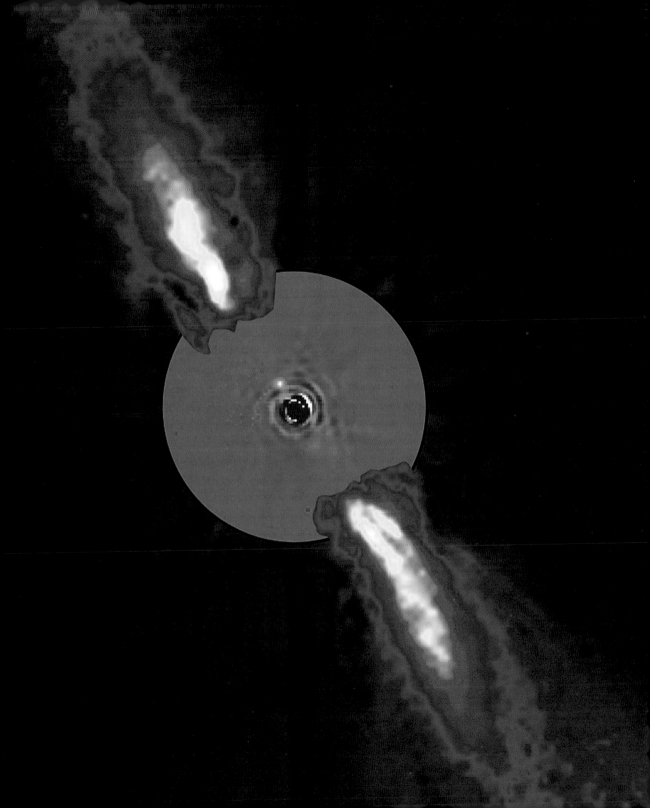

Voyager 2 at Uranus

During its flyby of Saturn in 1981, NASA spacecraft engineers used the gravity of the ringed planet to steer the *Voyager 2* space probe onward to an encounter with the giant planet **Uranus** in January 1986. This was the third stop in *Voyager 2*'s Grand Tour of the outer solar system, and remains the only spacecraft encounter with Uranus and its bluish atmosphere, icy rings, and moons.

Discovered in 1781 by English astronomer William Herschel, Uranus is tilted on its side—the planet rolls rather than spins around the Sun. *Voyager 2* mission planners had to guide the spacecraft to within about 50,000 miles (81,000 kilometers) of the upper cloud layers of Uranus to gain the precise gravity assist needed to propel the spacecraft on to Neptune. The trajectory allowed a close flyby of the small inner moon **Miranda**, and more distant views of the geology of the four other large moons—**Ariel**, **Umbriel**, **Titania**, and **Oberon**.

The encounter was incredibly successful and yielded many new and enigmatic discoveries about the Uranian system. Planetary scientists discovered that Uranus has a strong magnetic field, comparable to Saturn's but weaker than Jupiter's, and strangely tilted relative to the planet's sideways spin axis. Eleven small new moons were discovered in the imaging data, and detailed images were obtained of the nine dark rings that had been previously discovered from Earth-based telescopic measurements. *Voyager 2* data revealed that the aquamarine color of the Uranian cloud tops is caused by small amounts of methane, and analysis of density and other data indicated that beneath the hydrogen- and helium-dominated atmosphere, the planet has an icy mantle and an Earth-size rocky, metallic core. A lasting legacy of the *Voyager 2* encounters is the discovery that both Uranus and Neptune are ice giants, not gas giants.

For many, though, the highlight of the encounter was the spacecraft's close pass by the tiny moon **Miranda,** only 300 miles (480 kilometers) wide. The images revealed a patchwork-quilt pattern of giant fractures, ridges, and cliffs interspersed with bland, cratered, icy terrain—a landscape like no other seen before or since in our solar system. Miranda appears to have been ripped apart and then reassembled out of order, perhaps in the same giant impact event that may have toppled Uranus over.

SEE ALSO Uranus (c. 4.5 Billion BCE), Neptune (c. 4.5 Billion BCE), Discovery of Uranus (1781), Titania (1787), Oberon (1787), Ariel (1851), Umbriel (1851), Miranda (1948), Uranian Rings Discovered (1977), *Pioneer 11* at Saturn (1979), *Voyager* Saturn Encounters (1980, 1981).

Montage of Voyager 2 *images of the ice-giant planet Uranus from January 1986—our first and only encounter with the seventh planet. Counterclockwise from the foreground: the icy moons Ariel, Miranda, Titania, Oberon, and Umbriel.*

Supernova 1987A

If a star is massive enough—perhaps 8 to 10 times the Sun's mass—stellar evolution models indicate that once it converts all of its hydrogen to helium it will eventually end its life in a gigantic explosion called a supernova. Astronomers believe that a supernova explosion goes off in the Milky Way galaxy about once every 50 years or so. Most are too distant to notice, or are obscured by dust in the galactic plane. Chinese Astronomers recorded a number of "**Guest Stars**" over the centuries, including one in 185 and a "**Daytime Star**" supernova in 1054 that eventually formed the Crab Nebula. In 1572 **Tycho Brahe** obtained detailed observations of a supernova in the constellation Cassiopeia, and in 1606 Johannes Kepler wrote an entire book about a bright supernova in Ophiuchus that had occurred two years earlier. Kepler's supernova of 1604 is still the most recent known stellar explosion in our galaxy.

Modern astronomers finally got the opportunity to study a supernova "up close" when the blue supergiant star Sanduleak −69° 202 suddenly exploded on February 23, 1987, becoming Supernova 1987A. The explosion actually occurred 168,000 years earlier in the Large Magellanic Cloud, one of the Milky Way's dwarf satellite galaxies, but it took that long for the light to arrive at Earth. The star increased in brightness by a factor of about 4,000, becoming a naked-eye object to observers all over the world for more than six months before it faded.

Astronomers used Supernova 1987A as a grand cosmic experiment to understand stellar evolution and high-energy processes. Optical and infrared telescopes around the world, and ultraviolet, X-ray, optical, and infrared telescopes in space observed the event and its aftermath. Just three hours before the visible explosion, **Neutrinos** were detected at multiple observatories, confirming the core-collapse model for supernova explosions. In the past few years astronomers have watched the shock waves from the main explosion crash into a shell of previously ejected gas from the dying progenitor star.

Some stars die spectacular, violent deaths. One wonders whether any planets and their inhabitants were also destroyed in these cataclysmic events, and when the next supernova might go off in our neck of the galaxy. We seem overdue.

SEE ALSO Astronomy in China (c. 2100 BCE), Chinese Observe "Guest Star" (185), "Daytime Star" Observed (1054), Brahe's "Nova Stella" (1572), Main Sequence (1910), Neutrino Astronomy (1956).

Hubble Space Telescope photo of the bright speckled ring of light surrounding the remnants of Supernova 1987A (center). The ring is caused by powerful shock waves traveling outward from the exploding star. The two bright bluish stars in the foreground are unrelated to the supernova.

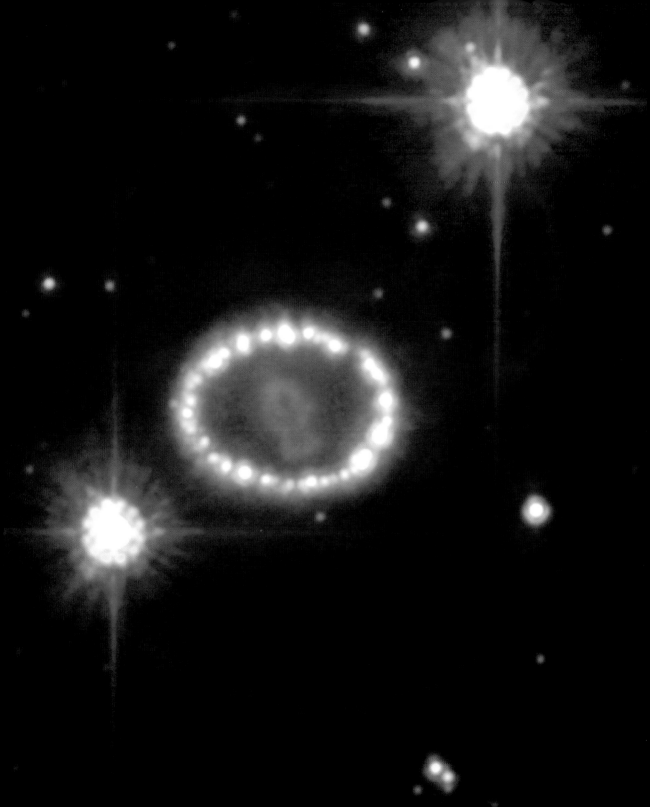

Light Pollution

To our ancestors, the night sky was a source of reverence, inspiration, and wonder. On a clear, moonless night it was possible even from the cities to see thousands of stars with the unaided eye, including the grand, sweeping arch of the **Milky Way**. But the advent of modern civilization, and especially the growth of major cities and urban centers and their widespread use of electricity for artificial illumination, have significantly changed our relationship with the night sky. Rather than thousands of stars, most people in industrialized countries can now usually see only hundreds of stars on a clear night; residents of major cities might be lucky to see just 10 or 20 stars and a slew of airplanes, but certainly not the Milky Way. The night sky has lost its wonder for most people, becoming instead a dull, faintly glowing, and featureless part of the background.

The culprit in this nocturnal cosmic dulling is light pollution, the alteration of natural outdoor light levels by artificial light sources. Light pollution obscures fainter stars for people living in cities or suburbs, interferes with astronomical observations of faint sources, and can even have an adverse effect on the health of nocturnal ecosystems. It's also economically inefficient—the point of lighting a home or building at night is to light the home or building, not to spend money and kilowatts lighting up the night sky.

In recognition of the growing global problem of light pollution, in 1988 a group of concerned citizens formed an organization known as the International Dark-Sky Association (IDA), with the mission "to preserve and protect the nighttime environment and our heritage of dark skies through quality outdoor lighting." IDA now has about five thousand members worldwide who work with city and local governments, businesses, and astronomers to raise awareness about the value of dark skies and to help implement lighting solutions that are more energy efficient and economical and that lead to less light pollution.

Despite some notable successes establishing ordinances and building codes that are decreasing light pollution, the effect on astronomy continues to limit the utility of major observatories near large cities (such as the Mount Wilson Observatory, perched above Los Angeles). New telescopes are now usually built in remote deserts or on isolated dark mountaintops to escape the night sky's growing glow.

SEE ALSO Milky Way (c. 13.3 Billion BCE), Stellar Magnitude (c. 150 BCE), First Astronomical Telescopes (1608).

A map of artificial night sky brightness for part of the Western Hemisphere, from the US Defense Meteorological Satellite Program. The reddest places, mostly in the eastern and western United States, are where light pollution makes the night sky nearly 10 times brighter than the natural sky.

Voyager 2 at Neptune

The launch of the *Voyager 1* and 2 missions in 1977 fortuitously occurred at a time when a rare alignment of **Jupiter**, **Saturn**, **Uranus**, **Neptune**, and **Pluto** could allow a space probe to gravitationally slingshot from one to the next and enable flybys of all five planets with little extra propulsion needed. Advocates for this **"Grand Tour"** of the outer solar system realized that such an opportunity would not occur again for 176 years.

Though the Pluto flyby option was given up in favor of a closer flyby of Saturn's moon **Titan** by *Voyager 1*, the *Voyager 2* probe was able to complete the initial reconnaissance of all four of the giant planets. Its last stop on the Grand Tour was in August 1989 at Neptune, where the spacecraft bent its path downward to a 24,000-mile (38,000-kilometer) flyby of the large icy moon **Triton**.

The *Voyager 2* Neptune encounter revealed much beauty and mystery in the Neptune system. The planet's hydrogen-, helium-, and methane-bearing atmosphere was much more dynamic than Uranus's had been in 1986, with dark- and light-blue belts and white clouds circling a giant cyclonic storm called the Great Dark Spot (analogous to Jupiter's **Great Red Spot**). Planetary scientists discovered that Triton's

relatively youthful surface contains active geysers of icy nitrogen, water, and carbon dioxide, erupting nitrogen into the moon's thin atmosphere of nitrogen, carbon monoxide, and methane. A new large moon, 250 miles (400 kilometers) across, was discovered close to the planet and was named Proteus, after the shape-shifting Greek sea god. Five other small moons were discovered and characterized during the flyby. Subsequent missions have gone on to orbit Jupiter (*Galileo*) and Saturn (*Cassini*), and plans are slowly being formulated to send orbiters to Uranus and Neptune to expand on Voyager's initial discoveries.

SEE ALSO Uranus (c. 4.5 Billion BCE), Neptune (c. 4.5 Billion BCE), Great Red Spot (1665), Discovery of Neptune (1846), Triton (1846), *Voyager* "Grand Tour" Begins (1977), Jovian Rings (1979), An Ocean on Europa? (1979), *Voyager* Saturn Encounters (1980, 1981), Rings around Neptune (1982), *Voyager 2* at Uranus (1986), Kuiper Belt Objects (1992), *Galileo* Orbits Jupiter (1995), *Cassini* Explores Saturn (2004).

LEFT: *Neptune's second-largest moon, Proteus, was discovered in images taken during the* Voyager 2 *Neptune flyby in August 1989.* RIGHT: *A composite view assembled from* Voyager 2 *images simulating a view of Neptune from the surface of the large icy moon Triton.*

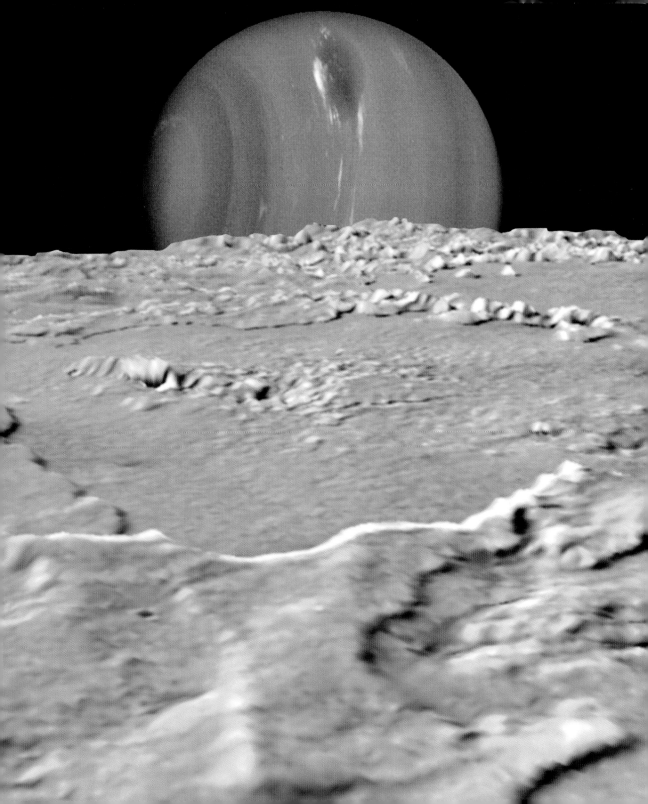

Walls of Galaxies

Margaret Geller (b. 1948), **John Huchra** (1948–2010)

Advances in telescopes, spectrometers, and photographic plates allowed early-twentieth-century astronomers such as Vesto Slipher and Edwin Hubble to determine the **Doppler Shifts** of distant galaxies in our expanding universe. Using **Hubble's Law**, the distances to those red-shifted galaxies could be estimated. It was slow going—only around 600 galaxies had estimated distances by the 1950s. By the 1970s and 1980s, however, larger telescopes, advances in digital detectors (such as charge-coupled devices, or CCDs), and the advent of dedicated all-sky galaxy surveys had enabled red shifts to be measured for more than 30,000 galaxies. Finally, the universe could begin to be mapped.

A pioneer in galaxy mapping has been the American astronomer Margaret Geller, who, along with fellow Harvard-Smithsonian Center for Astrophysics (CfA) astronomer John Huchra, has been using the results of several large-scale galaxy red-shift surveys to discover the structure of the cosmos. The first CfA red-shift survey ran from 1977 to 1982, and the second from 1985 to 1995. In 1989 Geller and Huchra reported that the

distribution of galaxies is far from uniform. Instead, galaxies are clumped into enormous filamentary structures surrounding giant voids containing few galaxies, in a structure now called the cosmic web. One of the structures found by Geller and Huchra was called the "Great Wall"; at over 500 million light-years long and 300 million light-years wide, the Great Wall is one of the largest known structures in the universe.

Astrophysicists have a powerful working model for the formation of these vast patterns in the universe: galaxies and larger patterns grow from very small irregularities in the matter distribution in the early universe. In some versions of this model, these irregularities can arise from a very early (during the first 10^{-32} seconds after the **Big Bang**!) epoch of inflation, which smeared matter into weblike filaments that may have clumped into galaxies.

Since the initial CfA studies, other more ambitious galaxy surveys have followed, such as the Sloan Digital Sky Surveys (starting in 2000), which despite more than a decade of work have only mapped out 1/10,000th of the visible universe so far!

SEE ALSO Big Bang (c. 13.7 Billion BCE), Milky Way (c. 13.3 Billion BCE), Doppler Shift of Light (1848), Hubble's Law (1929), Dark Matter (1933), Astronomy Goes Digital (1969).

LEFT: *Harvard astronomer Margaret Geller.* RIGHT: *A slice through part of a 3-D map of the structure of distant galaxies, from the Sloan Digital Sky Survey. Earth is at the center right, and the outer circle at left is 2 billion light-years away. Each dot is a galaxy, with redder dots representing galaxies containing older stars.*

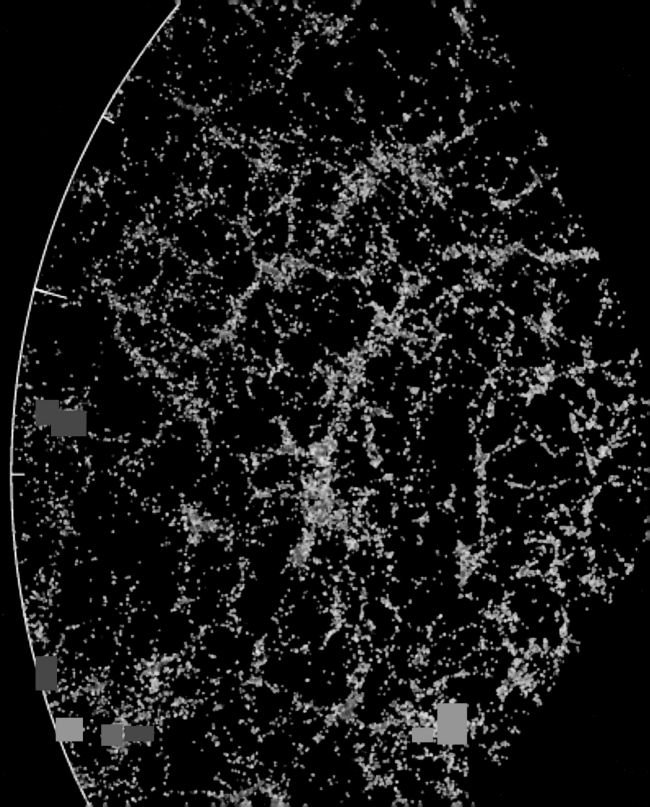

Hubble Space Telescope

Lyman Spitzer (1914–1997)

The **First Astronomical Telescopes**, developed in the early seventeenth century, opened the skies to astronomers; they and subsequent larger and more advanced instruments enabled amazing discoveries about the solar system, galaxy, and universe. But astronomers have always known that even the largest telescopes on Earth are fundamentally limited in two important ways: first, the unavoidable shimmering and twinkling of our atmosphere limit the resolution to much less than a large telescope's theoretical limit; and second, our atmosphere blocks many parts of the spectrum—especially in the ultraviolet and infrared ranges—making ground-based observations difficult or impossible at key wavelengths.

With the advent of space satellites in the 1960s, astronomers began advocating within NASA for a dedicated, space-based telescope to overcome those limitations. A chief champion of an orbiting space telescope was the American astronomer Lyman Spitzer, who led a critical grassroots lobbying campaign for the necessary support and funding of the project. After many bureaucratic hurdles and a forged partnership with the European Space Agency, the Large Space Telescope, later named the Hubble Space Telescope (HST) after astronomer Edwin Hubble, was approved in 1978. HST was eventually launched into low Earth orbit—about 350 miles (570 kilometers) above the surface—by the **Space Shuttle** *Discovery* in April 1990.

Shortly after launch, HST was discovered to have a major flaw in its primary mirror design. Fortunately, the telescope could be serviced by space shuttle astronauts, and five shuttle missions between 1993 and 2009 fixed the telescope and upgraded key instruments and components. As a result, HST serves as a cosmic time machine that uses CCD imaging and **Spectroscopy** to determine the nature and even the **Age of our Universe**. By today's giant telescope standards, HST is only a medium-size telescope, but its constant clear-sky and full-spectrum view of the cosmos have enabled it to realize the dreams of Spitzer and other early supporters, and fundamentally revolutionize modern astronomy and astrophysics.

SEE ALSO First Astronomical Telescopes (1608), Hubble's Law (1929), Space Shuttle (1981), Age of the Universe (2001).

The Hubble Space Telescope floats freely about 350 miles (560 kilometers) above Earth's surface after being released by the space shuttle Discovery during a servicing mission in February 1997. The telescope is about 8.2 feet (2.5 meters) in diameter and about 43 feet (13.1 meters) long, or slightly longer than an average school bus.

Venus Mapped by *Magellan*

When Galileo peered through his early astronomical telescope in 1610, he was the first person to realize that **Venus** has phases. In contrast to his studies of the Moon and Jupiter, however, Galileo's observation of Venus did not reveal any specific features or markings—and neither did those of any other astronomers. That's not surprising: recent telescopic observations and space missions, such as the Soviet Venera orbiters and landers, have showed that the surface of Venus is shrouded from view by a thick atmosphere of carbon dioxide and featureless sulfuric acid clouds and hazes.

Fortunately, radio waves can see through clouds and hazes. Indeed, radar weather mapping instruments used by terrestrial meteorologists exploit the fact that radio waves pass through clouds but bounce off raindrops and snowflakes. In the 1960s, astronomers figured out that radio waves could penetrate the clouds of Venus by bouncing radar signals off the planet using the **Arecibo Radio Telescope**. Some surface markings were detected this way, allowing the very slow (243 Earth days), backward rotation of the planet to be discovered. The Arecibo findings helped justify the development of orbital radar mapping missions to Venus. The first successful missions were the Soviet *Venera 15* and *16* orbiters in 1983–1984, which mapped about 25 percent of Venus's northern hemisphere, revealing mountains, ridges, faults, volcanoes, and other landforms.

Venera's results helped spur an even more ambitious Venus radar mapper: the NASA *Magellan* mission. Launched in 1989 by the **Space Shuttle** *Atlantis*, *Magellan* arrived in 1990 and systematically mapped 98 percent of the surface from pole to pole. *Magellan* data reveal the full scope of Venus topography—from high mountains to deep valleys—and provided geologists with images of a spectacular variety of volcanic, tectonic, impact, and erosional landforms. *Magellan* scientists discovered vast lava plains, pancake-shaped volcanic domes, and large Hawaiian-style shield volcanoes. They also discovered channels many thousands of miles long carved by very low-viscosity molten rock; vast networks of ridges and troughs, suggesting tectonic activity but not Earthlike plate tectonics; and very few impact craters, suggesting that most of the planet may have been resurfaced by massive outpourings of lava some 500 to 750 million years ago. While certainly not Earth's twin, Venus turns out to have been (and still is?) just as active over geologic time.

SEE ALSO Venus (c. 4.5 Billion BCE), Greenhouse Effect (1896), Arecibo Radio Telescope (1963), Microwave Astronomy (1965), *Venera* 3 Reaches Venus (1966), *Venera* 7 Lands on Venus (1970).

A colorized elevation map (reds and whites are higher; greens and blues are lower) of Venus, using radar data from the NASA Magellan mission and the Arecibo radio telescope in Puerto Rico.

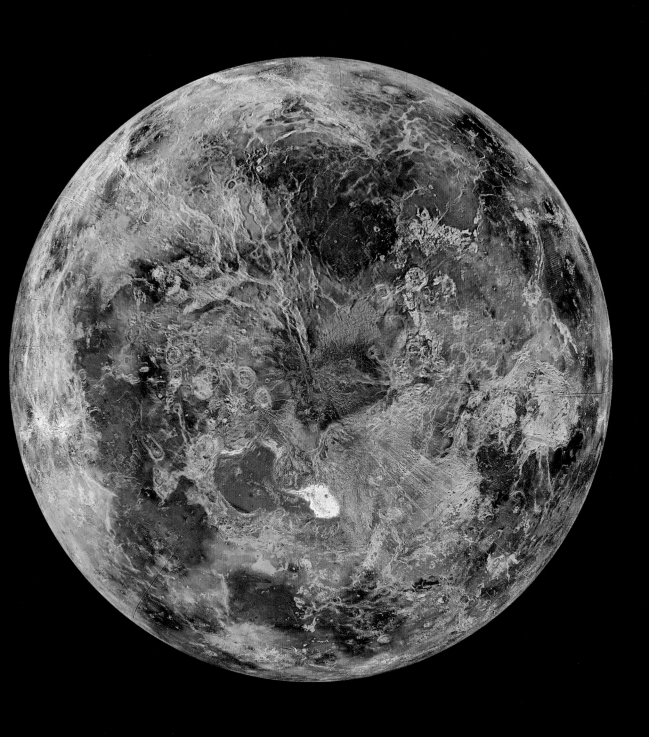

Gamma-Ray Astronomy

Arthur H. Compton (1892–1962)

In the 1970s, military scientists working with Earth orbital satellites designed to detect the telltale high-energy gamma-ray emission from nuclear explosions discovered that short, intense flashes of gamma radiation—called **Gamma-Ray Bursts**—were coming from all over the sky and from vast cosmic distances. Follow-on civilian space satellites verified this finding, but the origin of these unprecedented bursts of energy remained an enigma throughout the 1980s, partly because they fade away so quickly (within minutes, at most) and partly because the satellites didn't have enough resolution to associate the bursts with specific stars or galaxies at other wavelengths. Faster, sharper gamma-ray eyes were needed.

Those eyes were provided in 1991 with the launch of the Compton Gamma Ray Observatory (CGRO), the second of NASA's "great space observatories" to be launched, after the **Hubble Space Telescope**. Named after the American physicist and pioneer of gamma-ray studies Arthur H. Compton, CGRO scanned the skies for gamma-ray bursts for more than nine years. A main objective of the CGRO instruments was to constantly scan the sky for randomly occurring gamma-ray bursts, and when one was located, to quickly identify its energy and position to a very high level of accuracy. A network of ground-based telescopes was then alerted in near real time to conduct follow-on optical and infrared imaging and spectroscopic observations.

Instruments on CGRO detected more than 2,700 gamma-ray bursts (about one per day), and pinpointed the locations of more than 100 ultra-high-energy bursts that received quick follow-up from other telescopes. Measurements of the **Doppler Shifts** of the sources confirmed the extragalactic nature of the events, further underscoring their tremendous energies. CGRO data also reveal that there are two kinds of gamma-ray bursts: short duration (less than 2 seconds) and longer duration (up to a few minutes). The durations of the events are a clue that their origin involves physically small (compact) objects, and their enormous energies are a clue that the gamma rays are created in violent environments. Astronomers now believe that the longer events result from supernovas collapsing to form **Neutron Stars** or **Black Holes**, and that the shorter events are caused by the merger of pairs of binary neutron stars.

SEE ALSO Neutron Stars (1933), Nuclear Fusion (1939), Black Holes (1965), Pulsars (1967), Gamma-Ray Bursts (1973), Hubble Space Telescope (1990).

Photograph of the remote manipulator arm on the space shuttle Atlantis *releasing the Compton Gamma Ray Observatory into Earth orbit in April 1991.*

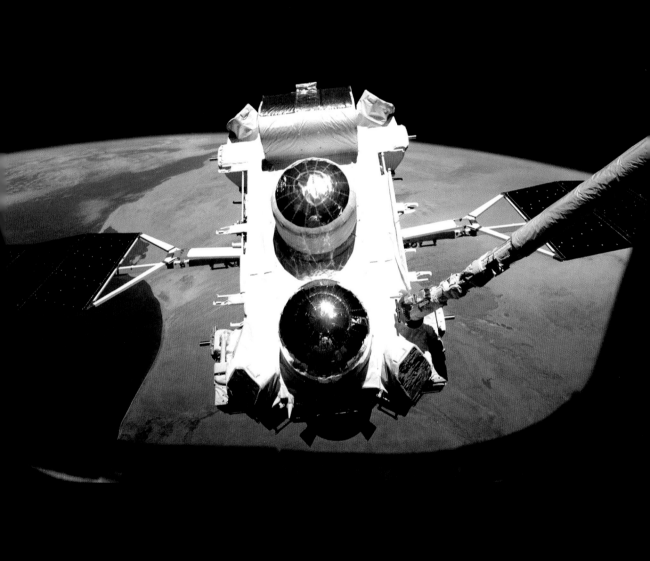

Mapping the Cosmic Microwave Background

The discovery of the **Cosmic Microwave Background** in 1964 had exciting implications because it showed that theories like the **Big Bang** could be tested with actual observations. A new generation of astronomers became interested in observational cosmology—the precise measurement of certain characteristics of the universe that could enable a detailed understanding of its origin and evolution. To reach the needed precision, though, these measurements had to be conducted from space.

NASA's continuing Explorer small satellite program, which had begun in 1958 with the *Explorer 1* discovery of the Van Allen **Radiation Belts**, was a perfect platform. Astrophysicists worked out the mission concepts in the 1970s, and in the 1980s NASA approved the Cosmic Background Explorer (COBE) mission, which was launched into Earth orbit in 1989. COBE used very sensitive infrared and microwave radiation detectors to slowly and accurately build up a map of the variations in cosmic background radiation across the entire sky.

In 1992 cosmologists announced that the initial map was completed, and that the results were exciting. The main variation that COBE detected was a weak so-called dipole signature, only about 1/1000th of the brightness of the sky, due to the Doppler shift from our galaxy's motion relative to the rest of the universe. Once that signal was removed, the next biggest COBE variation was from the faint microwave emission of our own Milky Way galaxy. Once that signal was removed, astronomers were delighted that there were still some small variations left, tiny fluctuations in background radiation at the scale of a few millionths of a degree.

Cosmologists believe that those tiny variations were formed during the inflation of the universe in the first 10^{-32} seconds after the **Big Bang**, concentrating normal and **Dark Matter** into the "seeds" that would eventually form galaxies and stars. COBE's results were confirmed and extended by the Wilkinson Microwave Anisotropy Probe (WMAP) Explorer mission in 2003, helping to provide new, accurate estimates of the **Age of the Universe**.

SEE ALSO Big Bang (c. 13.7 Billion BCE), Recombination Era (c. 13.7 Million BCE), Einstein's "Miracle Year" (1905), Dark Matter (1933), Earth's Radiation Belts (1958), Cosmic Microwave Background (1964), Walls of Galaxies (1989), Age of the Universe (2001).

Maps of the microwave radiation permeating the universe, as measured by the COBE satellite. The plane of the Milky Way goes through the "equator" of these maps. At the top is the total signal; in the middle is the map with the relative motion of the solar system removed; at the bottom is the map with the signal from the galaxy removed.

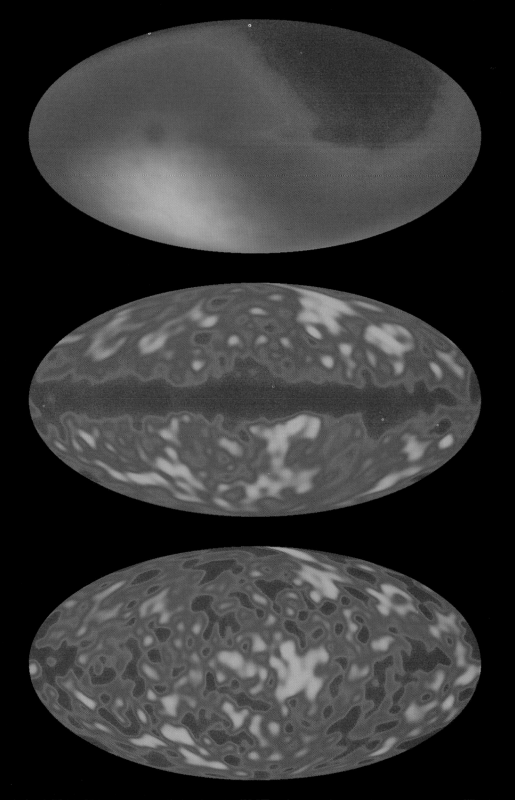

First Extrasolar Planets

Are there planets around other stars? For most of the history of astronomy, the question has been either too heretical to ask (Giordano Bruno was burned at the stake in 1600 because of it) or too technologically impossible to address. Recently, though, astronomers have discovered that the answer is a resounding *yes*.

By the late twentieth century, telescope and observational technology had advanced to the level where astronomers could detect the presence of planets around other stars using a variety of methods. One method exploits the fact that planets make their parent stars "wobble" in the sky; Jupiter's gravity, for example, makes our own Sun's path appear to wobble slightly as it orbits the center of the galaxy.

In 1992, a team of astronomers discovered that such slight wobbles could also be detected as very slight changes in the rotational speed of rapidly spinning **Neutron Stars** known as **Pulsars**. In 1990, astronomers using the **Arecibo Radio Telescope** discovered a millisecond pulsar named PSR B1257+12 in the constellation Virgo. Monitoring the pulses every 6.22 milliseconds (msec) from this collapsed neutron star's supernova remnant revealed small regular variations in the pulse rate. In 1992 researchers explained that this was caused by the gravitational pull of at least three planets in orbit around the pulsar. Mathematical modeling showed that two of the planets were likely around four times Earth's mass, and a third was around 2 percent of Earth's mass; all appear to orbit within 0.5 astronomical units of the pulsar.

This first confirmed evidence for the existence of extrasolar planets came as a surprise to most astronomers, because the expectation had been that planets would be found around other normal, main sequence stars like the Sun rather than around exotic objects like neutron stars. There is much speculation, then, about the nature of these particular pulsar planets. Perhaps they are the rocky and metallic cores of previous gas or ice giants that had their outer volatile layers stripped away by the supernova explosion that created the pulsar. Or perhaps they represent the results of a second round of **Solar Nebula** planet formation using remnant materials ejected by the supernova explosion.

Whatever the origin of these worlds, their detection appears to be robust, and so astronomers and planetary scientists now discovering and characterizing **Planets around Other Suns** must also consider an even wider range of ways that "extreme" extrasolar planets can form and evolve in a variety of environments.

SEE ALSO Solar Nebula (c. 5 Billion BCE), Bruno's *On the Infinite Universe and Worlds* (1600), First Astronomical Telescopes (1608), Neutron Stars (1933), Arecibo Radio Telescope (1963), Pulsars (1967), Planets around Other Suns (1995).

Artist's conception of the planetary system detected around the pulsar PSR B1257+12 (lower left).

Kuiper Belt Objects

Kenneth Edgeworth (1880–1972), **Gerard P. Kuiper** (1905–1973)

After the **Discovery of Pluto** in 1930, many astronomers began to wonder if the solar system really ended just beyond Neptune's orbit. In 1943 the Irish astronomer Kenneth Edgeworth hypothesized that Pluto might be one of many small trans-Neptunian bodies that failed to grow very large because of the large separation between early-growing planetesimals (kilometer-size clumps of dust and ice) and the resulting lower impact rate in the outer solar system. In the 1950s, the Dutch-American planetary scientist Gerard P. Kuiper studied planet formation in the outer solar system and similarly hypothesized the presence of a large disk of small bodies beyond Pluto. If Pluto was an Earth-size body, however (as was thought at the time), then Kuiper speculated that this disk would have been cleared out and scattered away by Pluto's gravitational influence.

The existence of what astronomers came to call the Edgeworth-Kuiper Belt (or just the Kuiper Belt) remained a subject of speculation for decades. With the advent of giant telescopes and super-sensitive charge-coupled device (CCD) detectors in the 1990s, however, it became possible to search for and ultimately detect small, faint, asteroid-like bodies at distances beyond Neptune's orbit. The first member of the Kuiper Belt (besides Pluto, **Charon**, and possibly former members **Triton** and **Phoebe**) was discovered in 1992. Named 1992 QB1, the object orbits between 40 and 46 astronomical units (AU; Neptune is at 30 AU) and is probably about 100 miles (160 kilometers) in diameter.

Since then, more than 1,000 Kuiper Belt Objects (KBOs) have been discovered by astronomers. Some, such as 136199 Eris, 136472 Makemake, and 136108 Haumea, are comparable in size to Pluto. In fact, Eris may be larger than Pluto, partly precipitating a decision in 1996 by the International Astronomical Union to **Demote Pluto** from planet to dwarf planet status. About 10 percent of known Kuiper Belt Objects have been discovered to have satellites (as Pluto does).

KBOs are primitive bodies containing comet-like mixtures of water, methane, and ammonia ices. The *New Horizons* probe will give us our first look at KBOs (Pluto and Charon) in 2015, and will go on to encounter more in the 2020s.

SEE ALSO Pluto and the Kuiper Belt (c. 4.5 Billion BCE), Triton (1846), Phoebe (1899), Discovery of Pluto (1930), Öpik-Oort Cloud (1932), Astronomy Goes Digital (1969), Discovery of Charon (1978), Demotion of Pluto (2006), Pluto Revealed! (2015).

Painting by space artist Michael Carroll depicting a heavily impacted, icy, trans-Neptunian object (TNO) or Kuiper Belt Object (KBO) far beyond the orbit of Pluto. Like about 10 percent of known KBOs, this one is part of a binary pair, with its companion just above and to the left of the faint, distant Sun.

Asteroids Can Have Moons

The first planetary satellites beyond our own Moon were discovered around Jupiter by Galileo Galilei in 1610. Since then, dozens more natural satellites have been discovered around all the other planets except Mercury and Venus. When a moon was found even around the small (dwarf) planet Pluto in 1978, many astronomers began to wonder whether there was a limit to how small an object could be in order to have a moon. Could asteroids have moons?

Telescopic searches in the 1970s and 1980s provided possible evidence for their existence. Definitive proof didn't come until 1992, however, when the NASA *Galileo* spacecraft flew past main belt asteroid 243 Ida en route to Jupiter. To everyone's surprise and delight, *Galileo*'s images revealed a small moon orbiting the asteroid. Mission scientists named the moon, which was one mile (1.6 kilometers) wide, Dactyl, after creatures that inhabited Mount Ida on the island of Crete in Greek mythology.

Buoyed by the certainty that even small bodies can have moons, planetary astronomers redoubled their efforts to find more. Using advances such as adaptive optics to increase the effective resolution of ground-based telescopes, as well as assets such as the **Arecibo Radio Telescope** and the **Hubble Space Telescope**, scientists have since discovered moons around more than 200 small bodies, including **Trojan Asteroids** of Jupiter and **Kuiper Belt Objects**. More than 220 minor planet moons have been found— some small bodies even have two or three moons of their own, and Pluto has five!

Planetary scientists have several leading theories about how asteroid-size or dwarf planet–size bodies can have moons. It may be that impacts into the primary asteroid knock off debris that stays and accumulates in orbit to form a satellite. The huge numbers of impact craters seen on the surfaces of asteroids such as 243 Ida and 433 Eros, the target of the NASA Near Earth Asteroid Rendezvous (**NEAR**) mission, support this idea. Alternatively, maybe asteroids can capture other asteroids during close encounters. Computer models show this to be possible for binary asteroids comparable in size and relatively close to each other, but less likely in situations like the Ida-Dactyl system, where one body is very large compared to the other.

SEE ALSO First Astronomical Telescopes (1608), Jupiter's Trojan Asteroids (1906), Arecibo Radio Telescope (1963), Discovery of Charon (1978), Hubble Space Telescope (1990), Kuiper Belt Objects (1992), *Galileo* Orbits Jupiter (1995), NEAR at Eros (2000).

In 1992 the NASA Galileo spacecraft flew past the asteroid 243 Ida while traveling through the main asteroid belt en route to its orbital mission at Jupiter. Images of the asteroid, which was 33 miles (53 kilometers) long, also revealed a tiny 1-mile (1.6-kilometers) companion moon, which was named Dactyl.

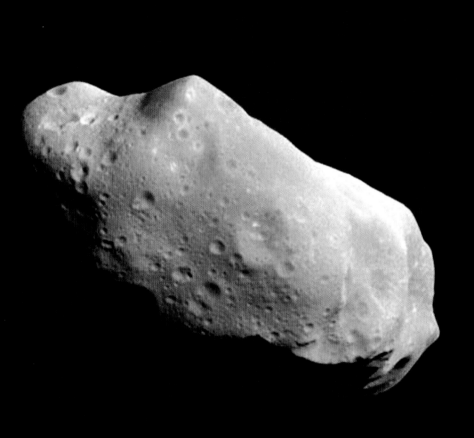

Giant Telescopes

The frontier of modern astronomy has been defined during the last few centuries by the limits of telescope and instrument technology. The need to observe extremely faint sources (oftentimes adjacent to much brighter sources) at extremely fine resolutions, to acquire spectroscopic observations over a wide range of the electromagnetic spectrum, and to accurately record observations for later detailed analysis have all pushed astronomy, and astronomers, to the cutting edge of optics, engineering, electronics, and software design. Perhaps nowhere is this more visible than in the recent proliferation of truly enormous optical telescopes at premier observing sites around the world.

When the 200-inch (5-meter) telescope mirror was cast for the Hale telescope at Mount Palomar in the late 1940s, it was a miracle of then-modern engineering. In the decades since, materials and mechanical engineering methods have improved dramatically, but still the largest practical single-mirror telescope size has only grown to about 320 inches (8 meters) because of gravity and materials' strength limitations. To substantially increase the light-collecting area has required engineers and opticians to innovate, and a key innovation has been the development of segmented mirrors—smaller, more practical mirrors that, when assembled together, simulate a single mirror of much larger size.

The first large-scale experiment in segmented-mirror telescopes was the twin 400-inch (10-meter) telescopes of the W. M. Keck Observatory atop Mauna Kea, Hawaii. Each telescope's mirror consists of 36 hexagonal, 70-inch (1.8-meter) segments that are individually and dynamically computer controlled to form a near-perfect giant parabolic reflector. Keck I went online in 1993; Keck II in 1996. Both have enabled incredible scientific discoveries ever since.

Two Keck telescopes were built to try to exploit another innovation in giant telescope design—using two or more telescopes, it is possible through electronics and software to combine the data from each telescope to yield angular resolutions equivalent to a single enormous telescope as large as the separation between the individuals. This process is known as interferometry. The Keck telescopes are just one example of a growing number of large optical interferometer telescopes that are now defining the newest frontier in ground-based astronomy.

SEE ALSO First Astronomical Telescopes (1608), Arecibo Radio Telescope (1963), Hubble Space Telescope (1990).

The twin telescopes at the Keck Observatory, near the summit of the extinct Mauna Kea volcano (elevation 13,600 feet [4,145 meters]) in Hawaii. Each telescope is 10 meters (33 feet) in diameter.

Comet SL-9 Slams into Jupiter

Eugene Shoemaker (1928–1997), **Carolyn Shoemaker** (b. 1929), **David Levy** (b. 1948)

Impacts have been a fundamental force in shaping the surfaces and atmospheres of the planets, and have even influenced the history of climate and life in our own world. But impacts are rare and unpredictable events in the solar system and thus have been impossible to study directly—or at least they were until a group of astronomers discovered a year in advance exactly when a comet would smash into the planet **Jupiter**.

In the summer of 1993, the American astronomers Eugene Shoemaker and Carolyn Shoemaker and the Canadian astronomer David Levy reported finding a strange "string of pearls" comet that their observations showed had recently passed so close to Jupiter that it had been broken apart into dozens of fragments. Incredibly, their data showed that the fragments, now collectively known as comet Shoemaker-Levy 9 (SL-9), were on return trajectories to Jupiter, and that they would crash into the planet in July 1994.

This was the first time that an impact event was known in advance, and astronomers worldwide mobilized telescopes and space observatories to monitor the events. As predicted, 21 cometary fragments 0.6–1.2 miles (1–2 kilometers) across slammed

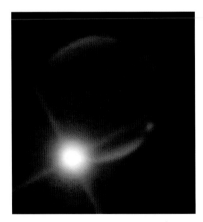

into Jupiter at speeds of 134,000 miles per hour (60 kilometers per second) between July 16 and July 22. The results were spectacular and surprising—that such small, icy objects could produce huge fireballs, plumes, and Earth-size blemishes lasting for months was not widely predicted. Indeed, some astronomers had predicted that Jupiter would simply swallow up the tiny fragments without any noticeable effect at all. How wrong they were!

The SL-9 impacts into Jupiter were a humbling reminder of the destructive power of small bodies traveling at very high speeds. They were also a public and media sensation, shared worldwide in near real time via a relatively new communication medium known as the Internet.

SEE ALSO Jupiter (c. 4.5 Billion BCE), Arizona Impact (c. 50,000 BCE), Great Red Spot (1665), Halley's Comet (1682), Tunguska Explosion (1908).

LEFT: *Fireball from the impact of SL-9 fragment G on July 18, 1994, as seen from Mount Stromlo Observatory, Australia.* RIGHT: *Earth-size blotches—scars from the impacts of fragments of comet SL-9—dot the midlatitudes of Jupiter south of the Great Red Spot in this July 1994 Hubble Space Telescope photo.*

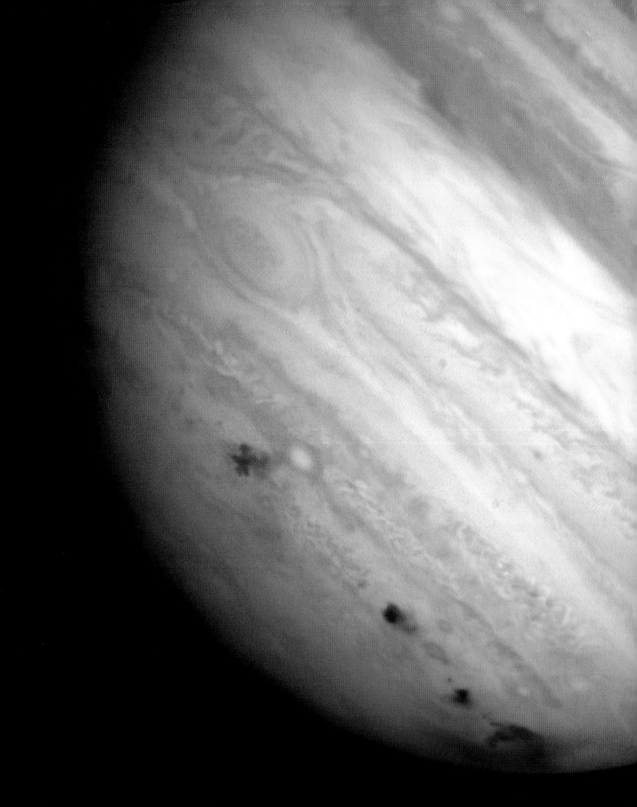

Brown Dwarfs

1994

Stars on the **Main Sequence** come in many colors and sizes, but they all share the common attribute of central core temperatures and pressures high enough to enable **Nuclear Fusion** of hydrogen into helium. Indeed, the lowest possible stellar mass that enables nuclear fusion is about 7–9 percent of the mass of the Sun, or about 75–80 times the mass of Jupiter. In the 1970s, astronomers hypothesized that there might be a population of substellar objects too large to be considered giant planets but too small to be considered stars. These objects were dubbed brown dwarfs because they would be smaller than stars, would radiate infrared heat from gravitational contraction, but would not emit their own visible light from nuclear fusion. The hunt was on to find and characterize these potentially important links between giant planets and small stars.

Potential candidates were found in the late 1980s and early 1990s, but it turned out to be difficult to figure out where the line between small stars and brown dwarfs was crossed. It wasn't until 1994 that a faint infrared source was identified as a possible brown dwarf near the star Gliese 229, a small red dwarf star about 19 light-years from our solar system. Follow-on observations from the **Hubble Space Telescope** and other observatories showed that the fainter object is indeed in orbit around the star, earning it the name Gliese 229B (GL229B).

GL229B has a luminosity and temperature (950 kelvins), well below those of even the smallest main sequence stars, but much higher than those of a gas-giant planet orbiting some 30 astronomical units (AU) from a red dwarf star. The telltale proof that GL229B is not a low-mass star came with the discovery of methane—a gas that is not stable in stellar atmospheres—in its spectrum. Current estimates are that GL229B is a brown dwarf of about 20–50 Jupiter masses.

Some of the largest of the newly discovered planets around other stars have masses around 20–50 times **Jupiter**'s mass. Does that mean that they are brown dwarfs rather than planets? The low mass line between a giant megaplanet and a low-mass star appears to be fuzzy. Astronomers often use density, infrared luminosity, or the presence of X-rays to decide, but when in doubt, they fall back on the "official" definition that a brown dwarf is above 13 Jupiter masses.

SEE ALSO Jupiter (c. 4.5 Billion BCE), Nuclear Fusion (1939), Hubble Space Telescope (1990), First Extrasolar Planets (1992).

False-color Hubble Space Telescope image of the first directly detected brown dwarf (near center of the image), orbiting the nearby star Gliese 229. Called GL229B, the substellar companion is some 20 to 50 times the mass of Jupiter–just a little too small to fuse hydrogen into helium in its core.

Planets around Other Suns

The discovery of the **First Extrasolar Planets** around pulsar B1257+12 in 1992 compelled astronomers to search even harder for evidence of extrasolar planets around "normal" **Main Sequence** or Sun-like stars. For decades it was known that stars in binary systems can show "wobbles" in their **Proper Motion** across the sky because both stars are actually orbiting the system's center of mass. In theory, the same kinds of wobbles—though much smaller—should be seen if a giant Jupiter-like (or larger) planet is in orbit around a single star. A breakthrough came when astronomers realized that they didn't need to measure the precise position of the star over time; instead, they could use the **Doppler Shift** of the star's spectrum to deduce its wobbling motion from the star's radial velocity—the part of that wobble directed toward or away from Earth.

In 1995, using this method, the first extrasolar planet was "found" orbiting the nearby Sun-like star 51 Pegasus (51 Peg). Based on the amount of wobble it induced in 51 Peg, and the timing of the Doppler-shifted spectral variations, planet 51 Peg b was inferred to be a gas giant many times the size of Jupiter, and orbiting very close—only 0.05 astronomical units—from its star. More than five hundred other planets around other nearby stars have been found since then using the radial velocity method. Most of these planets are known as "hot Jupiters" because they are large and they also orbit extremely close to their parent stars. Big planets, close in, are the easiest to find using the radial velocity method, so hot Jupiters may not be typical.

Other ways to find extrasolar planets, besides radial velocity and pulsar timing, are to watch for them passing in front of (transiting) their parent stars (the goal of NASA's *Kepler* Mission), to detect them by gravitational lensing, or to just directly image them through the glare of their parent stars. To date, most known extrasolar planets are still gas or ice giants. However, astronomers are now starting to find and catalog many Earth-size or super Earth-size worlds around nearby Sun-like stars using these methods. We are only at the tip of the iceberg of extrasolar planet discoveries!

SEE ALSO Solar Nebula (5 Billion BCE), First Astronomical Telescopes (1608), Proper Motion of Stars (1718), Doppler Shift of Light (1848), Gravitational Lensing (1979), First Extrasolar Planets (1992), *Kepler* Mission (2009).

Artist's conception of the "hot Jupiter" planet HD189733b, which Hubble Space Telescope observations show has methane and water vapor in its atmosphere.

Galileo Orbits Jupiter

The *Pioneer 10* and *11* and *Voyager 1* and *2* flybys of Jupiter revealed a beautiful, complex, and enigmatic mini solar system at the giant planet, including a dynamic atmosphere of zones, belts, and long-lived cyclonic storm systems, such as the **Great Red Spot**; a veritable zoo of icy and rocky satellites, including four that are essentially planets in their own right (**Io, Europa, Ganymede, and Callisto**), a dusty system of narrow rings; and an enormous magnetosphere that bathes much of the system in high-energy radiation. The discoveries during those brief flybys convinced planetary scientists that the next logical step would be to send a spacecraft to orbit Jupiter and stay for a while.

With congressional and international support, the Jupiter Orbiter and Probe mission, renamed *Galileo* in honor of the astronomer who first studied Jupiter and its moons through a telescope, was approved for funding in late 1977. Overcoming a number of technical and financial hurdles, *Galileo* was eventually launched on the space shuttle *Atlantis* in late 1989, and used gravity-assist flybys of Venus and Earth to slingshot out to Jupiter, where it arrived and entered orbit in December 1995. Along the way, *Galileo* performed the first close asteroid flybys (of 952 Gaspra and 243 Ida) while passing through the **Main Asteroid Belt**.

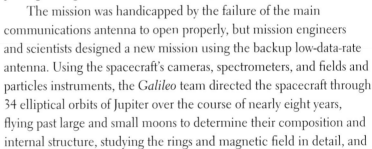

The mission was handicapped by the failure of the main communications antenna to open properly, but mission engineers and scientists designed a new mission using the backup low-data-rate antenna. Using the spacecraft's cameras, spectrometers, and fields and particles instruments, the *Galileo* team directed the spacecraft through 34 elliptical orbits of Jupiter over the course of nearly eight years, flying past large and small moons to determine their composition and internal structure, studying the rings and magnetic field in detail, and releasing a probe into the atmosphere that relayed direct measurements of composition, temperature, and pressure. The *Galileo* spacecraft is gone, crushed and vaporized by Jupiter's deep atmosphere at the end of its mission, but its scientific legacy will be long-lasting—living up to the larger legacy of the name Galileo.

SEE ALSO Jupiter (c. 4.5 Billion BCE), Great Red Spot (1665), Io (1610), Europa (1610), Ganymede (1610), Callisto (1610), Amalthea (1892), Himalia (1904), Jupiter's Magnetic Field (1955), *Pioneer 10* at Jupiter (1973), Active Volcanoes on Io (1979), Jovian Rings (1979), An Ocean on Europa? (1979).

A montage of Jupiter's Great Red Spot and the four Galilean satellites (Io, Europa, Ganymede, and Callisto), all studied in detail from 1995 to 2003 by the NASA Galileo orbiter (left).

Life on Mars?

The idea of **Mars** as a potential abode of life was popularized in the late nineteenth and early twentieth centuries by telescopic observers such as the American businessman and astronomer Percival Lowell, who claimed to see evidence for vast networks of linear canals crisscrossing the parched Martian plains. The media picked up on the notion, exploiting the idea of Martian inhabitants through fantastic (and highly rated) science fiction stories such as the 1938 radio adaptation of H. G. Wells's *The War of the Worlds*.

The search for evidence of life on Mars permeated much of twentieth-century exploration of the Red Planet as well, culminating in the design and operation of a number of highly sensitive organic and biologic detection experiments on the two NASA Viking landers in 1976. While the results of those experiments are widely considered to have been negative (there was no evidence for organic molecules at either landing site, even at the parts-per-billion level), it's not clear that the uppermost sampled surface materials, which have been exposed to harsh solar UV radiation for perhaps billions of years, should contain any organic molecules. The experiments were limited.

It was against that backdrop that in 1996 a team of NASA scientists studying samples from a meteorite called ALH84001, which was blasted off Mars and eventually fell in Antarctica, made an extraordinary claim. Inside that piece of ancient Martian rock, they claimed, was chemical, mineral, and geologic evidence for fossilized microbes—there had been life on Mars.

The astronomer Carl Sagan was fond of saying that extraordinary claims require extraordinary evidence. In the case of ALH84001, most of the scientific community does not believe that the evidence in support of life is extraordinary—rather, there are nonbiologic explanations for each of the pieces of evidence presented by the NASA researchers. Still, the original team remains convinced. Ultimately, though, it may not matter whether there is convincing evidence of life inside ALH84001. More important may be the fact that most scientists agree that the requirements for life—liquid water, heat and energy sources, and organic molecules—were all present in the environment when this rock was on Mars. That is, ALH84001 helps us know that it was at least habitable.

SEE ALSO *Mars and Its Canals* (1906), SETI (1960), *Vikings* on Mars (1976), *Cosmos: A Personal Voyage* (1980).

High-resolution scanning electron micrograph image of segmented, tubelike structures—fossilized microbes?—in a chip from the Martian meteorite ALH84001. The longest structure here is only about 100 nanometers wide, or about 1/1000th the width of a human hair and half the size of the smallest known living cells on Earth.

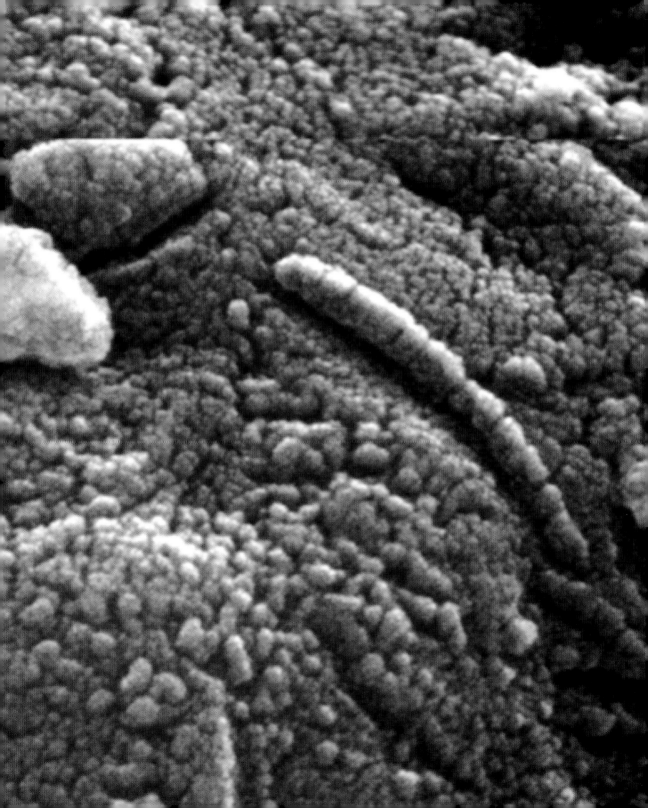

"Great Comet" Hale-Bopp

Comets—small rocky and icy bodies that evaporate in spectacular fashion when they get close to the Sun—have excited and even alarmed people for ages. Some are visible in the sky at predictable times, like **Halley's Comet**, which returns every 76 years. Many others, however, appear suddenly and at unpredictable times. Perhaps the most spectacular of these unexpected comets was comet Hale-Bopp, which was visible to most of the world for many months in the early evening skies during the spring of 1997.

Comet Hale-Bopp was discovered by the American amateur astronomers Alan Hale and Thomas Bopp in July 1995 as it was heading toward the Sun, but when it was still way out past the orbit of Jupiter. Tracking of the comet's orbit revealed that it is on a long, elliptical path that takes more than 2,500 years to circle the Sun. At its farthest distance, the comet is more than 370 times farther from the Sun than the Earth is. Astronomers calculated that prior to 1997 the last time Hale-Bopp had passed by the Earth was in the summer of the year 2215 BCE. They classify this comet as "young" because its complement of highly volatile ices shows that it has spent most of its time in the cold outer solar system, rather than in the warm environs closer to the Sun.

Several years' advance warning of the comet's pass by Earth gave astronomers a rare opportunity to study such a young comet up close for the first time. Using **Spectroscopy** and other methods, astronomers discovered rocky dust, water ice, sodium, and other molecules in the comet's ion and dust tails—including some complex organic molecules never before detected in comets. The brightness of the head, or coma, of Hale-Bopp implied that the rocky and icy nucleus must be big for a comet—about 37 miles (60 kilometers) in diameter, or about six times the size of Halley's comet. While Hale-Bopp is not a direct threat to us, a comet that size impacting the Earth at a velocity of more than 50 kilometers per second would wipe out civilization and potentially kill most life on our planet.

Billions of people saw Hale-Bopp because, for several months, it was bright enough to see with the naked eye just after sunset. There was a lot of media coverage about the "Great Comet of 1997," and there was even some unfounded speculation and hype about possible contamination of the Earth by comet debris. Beautiful heavenly bodies or possible portents of doom? Comets have always been a bit of both.

SEE ALSO Pluto and the Kuiper Belt (c. 4.5 Billion BCE), Halley's Comet (1682), Öpik-Oort Cloud (1932), *Deep Impact*: Tempel-1 (2005), Comet Hartley-2 (2010).

An April 4, 1997, wide-field camera photo of comet Hale-Bopp, taken by astronomers at the Johannes Kepler Observatory in Linz, Austria. This 10-minute exposure shows the comet's beautiful blue ion tail pointing away from the Sun, and the yellow-white dust tail pointing back along the comet's curved trajectory.

253 Mathilde

The NASA *Galileo* mission's flybys of the **Main Belt Asteroids** 951 Gaspra and 243 Ida in 1991 and 1993 revealed that even small planetary bodies can have interesting surface features (and moons of their own), spurring interest in a dedicated space mission to study asteroids up close. The first such project, NASA's Near Earth Asteroid Rendezvous (NEAR) mission, was launched in 1996 to conduct a study of the Earth-approaching asteroid 433 Eros. Along the orbital path toward its encounter with Eros, NEAR acquired some "bonus" science in 1997 by flying past the large main belt asteroid 253 Mathilde.

Mathilde, discovered in 1885 by the prolific Austrian astronomer and asteroid hunter Johann Palisa, orbits along an elliptical path between Mars and Jupiter. More recent telescopic observations show that Mathilde is quite dark—almost black as coal—reflecting only about 4 percent of the incident sunlight. This low albedo (ratio of reflected light to incident sunlight), along with its relatively gray color and featureless spectrum, led astronomers to classify Mathilde as a C-type (comparable to carbon-bearing meteorites) asteroid, in a different class from the S-type (more like stony meteorites) asteroids Gaspra and Ida. Mathilde's different spectral class and the fact that it rotates very slowly, taking more than 17 days to spin once on its axis, created some excitement about what NEAR might see.

The NEAR flyby of 253 Mathilde confirmed the asteroid's low albedo and also revealed a number of surprises. Even though only about 60 percent of the asteroid was photographed, a good estimate of the asteroid's volume combined with an estimate of its mass allowed scientists to estimate Mathilde's density at about 1.3 grams per cubic centimeter. This surprisingly low value is much less than typical rock densities. Since the asteroid is likely too close to the Sun (too warm) to contain ice, the best alternative was to deduce that there is up to 50 percent empty space inside Mathilde, between large chunks of rocky material. Mathilde seems to be a porous, rubble-pile asteroid.

NEAR's discovery of half a dozen giant impact craters on Mathilde support this idea. If the asteroid were a coherent rocky body, then any one of those impacts should have destroyed it; however, interior pore space absorbs impact energy, enabling the asteroid to remain intact in the face of otherwise devastating blows.

SEE ALSO Main Asteroid Belt (c. 4.5 Billion BCE), Jupiter's Trojan Asteroids (1906), Asteroids Can Have Moons (1992), NEAR at Eros (2000).

Ghostly view of the large C-type main belt asteroid 253 Mathilde, photographed by the NASA Near Earth Asteroid Rendezvous (NEAR) spacecraft during a June 1997 flyby. Mathilde is about 36 miles (60 kilometers) across, with a dark surface dominated by many large impact craters.

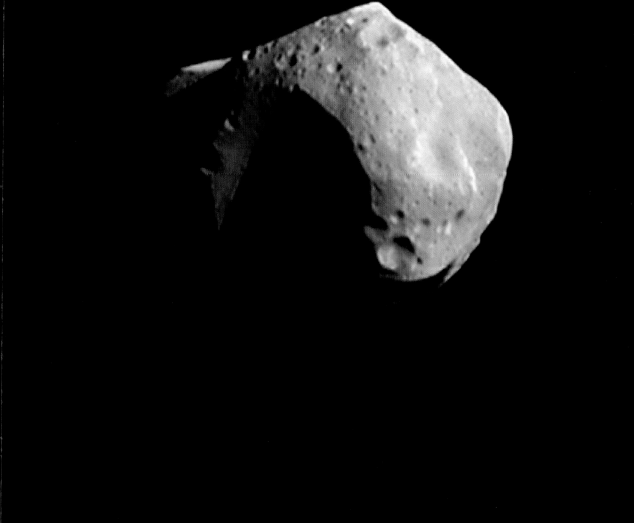

First Rover on Mars

The NASA *Viking* lander missions were unquestionable successes in the exploration of **Mars**, but they provided at least two important lessons for the future: first, it would be scientifically better to have mobility on the planet's surface, rather than having to stay at one fixed landing site; and second, it would be better economically to be able to explore a planetary surface for much less than the multibillion-dollar cost of a flagship-class mission such as Viking.

Both those lessons were taken to heart as part of NASA's third attempted Mars landing mission, called Mars Pathfinder. Launched in 1996, Mars Pathfinder was one of the first of NASA's new "better, faster, cheaper" Discovery-class missions, with a cost target of only 10–20 percent of the cost of the original *Viking* missions. Not only was the team tasked with landing a spacecraft on Mars for this lower cost, they also had to accommodate and operate a small rover as part of the mission, to demonstrate the advantages of having mobility on the Red Planet.

Using a novel (and risky) airbag-assisted landing system, *Pathfinder* successfully landed on Mars on July 4, 1997, and deployed a small rover named *Sojourner* (after the nineteenth-century African American abolitionist and women's rights advocate Sojourner Truth) shortly thereafter. For nearly three months the *Pathfinder* lander and *Sojourner* rover obtained images and chemical data on rocks and soils at the mission's landing site in Ares Vallis, an ancient Martian outflow channel that is thought to have been the site of a catastrophic flood early in the planet's history. Evidence for the region's watery past was provided by a variety of geologic and geochemical clues obtained during the mission.

The *Sojourner* rover, moving at a top speed of about 0.022 miles per hour (1 centimeter per second), only traveled about 330 feet (100 meters) along a circuitous path around the lander during the mission's 83-Martian-day duration. Still, it was a remarkable demonstration of the value of mobility in robotic exploration, as the rover was able to photograph and chemically sample a much wider variety of rocks and soils at the site than the lander alone could have. The successful basic design and operation principles for Mars roving demonstrated by *Sojourner* were essentially scaled up by about a factor of three to create the next-generation Mars Exploration Rovers, **Spirit and Opportunity**, which landed on Mars in 2004.

SEE ALSO Mars (c. 4.5 Billion BCE), *Vikings* on Mars (1976), Life on Mars? (1996), Mars Global Surveyor (1997), *Spirit* and *Opportunity* on Mars (2004), Mars Science Laboratory *Curiosity* Rover (2012).

NASA's Sojourner rover, about the size of a microwave oven, nuzzles up to a rock called Yogi to acquire a measurement of the rock's elemental chemistry during the summer of 1997.

Mars Global Surveyor

Observations from the **First Mars Orbiters** in the early 1970s and then the ***Viking Orbiters*** in the late 1970s and early 1980s provided tantalizing evidence that ancient **Mars** was significantly different—and perhaps more Earthlike—than current Mars. More extensive observations were needed, so in 1992 NASA launched the *Mars Observer* orbiter to conduct that study. Unfortunately, communications were lost with *Mars Observer* just three days before it arrived at Mars, probably due to a fuel line rupture.

Much of the science lost from *Mars Observer* in 1993 was recovered in 1997 with the successful arrival of NASA's Mars Global Surveyor (MGS) orbiter. MGS carried cameras, an infrared spectrometer, a laser altimeter, and a magnetometer to map the geology, mineralogy, topography, and magnetic properties of the planet from pole to pole over the course of nine Earth years (about four Mars years).

MGS measurements revolutionized our understanding of the surface and atmosphere of Mars, much as *Viking* measurements had done more than a decade earlier. For example, high-resolution (meter-scale) images revealed fine details of channels, gullies, and even deltas formed by the persistent flow of liquid water during a

time in early Mars history when the climate must have been warmer and wetter. Volcanic and potentially water-formed minerals were mapped globally, and the topography was measured better than it had been for our own planet. Evidence was found that Mars used to have a strong global magnetic field, perhaps from a time when the core was partially molten and the interior more geologically active.

MGS images, topography data, and mineral discoveries provided the primary means to select the most compelling potential landing sites for the 2004 Mars Exploration Rovers ***Spirit*** and ***Opportunity***, and helped to guide the selection of next-generation cameras and spectrometers for the 2003 European Space Agency *Mars Express* orbiter and the 2006 NASA *Mars Reconnaissance* orbiter. Contact with MGS was lost in late 2006, but other orbiters and rovers are still carrying on its legacy.

SEE ALSO Mars (c. 4.5 Billion BCE), First Mars Orbiters (1971), *Vikings* on Mars (1976), Life on Mars? (1996), First Rover on Mars (1997), *Spirit* and *Opportunity* on Mars (2004), Mars Science Laboratory *Curiosity* Rover (2012).

LEFT: *Artist's rendering of the Mars Global Surveyor (MGS) orbiter.* RIGHT: *An MGS Mars Orbiter Camera photo from 2002 showing a fan-shaped set of landforms that have been interpreted as the eroded remains of a shallow water delta within Eberswalde crater. Features like this suggest that liquid water was persistent on early Mars.*

2 km

International Space Station

Early-twentieth-century rocket pioneers such as Konstantin Tsiolkovsky and Robert Goddard were among the first to work out the technical details of orbiting stations and habitats in space. For most of the century, however, the idea of a human outpost in Earth orbit was only realized in science fiction books, magazines, TV shows, and movies. In the 1970s the Soviet Union launched the first of nine long-duration *Salyut* space research modules, followed up in the 1980s by the orbital assembly of their *Mir* space station—the first long-duration, multicrew outpost in space.

NASA's plans to launch a US space station (called *Freedom*) in the 1980s never materialized, due to cost overruns and technical delays. The fall of the Soviet Union in 1991, technical problems with the *Mir* station, and the high cost of launching and operating space vehicles in general all compelled NASA, Russia, and other space-faring nations to pool resources toward the design and operation of a joint International Space Station (ISS), begun in 1993.

The first component of the new ISS was a Russian electrical power, propulsion, and storage module called *Zarya*, launched into low Earth orbit (about 230 miles [370 kilometers] above the surface) on a Russian Proton rocket in November 1998. The second component, a US docking, airlock, and research module called *Unity*, was launched and connected to *Zarya* a few weeks later by the crew of the **Space Shuttle** *Endeavour*. Fifteen more launches of shuttles and Russian Proton and Progress rockets over the next 13 years added additional solar panels, living quarters, laboratories, airlocks, and docking ports. Completed in 2011, the ISS now spans the area of a US football field, with a total mass of more than 920,000 pounds (420,000 kilograms), making it the largest artificial satellite ever constructed. In addition to the United States and Russia, the European, Japanese, and Canadian space agencies are also key partners.

The ISS is primarily an international research laboratory designed to take advantage of its unique microgravity, orbital environment to enable space-related medical, engineering, and astrophysical research. But it also serves an important role as an outpost for a permanent human presence in space, a place where we can learn how to live and work there, and how best to prepare to venture further beyond low Earth orbit for deep space voyages of exploration.

SEE ALSO Liquid-Fueled Rocketry (1926), Space Shuttle (1981).

The International Space Station orbits about 190 miles (305 kilometers) above Earth's surface. Assembly of the space research outpost began in 1998; this 2009 photo taken by the crew of the space shuttle Discovery *shows the station's solar panels, trusses, and pressurized modules.*

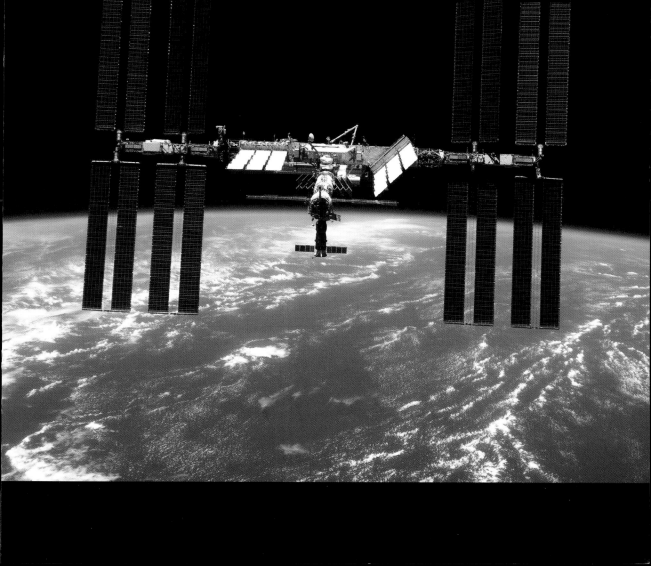

Dark Energy

Albert Einstein (1879–1955), **Edwin Hubble** (1889–1953)

When physicist **Albert Einstein** developed his theory of general relativity early in the twentieth century to explain the relationship between space, time, and matter in the presence of gravitational fields, it was in the context of what astronomers believed at the time to be a static, or unchanging, universe. To make his theory work, Einstein had to invent an as-yet-unseen force, which he called the cosmological constant, to counteract the attractive force of gravity and enable the universe to remain static. When **Edwin Hubble** discovered in 1929 that space is actually expanding, Einstein figured that his cosmological constant wasn't needed after all. Everything seemed to fit nicely.

But astronomers' detailed study of **Spiral Galaxies** and galactic cluster motions over subsequent decades led to the surprising discovery of an unseen but gravitationally attractive material known as **Dark Matter**, and the humbling realization that most of the universe consists of matter that we cannot observe directly. Even more surprising was the discovery by astronomers in 1998 that the expansion of the universe appears to be accelerating with time. That is, galaxies nearer to us, which we observe in the relatively "modern" era of the universe, are moving away from each other faster than extremely distant galaxies, which we observe from much earlier in the universe's history. One possible explanation is that there is some sort of unseen energy force or pressure that permeates the vacuum of space and acts opposite to gravity, helping to accelerate the normal expansion of space from the **Big Bang**. Cosmologists call this hypothesized force dark energy. Maybe Einstein was right about his cosmological constant after all.

The true nature (or existence) of dark energy is impossible to study directly by traditional telescopic observations. As with dark matter, its presence can only be inferred indirectly, by studying its gravitational effects on normal matter. If dark energy proves to be real, however, it would lead to an even more stunning and humbling realization about our universe: dark energy and dark matter, things we currently have no way to measure or characterize, comprise 96 percent of the energy in our universe, while ordinary matter—galaxies, stars, planets, us—accounts for only 4 percent of the cosmos!

SEE ALSO Big Bang (c. 13.7 Billion BCE), Recombination Era (c. 13.7 Billion BCE), Einstein's "Miracle Year" (1905), Hubble's Law (1929), Dark Matter (1933), Spiral Galaxies (1959), Gravitational Lensing (1979), Walls of Galaxies (1989), Hubble Space Telescope (1990), Mapping the Cosmic Microwave Background (1992).

Clusters of galaxies, as in this 2002 Hubble Space Telescope photo of Abell 1689, represent concentrations of normal matter (galaxies), dark matter (colored blue based on astronomers' inferences from gravitational lensing), and the hypothetical force known as dark energy.

Earth's Rotation Speeds Up

Our home planet spins on its axis once per day. Despite some impressively accurate determinations of the length of the day by early Egyptian, Arabic, and Indian astronomers, for most of the history of astronomy it has been sufficient to know that the rotation rate of our planet relative to the distant "fixed" stars is about 23 hours and 56 minutes. The fact that the Earth spins at this rate about 365 and a quarter times during every trip around the Sun has led to a variety of creative ways to add leap years to the **Julian Calendar**, culminating in the modern leap-year method developed during the **Gregorian Calendar** reform of 1582.

In our modern era of digital computers, global positioning system satellites, and interplanetary space probes, it has become much more important to be able to record time, including the Earth's rotational rate, to a much higher degree of precision. Atomic clocks began to be used in the 1950s and 1960s to precisely reckon the passing of time, using the frequency of stable atomic-energy-level transitions in elements such as cesium. An internationally agreed-upon timekeeping system called Coordinated Universal Time (UTC) was developed based on these atomic clocks. Using modern technology, it is now possible to measure the length of the day to almost one part in 10 billion.

The problem for astronomers and timekeepers, however, is that the Earth's rotation is not constant. Tidal friction with the Moon and Sun are slowing down our planet's spin ever so slightly each year, and very slight changes in the distribution of mass on the Earth's surface and interior can also have tiny effects on our planet's spin rate. Thus, since 1972, to keep UTC time precisely aligned with the passage of time as reckoned by the motion of the Sun in the sky, an organization known as the International Earth Rotation and Reference Systems Service has had to occasionally add extra leap seconds to UTC time.

From 1972 to 1998, 21 leap seconds were added to keep UTC in sync with the Earth's slowing spin. In 1999, however, the Earth sped up a tiny amount, and only 2 leap seconds have had to be added since then. What caused Earth's day to become about 1 millisecond shorter in 1999 remains a mystery; geologists and timekeepers are continuing to try to understand what makes our planet tick.

SEE ALSO Egyptian Astronomy (c. 2500 BCE), Earth Is Round! (c. 500 BCE), First Computer (c. 100 BCE), Julian Calendar (45 BCE), *Aryabhatiya* (c. 500), Early Arabic Astronomy (825), Gregorian Calendar (1582), Origin of Tides (1686), Foucault's Pendulum (1851).

The astronomical clock in the town square of Prague, Czech Republic. Analog clocks such as this one, tracking hours and minutes and the motions of the Sun and Moon, have been replaced in astronomical research by precise digital clocks and an internationally regulated system of timekeeping.

Torino Impact Hazard Scale

While evidence for all but a few hundred terrestrial impact craters has been erased by our planet's dynamic geology and hydrology, we can tell just by looking at the ancient, heavily scarred surface of our planetary neighbor—the Moon—that significant numbers of asteroids and comets have hit Earth in the past. These high-speed impact events released huge amounts of energy, and geologic evidence and the fossil record suggest that they occasionally significantly altered the planet's climate and biosphere.

The impact rate has decreased exponentially with time over Earth's history, but even in modern times that rate is not zero—consider, for example, the 1908 explosion of a comet or asteroid in the atmosphere above Siberia (the **Tunguska** event) and the observation of several large atmospheric fireball explosions every year by military and civilian planetary monitoring satellites.

Fueled by public and political interest in understanding the risks associated with cosmic impacts, the rate of discovery of small asteroids and comets, especially within the population known as near-earth objects (NEOs), has increased over the past few decades. Dedicated telescopic surveys have identified more than half a million **Main Belt Asteroids** and nearly 1,000 NEOs. The few hundred NEOs that could potentially cause a threat to life on our planet get a special acronym: PHAs, for potentially hazardous asteroids.

As the rate of PHA discovery increased, it became clear that there was no systematic or simple way to understand and communicate the risk of PHA impacts; indeed, there was much potential for confusion or even unfounded panic about this issue. So, in 1999, a group of planetary astronomers developed an index called the Torino Impact Hazard Scale to quantify the risks. Torino values for newly discovered PHAs range from 0 (no chance of impact) to 10 (certain impact with likely catastrophic consequences).

Most PHAs have Torino values of 0. About a dozen have had nonzero values (most were downgraded to 0 with follow-on observations), with the record so far being an original value of 4 (a 1 percent or greater chance of collision) for the asteroid **99942 Apophis**, which will pass very close to the Earth on April 13, 2029. The Apophis risk has since been downgraded to a 0, but astronomers still monitor it carefully.

SEE ALSO Main Asteroid Belt (c. 4.5 Billion BCE), Dinosaur-Killing Impact (65 Million BCE), Arizona Impact (c. 50,000 BCE), Ceres (1801), Vesta (1807), Tunguska Explosion (1908), Asteroids Can Have Moons (1992), Comet SL-9 Slams into Jupiter (1994), 253 Mathilde (1997), Apophis Near Miss (2029).

A dark, cratered, near-Earth asteroid—perhaps like 99942 Apophis—approaches Earth in this painting by planetary scientist and artist William K. Hartmann.

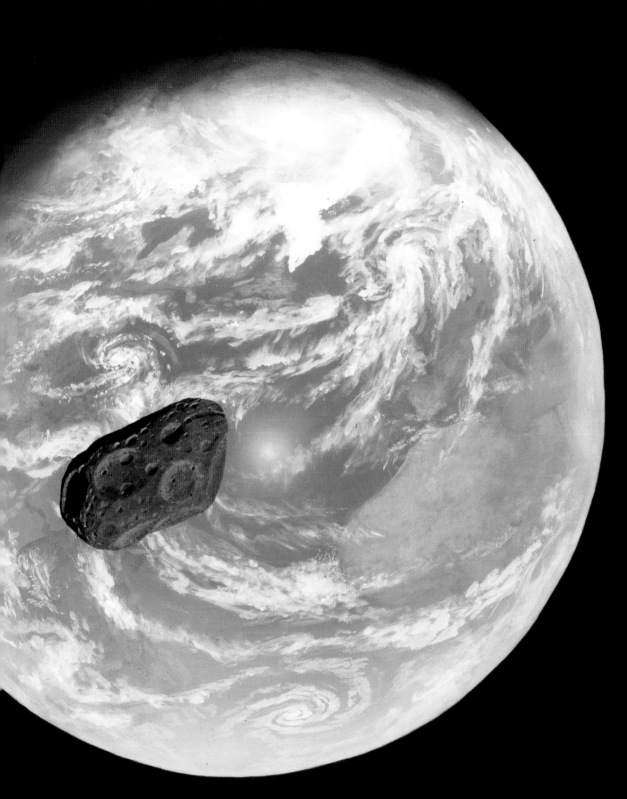

Chandra X-ray Observatory

In 1895 the German physicist Wilhelm Röntgen discovered a mysterious form of radiation created in high-voltage cathode-ray-tube experiments. He dubbed the unknown type of radiation X-rays. Twentieth-century physicists eventually learned that X-rays can be created by electrons accelerated to high velocities in laboratory experiments or in high-energy astrophysical events, such as supernova explosions. The astronomical study of X-ray sources was severely limited, however, by the fact that the Earth's atmosphere absorbs most X-rays generated by cosmic events. A space-based platform was needed.

In 1978 NASA launched the Einstein X-ray imaging satellite to conduct the first space-based observations of high-energy cosmic X-ray sources. For nearly three years Einstein surveyed the sky, studying the details of supernova explosions and identifying new X-ray sources. The success of Einstein compelled astronomers to propose a more sensitive, higher-resolution X-ray space telescope mission as part of NASA's Great Observatories program—four space telescopes designed to allow astronomers to make measurements that are impossible to make from Earth-based telescopes.

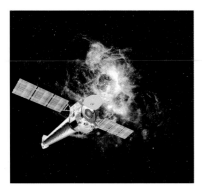

After more than 20 years of development effort, NASA's Advanced X-ray Astrophysics Facility space telescope, renamed Chandra after the Indian-American astrophysicist Subrahmanyan Chandrasekhar, was launched in 1999. Expected to last only five years, Chandra has now operated for more than a dozen years, taking data on supernovae, pulsars, gamma-ray bursts, supermassive black holes, brown dwarfs, and dark matter.

Like its Great Observatory cousins the **Hubble Space Telescope**, Compton Gamma Ray Observatory, and the **Spitzer Space Telescope**, Chandra has completely revolutionized an entire subfield of astronomy and astrophysics, opening an orbital telescopic window into violent, high-energy environments in the cosmos that would otherwise have been impossible to study.

SEE ALSO "Daytime Star" Observed (1054), Brahe's "Nova Stella" (1572), White Dwarfs (1862), Dark Matter (1933), Black Holes (1965), Pulsars (1967), Hubble Space Telescope (1990), Gamma-Ray Astronomy (1991), Spitzer Space Telescope (2003).

LEFT: *Artist's rendering of the Chandra X-ray Observatory.* RIGHT: *A composite Hubble Space Telescope (pink) and Chandra X-ray Observatory (green and blue) image of supernova remnant 0509-67.5, from a stellar explosion 160,000 light-years away in the Large Magellanic Cloud.*

An Ocean on Ganymede?

Between 1995 and 2003, NASA's *Galileo* **Jupiter Orbiter** conducted a grand orbital investigation of the giant planet's atmosphere, magnetic field, moons, and rings. The spacecraft's trajectory was designed so that it would make close flybys of the Galilean satellites **Io**, **Europa**, **Ganymede**, and **Callisto** in order to study their surface features up close, but also to measure their masses and gravity fields by monitoring the way that they slightly deflected *Galileo*'s orbit.

Over the course of the mission, *Galileo* made six close flybys of **Ganymede**, the solar system's largest moon. Gravity data indicated that the interior of Ganymede is differentiated—that is, segregated into a dense core of rock and iron, a lower-density (presumably icy) mantle, and an outer, icy crust. The biggest surprise from the flybys, however, was the discovery that Ganymede has its own magnetic field, embedded within the strong magnetosphere of Jupiter. Ganymede is the only moon in the solar system known to have its own magnetosphere.

Ganymede's magnetic field is thought to be generated by the same process that generates Earth's magnetic field: a spinning, partially molten core of conductive iron. Decay of radioactive elements in the satellite's interior, and some tidal heating from gravitation interactions with Europa and Io, provide the heat to keep the core molten. By studying Ganymede's magnetic field, *Galileo* scientists were able to determine that the satellite's mantle is also electrically conductive. The simplest explanation, given the likely icy composition of the mantle and the strong internal heat sources, is that Ganymede has a deep layer of salty, liquid water—a subsurface ocean—about 125 miles (200 kilometers) beneath the icy crust. Similar *Galileo* observations of Callisto also revealed tentative evidence for a subsurface liquid-water ocean, though less deep, on that moon as well. Could these oceans, or Europa's subsurface ocean, harbor life?

While compelling, the *Galileo* results don't prove that there is an ocean on Ganymede or Callisto. That proof, as well as proof of an **Ocean on Europa**, would have to come from future missions, perhaps future orbiters and landers, that will have to use radar sounding and perhaps deep drills to settle the issue for sure.

SEE ALSO Io (1610), Europa (1610), Ganymede (1610), Callisto (1610), Radioactivity (1896), Jupiter's Magnetic Field (1955), Jovian Rings (1979), An Ocean on Europa? (1979), *Galileo* Orbits Jupiter (1995).

Artist's conception cutaway view of the interior of Jupiter's largest moon, Ganymede, showing the hypothesized deep subsurface ocean of liquid water above a lunar-size core of rock and metal. The ocean is hypothesized to exist based on gravity and magnetic field data from NASA's Galileo Jupiter orbiter mission.

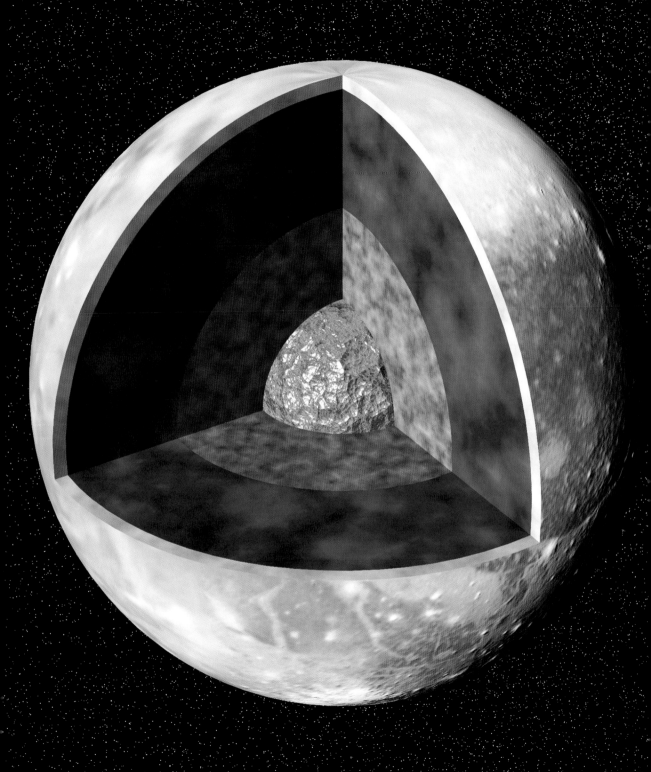

NEAR at Eros

Two centuries of observations since the discovery of **Ceres**, the first known asteroid, have led to the discovery so far of more than half a million minor planets orbiting throughout the solar system. Many orbit in the **Main Asteroid Belt** between Mars and Jupiter, many orbit in the **Trojan Asteroid** clouds ahead of and behind Jupiter's orbit, and many—in orbits that approach or cross that of our own planet—orbit much closer in. Members of this latter class are known as near-Earth asteroids (NEAs); about nine thousand have been found, 10 percent of which are larger than 0.6 mile (1 kilometer).

The first NEA discovered was 433 Eros, found in 1898 by German astronomer Gustav Witt and French astronomer Auguste Charlois. Eros occasionally comes close to Earth; in fact, its proximity enabled the use of parallax to make some of the first direct estimates of the astronomical unit (AU), the average distance between the Earth and the Sun. It is also one of the largest known NEAs, and while it is not currently a threat to impact the Earth, future perturbations in its orbit could make it a threatening object.

In order to learn more about Eros and NEAs in general, NASA launched a robotic mission in 1996 called NEAR—the Near Earth Asteroid Rendezvous—to fly by and then spend a year in orbit around Eros, studying the asteroid closely with CCD imaging and **Spectroscopy**. After a flyby of **253 Mathilde**, the spacecraft went into Eros orbit in 2000 and was renamed NEAR-Shoemaker, in honor of the planetary geologist and asteroid and comet hunter Eugene M. Shoemaker.

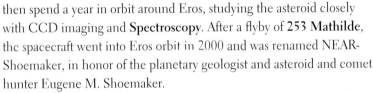

Eros is a big NEA—about the size of Manhattan island—and has a rocky density of about 2.7 grams per cubic centimeter. Its ancient surface is heavily cratered, and its color and spectrum suggests that it is made of the same primitive material as ordinary chondrite meteorites, the building blocks of the Earth and other planets.

At the end of its year-long mapping mission, *NEAR-Shoemaker* was guided to a gentle, low-gravity touchdown on the surface of Eros, where it rests today as a monument and a testament to early-twenty-first-century planetary exploration.

SEE ALSO Main Asteroid Belt (c. 4.5 Billion BCE), Late Heavy Bombardment (c. 4.1 Billion BCE), Arizona Impact (c. 50,000 BCE), Ceres (1801), Vesta (1807), Jupiter's Trojan Asteroids (1906), 253 Mathilde (1997), Torino Impact Hazard Scale (1999), *Hayabusa* at Itokawa (2005), *Rosetta* Flies by 21 Lutetia (2010).

An orbital view of the near-Earth asteroid 433 Eros photographed by the Near Earth Asteroid Rendezvous (NEAR) *spacecraft (left) from a distance of only 125 miles (201 kilometers). The asteroid is about 21 miles (34 kilometers) long and exhibits impact, tectonic, and erosional (landslide) features on its surface.*

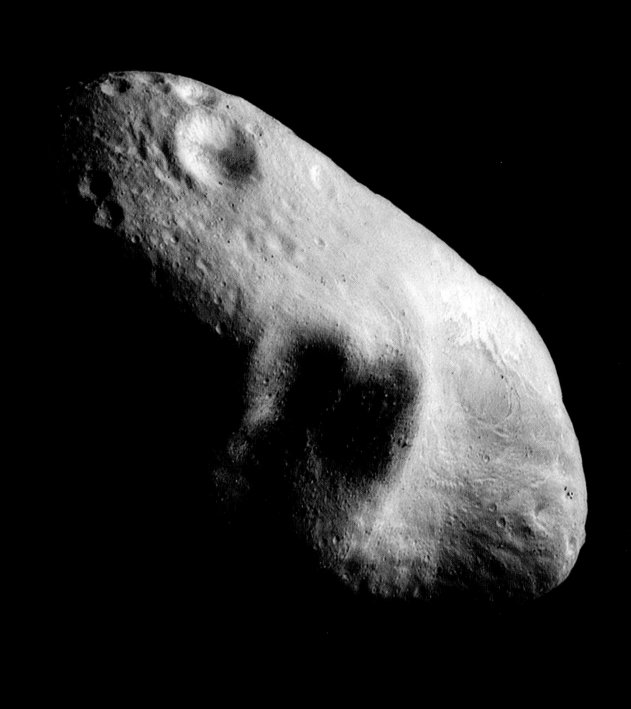

Solar Neutrino Problem

Raymond Davis, Jr. (1914–2006), **Masatoshi Koshiba** (b. 1926)

The so-called standard model of twentieth-century physics provides a theory that links matter and energy through the interactions of elementary particles and forces. Familiar elements of the standard model include the electron, a fundamental particle of charge, and the photon, a fundamental particle of light. The discovery of the neutron in 1933 led to the prediction and discovery of the neutrino in 1956, and the hypothesis that neutrinos are massless elementary particles that travel at the speed of light (like photons) and that are created in three "flavors" (called electron, muon, and tau neutrinos) depending on the process and environment. **Neutrino Astronomy** became an active way to probe high-energy processes in inaccessible places, like the interior of the Sun.

Large neutrino detectors started coming online in the 1960s in order to detect elusive neutrino particles generated in the Sun or during high-energy cosmic events such as supernova explosions. The standard model predicted that fusion of hydrogen to helium inside the Sun should create electron neutrinos that these detectors could find. Indeed, the predicted solar neutrinos were discovered, but at a rate only about one-third of what the model predicted. The "missing" neutrinos created what particle physicists called the solar neutrino problem.

Finding the solution to the solar neutrino problem became a high priority for physicists, because if it could not be solved, then the standard model might not actually be correct. Theorists began to wonder if neutrinos could change flavors, or oscillate between their different types. New higher-resolution detector experiments were built in the 1990s to improve the fidelity of the measurements, and in 2001 the new data revealed a surprise: neutrinos are not massless particles. They appear to have very small masses and travel at just slightly below the speed of light. Most important, experiments showed that they can oscillate between electron, muon, and tau flavors, and that about two-thirds of the electron neutrinos created inside the Sun do eventually convert to other flavors.

Solving the solar neutrino problem led to the Nobel Prize in Physics for Raymond Davis, Jr. and Masatoshi Koshiba in 2002. More importantly, it led to critical revisions of the standard model so that it now accounts for oscillations by neutrons and other particles.

SEE ALSO "Daytime Star" Observed (1054), Neutron Stars (1933), Nuclear Fusion (1939), Neutrino Astronomy (1956).

Sunspots photographed from the Swedish Solar Telescope at La Palma, Chile. A spectrum of electromagnetic radiation and a wide variety of elementary particles such as neutrinos are produced by nuclear-fusion reactions in the Sun. The scene here spans nearly five times the diameter of Earth.

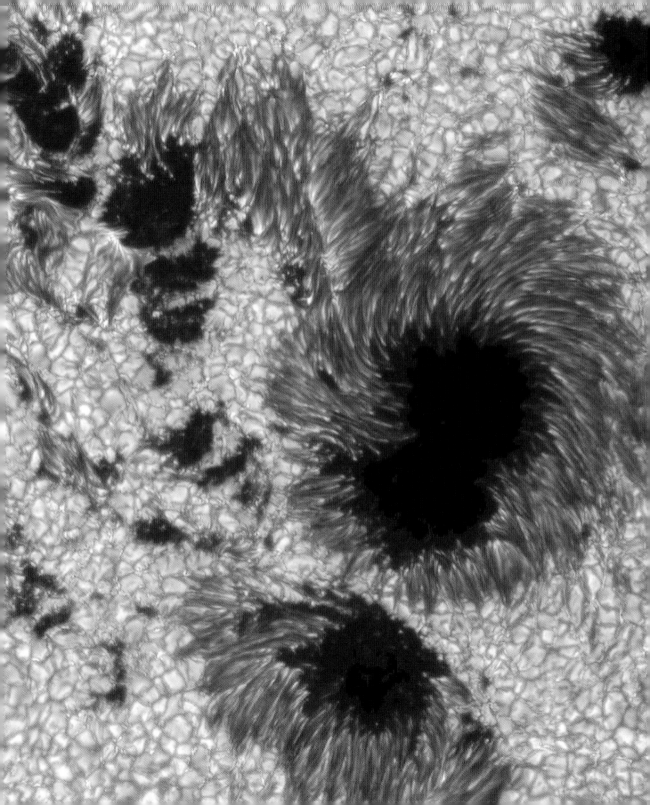

Age of the Universe

David T. Wilkinson (1935–2002)

The **Hubble Space Telescope** (HST) was essentially built to be a time machine. By gazing deeper into space, it also peers farther back in time, detecting light that was emitted by stars and galaxies many billions of years ago. Using standard candles, such as **Cepheid Variable Stars**, or certain types of supernova explosions in far-distant galaxies, astronomers have been able to use HST to refine **Hubble's Law**, and to determine the rate of the expansion of the cosmos. After a decade of observations, in 2001 HST scientists announced that by turning the clock backward on their derived expansion rate, they could deduce that the **Big Bang** had occurred approximately 13.7 billion years ago.

Around the same time, a new space satellite was launched, designed to measure the details of the **Cosmic Microwave Background** radiation from the initial expansion of the early universe. Dauntingly named the Wilkinson Microwave Anisotropy Probe (WMAP) after the American cosmologist David T. Wilkinson, the satellite took high-resolution images of variations (anisotropies) in the 3-kelvins background radiation that is the remnant of the universe's expansion following the first few hundred thousand years after the Big Bang, at the beginning of the so-called dark ages but before the birth of the **First Stars**. Amazingly, independent of HST or other methods, the WMAP data lead to an estimate of the age of the universe of 13.7 billion years.

By combining data from HST, WMAP, and other studies, cosmologists today claim that the Big Bang happened 13.75±0.11 billion years ago—a stunning level of precision! The Big Bang frames the standard model of cosmology, with hydrogen, helium, and a few other light elements created in the Big Bang, and heavier elements—the eventual building blocks of life—subsequently created in stars and supernova explosions. Explosion, rapid inflation, deionization, reionization, slow expansion, and accelerating expansion are all now key milestones in cosmologists' understanding of the universe.

Knowing when the Big Bang occurred raises more questions, though. Why did it occur? What was there "before" space and time were created? How will it end? Such deep questions push the boundaries of what can be studied by modern science.

SEE ALSO Big Bang (c. 13.7 Billion BCE), Recombination Era (c. 13.7 Billion BCE), First Stars (13.5 Billion BCE), Hubble's Law (1929), Cosmic Microwave Background (1964), Hubble Space Telescope (1990), Mapping the Cosmic Microwave Background (1992), Dark Energy (1998), How Will the Universe End? (The End of Time).

A cartoon representation of the standard model of the universe from WMAP data, dating back 13.7 billion years to the Big Bang. After a rapid moment of inflation, the ensuing gradual expansion of the universe has recently begun to speed up, potentially due to the effects of dark energy.

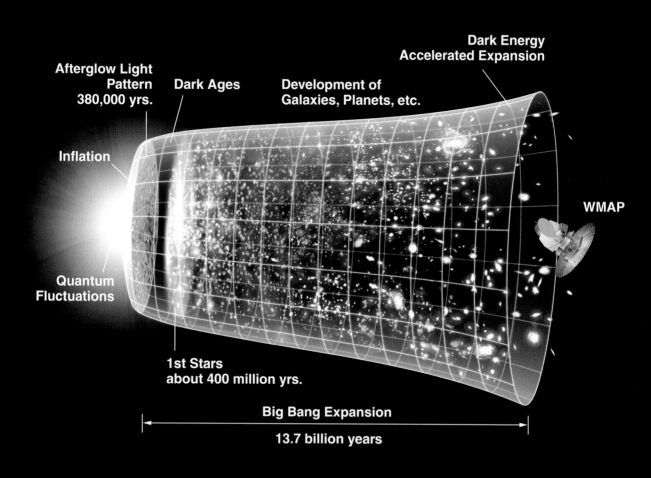

Afterglow Light
Pattern
380,000 yrs.

Dark Ages

Development of
Galaxies, Planets, etc.

Dark Energy
Accelerated Expansion

Inflation

Quantum
Fluctuations

1st Stars
about 400 million yrs.

WMAP

Big Bang Expansion

13.7 billion years

Genesis Catches Solar Wind

The Sun, like all stars, is powered by a dynamic balance between the gravitational contraction of an enormous concentration of mass and the outward pressure of intense **Nuclear Fusion** reactions occurring in the deep interior. Radiation escapes from the Sun, providing heat and light to the planets, but highly energetic particles also emerge from the Sun's photosphere (effectively, its "surface") and chromosphere (atmosphere) and are spewed out at high speeds into interplanetary space. This stream of particles is called the solar wind.

In 1995 the European Space Agency (ESA) and NASA teamed up to launch a satellite called the Solar and Heliospheric Observatory (SOHO), which was specifically designed to study the dynamic environment of the Sun, including the solar wind, using high-resolution ultraviolet images and **Spectroscopy**. SOHO continues to provide spectacular time-lapse views of our dynamic star.

Solar astronomers' discoveries about the solar wind from SOHO and other studies generated substantial interest in a dedicated space mission to try to collect some of these far-flung pieces of the Sun. Planetary scientists, too, are keenly interested in the Sun's composition, as it represents about 99.9 percent of the mass of the solar system and is the starting composition from which the planets formed. This interest ultimately led to a NASA mission called Genesis, which was designed to collect and return samples of the solar wind. The mission was launched in 2001 on a looping orbit around one of Earth's gravitational **Lagrange Points** (L1), where a specially designed sample-return canister was exposed to solar wind particles until early 2004.

Despite the 2004 crash of the canister in the Utah desert because of a failed reentry parachute, many of the Genesis samples survived intact and have been intensely studied by geochemists and solar astronomers. Particles from the three types of solar wind (fast, slow, and coronal mass ejections) were collected and analyzed, and the results are providing important new data—some of it surprising—about the Sun's composition, helping to refine our understanding of processes at work inside the Sun and other stars.

SEE ALSO Birth of the Sun (c. 4.6 Billion BCE), Lagrange Points (1772), Eddington's Mass-Luminosity Relation (1924), Nuclear Fusion (1939).

A giant coronal mass ejection from the Sun, photographed by the SOHO satellite in January 2002. Billions of tons of material are ejected by solar events such as this, creating a "solar wind" that interacts with the planets as it passes by. Samples of the solar wind were collected by the Genesis spacecraft (left).

Spitzer Space Telescope

Galaxies, stars, planets, moons, asteroids, comets, and even cosmic dust grains all emit thermal infrared energy that is dependent on temperature, composition, and environment. During recent decades astronomers have used sensitive new infrared detectors to study the thermal energy from these objects. Major advances came in 1983 with the launch of the Infrared Astronomical Satellite (IRAS), which conducted the first all-sky survey of infrared heat energy emitted by cosmic objects, and then in 1995 with the launch of the Infrared Space Observatory (ISO) satellite, which followed up IRAS discoveries with higher-resolution imaging and spectroscopy observations until early 1998. IRAS and ISO made important discoveries about **Circumstellar Disks** and planet formation processes, as well as star formation and galaxy evolution.

Motivated by these discoveries, NASA dedicated its fourth and final Great Observatory satellite to an IRAS/ISO follow-up mission initially called the Space Infrared Telescope Facility, and later renamed the Spitzer Space Telescope after the American astronomer and long-time space telescope advocate Lyman Spitzer. Spitzer was launched in 2003 and was "parked" in a heliocentric orbit close enough to Earth to allow frequent, high-bandwidth communications but far enough away to avoid interference from Earth's thermal background signature.

Spitzer used liquid helium to cool its instruments below 4 kelvins, making them supersensitive to extremely faint cosmic sources of thermal energy. Astronomers exploited the ability to see through optically thick dust in the infrared field to study star-forming regions such as the Orion Nebula. Major discoveries were also made studying **Quasars**, galaxies, protoplanetary disks, hot young stars, **Extrasolar Planets**, and our own solar system. The telescope's supply of helium ran out in 2009, but Spitzer continues to make less sensitive but still unique measurements of many infrared sources, and is expected to operate for many more years.

SEE ALSO Star Color = Star Temperature (1893), Circumstellar Disks (1984), Hubble Space Telescope (1990), Gamma-Ray Astronomy (1991), Planets around Other Suns (1995), Chandra X-ray Observatory (1999).

A false-color infrared composite photo, taken by the NASA Spitzer Space Telescope (seen in an artist's conception, left, of a star formation in the heart of the Orion Nebula (a region known as the Trapezium), merged with a 2-micron all-sky survey background photo of the stellar neighborhood.

Spirit and *Opportunity* on Mars

More than three decades of successful orbital and landed investigations of **Mars** by scientists involved in the Mariner and Viking missions painted a compelling picture of major past climate changes on the Red Planet. The Martian surface today is extremely cold, bone dry, and inhospitable to life as we know it. But ancient Mars, as revealed by these missions, appears to have been a warmer, wetter, and potentially more Earthlike place. If so, then early Mars (during the first billion years or so after its formation) may have been a habitable environment where, as on our own planet, life could have thrived.

Planetary scientists wanted to move beyond photographic evidence of a potentially habitable early Mars, however, and make quantitative geologic, geochemical, and mineralogic measurements that could provide smoking-gun proof. Experience gained from the 1997 Mars Pathfinder mission proved the value of mobility in doing geologic field work with robots in distant locations, leading to the choice to embark on an even longer-range rover mission. Because of two Mars missions failures in 1999, NASA decided to reduce its risk: instead of just one rover, it would launch twin rovers—named *Spirit* and *Opportunity*—in 2003.

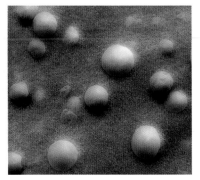

Both rovers landed safely in early 2004 and began their separate adventures on opposite sides of the planet: *Spirit* in an ancient crater named Gusev, which may once have hosted a lake, and *Opportunity* in the cratered area Meridiani Planum, where **Mars Global Surveyor** data showed evidence for water-formed minerals. After several years of virtually roving around Gusev with *Spirit*, mission scientists discovered evidence of water-bearing minerals in an ancient hydrothermal system that provided smoking-gun evidence for past habitability in Gusev. At Meridiani, the team immediately found other water-bearing minerals and geologic clues that provided conclusive proof of past habitability there as well. *Spirit*'s last data came in early 2010, but as of mid-2012 *Opportunity* continues to roll on and make new discoveries.

SEE ALSO Mars (c. 4.5 Billion BCE), *Mars and Its Canals* (1906), First Mars Orbiters (1971), *Vikings* on Mars (1976), First Rover on Mars (1997), Mars Global Surveyor (1997), Life on Mars? (1996), Mars Science Laboratory *Curiosity* Rover (2012), First Humans on Mars? (~2035–2050).

A computer-generated version of the NASA Mars rover Opportunity *placed into an actual* Opportunity *Pancam mosaic of finely layered rocks inside Endurance crater. These rocks contain evidence of past liquid water on Mars, including millimeter-size iron-rich spheres (inset) called concretions.*

Cassini Explores Saturn

Centuries of telescopic observations and the successful flybys of *Pioneer 11* and *Voyager*s *1* and *2* revealed **Saturn** to be both a stunningly beautiful and scientifically interesting destination, with important similarities to and differences from the other giant planets. Following a long-term planetary exploration strategy of flyby, orbit, land, rove, and sample return, NASA began planning for a dedicated Saturn orbiter mission in the 1980s. The agency finally gained congressional approval (and funding) in the 1990s, and in 1997 launched a joint US-European spacecraft named *Cassini*, after the Italian-French mathematician and astronomer Giovanni Cassini, who had made some of the earliest scientific observations of Saturn, its rings, and moons.

After gravity-assist flybys of Venus, Earth, and Jupiter, *Cassini* finally went into orbit around Saturn in 2004. The orbiter has been returning incredible images (visible, infrared, and radar) and **Spectroscopic** measurements ever since, and mission scientists have made many important discoveries. Among the most exciting of these have been the discovery of seven new moons, including several that create strange three-dimensional wakes and other structures within Saturn's thousands of individual rings; the discovery

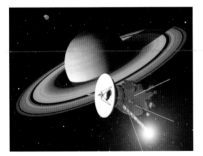

of active geysers spewing water vapor and organic molecules from the subsurface of the tiny moon **Enceladus**; the first close-up images of the moon **Phoebe**, a potentially captured Centaur object; detailed infrared and radar imaging and mapping of the haze-shrouded surface of **Titan**, revealing lakes of liquid methane, ethane, or propane on its surface; and the most detailed compositional and dynamical studies yet of Saturn's atmosphere and magnetic field.

Cassini also carried and successfully deployed a probe named *Huygens*, which landed on the surface of Titan in 2005, becoming the first successful mission to the surface of an outer-solar-system object. The *Cassini* orbiter is expected to continue studying the Saturn system in great detail until at least 2017.

SEE ALSO Saturn (c. 4.5 Billion BCE), Titan (1655), Saturn Has Rings (1659), Iapetus (1671), Rhea (1672), Tethys (1684), Dione (1684), Enceladus (1789), Mimas (1789), Phoebe (1899), *Pioneer 11* at Saturn (1979), *Voyager* Saturn Encounters (1980, 1981), *Huygens* Lands on Titan (2005).

A spectacular NASA Cassini mission photo from 2005, when the spacecraft (seen at left in an artist's rendering) crossed through the plane of Saturn's rings. The rings are so thin that they almost disappear when viewed edge-on, but their shadows can be clearly seen against Saturn's clouds.

Stardust Encounters Wild-2

Fred Whipple (1906–2004)

Comets can be spectacular events in the night sky, as witnessed during the many appearances of **Halley's Comet** over recorded history and the singularly dramatic occurrences of great one-time comets, including the appearance of comets Hyakutake in 1996 and **Hale-Bopp** in 1997. Most of the comets' spectacular display, however, comes from reflected sunlight or gaseous emissions from their tails. Little was known about the icy, rocky solid nucleus of comets until the 1986 flyby of Halley's comet by the European *Giotto* spacecraft, which helped to support the American astronomer Fred Whipple's idea of comet nuclei as small, irregular, "dirty snowballs."

The diversity of comet orbits and origins (some coming from the **Kuiper Belt**, others from the **Oort Cloud**, and still others with orbits that interact with and occasionally impact the planets, such as Shoemaker-Levy 9) motivated planetary scientists to propose more robotic missions to study them up close, and in 1999 a NASA mission called Stardust was launched to not only fly closely past the nucleus of comet Wild-2 (pronounced VILT-two) but also to collect tiny samples of dust and gas from the

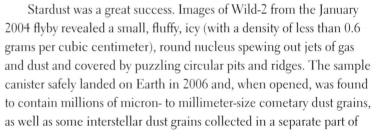

comet's tail and return them back to Earth in a special capsule.

Stardust was a great success. Images of Wild-2 from the January 2004 flyby revealed a small, fluffy, icy (with a density of less than 0.6 grams per cubic centimeter), round nucleus spewing out jets of gas and dust and covered by puzzling circular pits and ridges. The sample canister safely landed on Earth in 2006 and, when opened, was found to contain millions of micron- to millimeter-size cometary dust grains, as well as some interstellar dust grains collected in a separate part of the experiment. Analysis of the comet dust has revealed not only the expected water ice and silicate minerals but also a variety of organic molecules—some similar to the simple organics found in the primitive **Murchison Meteorite**, others having much more complex hydrocarbon chains comparable to those found in organic molecules. Comets, it would seem, are important potential contributors to the chemistry of life.

SEE ALSO Pluto and the Kuiper Belt (c. 4.5 Billion BCE), Halley's Comet (1682), Encke's Comet (1795), Miss Mitchell's Comet (1847), Öpik-Oort Cloud (1932), Organic Molecules in Murchison Meteorite (1970), Kuiper Belt Objects (1992), "Great Comet" Hale-Bopp (1997), *Deep Impact*: Tempel-1(2005).

The nucleus of comet Wild-2, photographed by the NASA Stardust spacecraft (seen at left in an artist's rendering) on January 2, 2004. The icy nucleus is approximately 2.5 miles (4 kilometers) across and covered by enigmatic circular features that could be impact craters or pits where jets of ice emerge from the interior.

Deep Impact: Tempel-1

Missions like the *Giotto* flyby of **Halley's Comet** in 1986 or the *Stardust* **Flyby of Wild-2** in 2004 revealed comet nuclei to be small, icy bodies. They are also very dark bodies, typically reflecting only about 3–5 percent of the sunlight that strikes them (about as dark as charcoal). It seems odd that icy bodies are so dark—but the reason is that ice is evaporated by the Sun's heat, leaving behind a lag of rocky and organic grains that darken the surface. Impacts or tidal forces can crack this surface lag, letting fresh ice escape from cracks into the interior.

If this model of a comet's surface is correct, reasoned some scientists in a mission proposal to NASA is 1999, then it should be possible—with a powerful enough impact—to design a mission to punch a hole in the surface lag and expose a comet's "pristine" icy materials for study. NASA accepted the bold mission idea, and the Deep Impact mission, carrying a 815-pound (370-kilogram) copper projectile probe, was launched toward comet Tempel-1 in early 2005.

As *Deep Impact* got close to Tempel-1, imaging of the comet's nucleus revealed it to indeed be a dark, irregular body about 5 × 3 miles, or 8 × 5 kilometers, in size, with surprisingly complex geologic landforms, including possible impact craters and lobes of strangely layered terrain. The spacecraft released the impactor for its July 4, 2005, crash landing, and the resulting impact generated some impressive cosmic fireworks. The projectile penetrated below the lag crust and into the comet's subsurface, releasing a huge spray of ice and dust (including clays, carbonates, and silicates). Images from the *Deep Impact* flyby spacecraft showed a spectacular flash and a giant debris cloud that hid the resulting impact crater from view. Later analysis of the flyby data showed Tempel-1 to have a low density (0.6 grams per cubic centimeter), indicating a rather porous, icy interior composition.

NASA's *Stardust* spacecraft, which had completed its sample return mission from comet Wild-2 in 2006, was redirected to a close flyby past Tempel-1 in February 2011. *Stardust* successfully imaged the 500-foot-diameter (150 meters) impact crater made by the *Deep Impact* probe's encounter six years earlier.

SEE ALSO Pluto and the Kuiper Belt (c. 4.5 Billion BCE), Halley's Comet (1682), Öpik-Oort Cloud (1932), Organic Molecules in Murchison Meteorite (1970), Kuiper Belt Objects (1992), "Great Comet" Hale-Bopp (1997), *Stardust* Encounters Wild-2 (2004).

A blinding shock wave of light emerges from the nucleus of comet Tempel-1 about 67 seconds after the Deep Impact *spacecraft's impact probe smashed into the comet at 23,000 miles per hour (10 kilometers per second). The collision excavated ice and dusty silicate minerals from the comet's subsurface.*

Huygens Lands on Titan

Saturn's moon **Titan** is the second-largest satellite in the solar system (larger than the planet Mercury) and the only satellite with a substantial atmosphere. Hopes were high that the surface would reveal interesting landforms and processes during the **Voyager flybys of Saturn** in 1980 and 1981; however, Titan turns out to be shrouded by a thick layer of haze that obscured the surface from view at the visible wavelengths of the *Voyager* cameras. **Spectroscopic Data** did show that Titan's atmosphere is made up of mostly nitrogen (N_2) with a small amount of methane (CH_4), and that it has a surface pressure 50 percent higher than Earth's pressure at sea level but a temperature only about 90 degrees above absolute zero.

Much was learned about Titan from Voyager data and subsequent telescopic observations, but much remained unknown, motivating a dedicated Titan exploration payload to be added to the **Cassini Saturn Orbiter**. When *Cassini* launched in 1997, it carried a Titan lander called *Huygens*, after Titan's discoverer, the Danish astronomer Christiaan Huygens.

Huygens completed a successful aerobraking and parachute-assisted landing on January 14, 2005, becoming the first lander on an outer planetary surface. During the descent the probe obtained spectacular photographs of river-like channel systems, shorelines, and dark, flat-floored plains that have been interpreted as lakes of liquid ethane, methane, or propane. The lander survived on the surface for about 90 minutes, taking photographs of the strange alien landscape and measuring the pressure, temperature, and chemistry.

Images of the surface of Titan from *Huygens* resemble images of the Martian surface in some ways, and Earth's surface in other ways. But there are substantial differences. For example, the "rocks" in the images are not made of silicates but are likely chunks of water or hydrocarbon ices (which act like rock at very low temperatures), and although the resulting landforms are surprisingly similar, Titan's river channels and shorelines are carved by liquid hydrocarbons, not liquid water.

SEE ALSO Titan (1655), Organic Molecules in Murchison Meteorite (1970), *Voyager* Saturn Encounters (1980, 1981), *Cassini* Explores Saturn (2004).

LEFT: *The 4- to 8-inch (10- to 20-centimeter) "rocks" in this artist rendering of the* Huygens *probe on Titan are pebbles of water and hydrocarbon ices, rounded by transport in streams or lakes of liquid hydrocarbons.* RIGHT: *A flattened (Mercator) projection of the probe's view of Titan from an altitude of 6 miles (10 kilometers).*

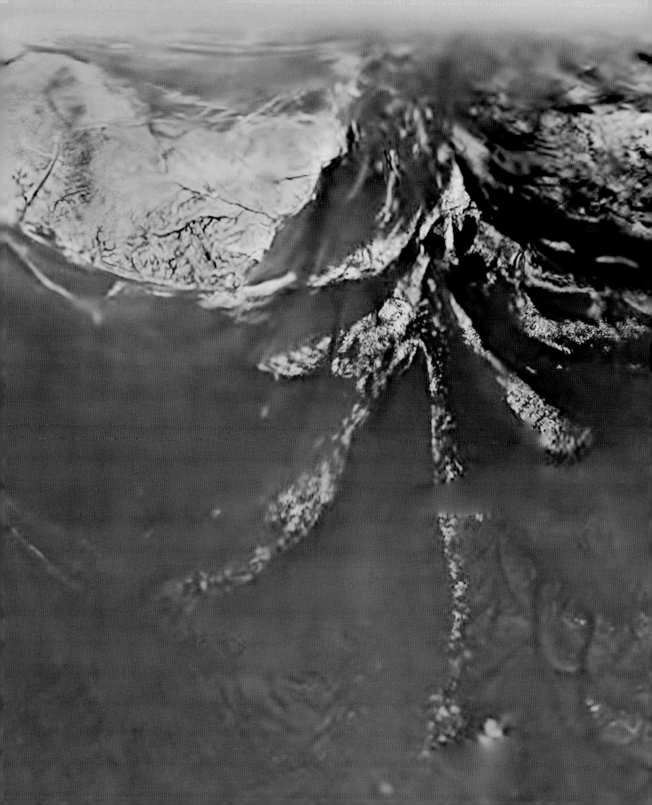

Hayabusa at Itokawa

Possible future impacts of Earth by asteroids or comets represent a threat to the entire human species. As such, studying small bodies and learning about their orbits and properties is appropriately an international endeavor. During the late twentieth and early twenty-first centuries, space missions by the US, European, and Russian space agencies to comets Halley, Borrely, Wild-2, and Tempel-1, and the asteroids Gaspra, Ida, Mathilde, Eros, and Lutetia, dramatically increased our understanding of kilometer-size (or tens-of-kilometers-size) bodies. But it wasn't until the mission of the Japanese *Hayabusa* spacecraft that a truly tiny asteroid was studied in detail.

Hayabusa, Japanese for "falcon," was the first asteroid mission by the Japanese Aerospace Exploration Agency (JAXA) and the first attempted robotic sample return from an asteroid. Launched in 2003, *Hayabusa* used novel ion propulsion engines to slowly and gently match its trajectory to that of a small, recently discovered near-Earth asteroid named 25143 Itokawa.

In September 2005 the spacecraft began its rendezvous alongside Itokawa (the asteroid's gravity was too low to allow the spacecraft to truly orbit), which turned out to be a gray, elongated, rocky object only about 1,750 feet (535 meters) long. Large boulders and strangely smooth areas were discovered covering the surface of the lumpy little world. Two months later *Hayabusa* was directed to approach even closer to the asteroid, and to gently touch down, release a small rover, collect some surface soil and rock samples, and then return a sample capsule to Earth. No problem, right?

In reality there were lots of problems. *Hayabusa* landed on the asteroid, briefly, but the rover deployment failed, and the spacecraft ascended out of control from the surface before it could be confirmed that it had gathered samples. By the time control was regained, it was too late to try again; the sample capsule had to be launched back to Earth without knowing whether or not it was empty.

The sample capsule made a safe parachute landing in Australia in June 2010, and after some tense weeks of careful unpacking and testing, scientists were delighted to find about 1,500 small dusty grains of Itokawa inside, with chemistry matching that of some primitive meteorites. Hayabusa was a success!

SEE ALSO Main Asteroid Belt (c. 4.5 Billion BCE), Lunar Robotic Sample Return (1970), Torino Impact Hazard Scale (1999), NEAR at Eros (2000).

Near-Earth asteroid 25143 Itokawa, photographed by the Japanese Aerospace Exploration Agency's Hayabusa *spacecraft in September 2005. Itokawa is a tiny lump of silicate rock, only about 1,750 feet (535 meters) long, or about the length of six New York City blocks.*

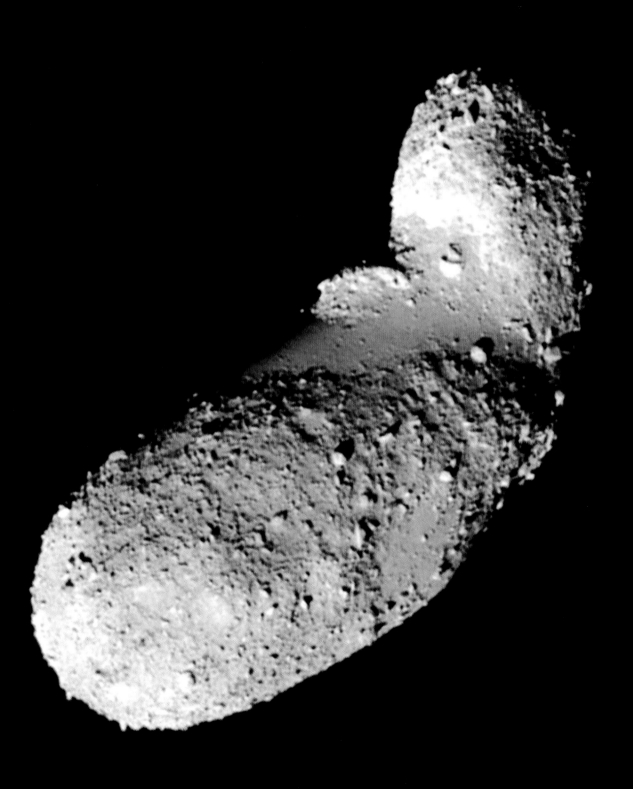

Shepherd Moons

In 1659 Christiaan Huygens recognized that a "thin, flat ring" surrounded Saturn, and then in 1675 Giovanni Domenico Cassini discovered a dark gap in the rings and realized that they must actually be a series of many narrower, separate rings. Subsequent detailed telescopic studies and especially spacecraft studies by the *Pioneer 11, Voyager 1 and 2,* and *Cassini* probes have since revealed that Saturn is surrounded by thousands of separate rings, each consisting of dust- to house-size chunks of ice.

Astronomers have grappled with many deep questions about Saturn's beautiful system of rings. Are they young or are they ancient? How did they form? And how do millions of separate "moonlets" manage to stay so sharply confined in their specific orbital paths? Important clues to the last question came from the 1990 discovery, in images taken during the 1981 *Voyager 2* Saturn flyby, of a moon 19 miles (30 kilometers) in diameter traveling within one of the gaps in the rings called the Encke gap. The moon's gravity keeps the gap clear and confines, or "shepherds," the edges of the surrounding ring particles. It was aptly named Pan, after the Greek god of shepherds.

In 2005 Cassini mission team members discovered another small shepherd moon,

this one embedded within the A ring, the outermost of the large, bright rings. Named Daphnis, after a mythological Greek shepherd, it is only about 5 miles (8 kilometers) wide, yet its gravitational effect creates three-dimensional wakes (like ripples on a pond) and other structures in the A ring as it moves past other ring particles. Two other small moons, Prometheus and Pandora, have also been recognized as shepherding Saturn's thin, outermost F ring.

While it is still uncertain how Saturn's rings formed and whether they are ancient or relatively young, *Cassini* photos reveal that they are nonetheless a beautiful, dynamic, and constantly changing natural laboratory within which to study complex gravitational interactions among the solar system's smallest bodies.

SEE ALSO Saturn (c. 4.5 Billion BCE), Saturn Has Rings (1659), *Pioneer 11* at Saturn (1979), *Voyager* Saturn Encounters (1980, 1981), *Cassini* Explores Saturn (2004).

LEFT: *A 3-D wake structure created in the rings by the tiny shepherd moon Daphnis.* RIGHT: NASA Cassini *Saturn orbiter photograph of Saturn's rings, showing the embedded "shepherd moon" Pan (in the wide Encke gap near center), which helps confine the rings' orbits.*

Demotion of Pluto

When **Pluto Was Discovered** in 1930, it was hailed as the solar system's ninth planet, partially because it was estimated to be about the size of the Earth. Follow-up observations over the decades, and the discovery of Pluto's moon **Charon**, eventually led to the realization that Pluto is a small world, only about 20 percent of the diameter and less than 1 percent of the mass of Earth. Still, with decades of inertia and countless textbooks touting Pluto as a cold and lonely outpost at the edge of the solar system, Pluto retained its status—and popularity—as a full-fledged planet.

In the 1990s, however, it became clear that the solar system does not end at Pluto. In addition to the long-known population of long-period comets that likely originate from the **Oort Cloud**, more than a thousand **Kuiper Belt Objects** (KBOs) have now been discovered orbiting well past the orbit of Neptune, and the expectation is that many tens to hundreds of thousands yet remain to be discovered. Many of them are comparable in size to Pluto, or even larger.

If Pluto is a planet, then the potential proliferation of Pluto-size worlds would of course dramatically increase the number of planets in the solar system. This possibility caused consternation among some astronomers and compelled the world's relevant governing agency, the International Astronomical Union (IAU), to reconsider the classification of Pluto-size worlds. In 2006 the IAU decided to formally "demote" Pluto and other large KBOs from planethood to a new class of object—dwarf planet—created to recognize the importance of this newly discovered class of small worlds but to distinguish them from the classical planets, which have significantly larger influences on their surroundings.

The demotion of Pluto caused significant public outcry, and even many astronomers and planetary scientists remain confused and deeply divided about the new formal definition of what it takes to be a planet. To many, any object that is large enough to have become roughly spherical under its own self-gravity, or that has had active internal processes that have differentiated it into core, mantle, and crust, could be considered a planet. And why not reclassify giant satellites such as Ganymede, Titan, and Europa (comparable to or larger than Mercury), as planets? Today, the debate continues to rage on about whether our solar system has 8 known planets, or perhaps 40 or more.

SEE ALSO Pluto and the Kuiper Belt (c. 4.5 Billion BCE), Discovery of Pluto (1930), Öpik-Oort Cloud (1932), Charon (1978), Kuiper Belt Objects (1992), Pluto Revealed! (2015).

Artist's rendition of some of the largest known Kuiper Belt Objects (KBOs), compared to the Earth (below) for scale. Many of these dwarf planets have moons, including the first known KBO, Pluto (distances of the moons from their primary planet are not shown to scale).

Largest known Kuiper Belt objects

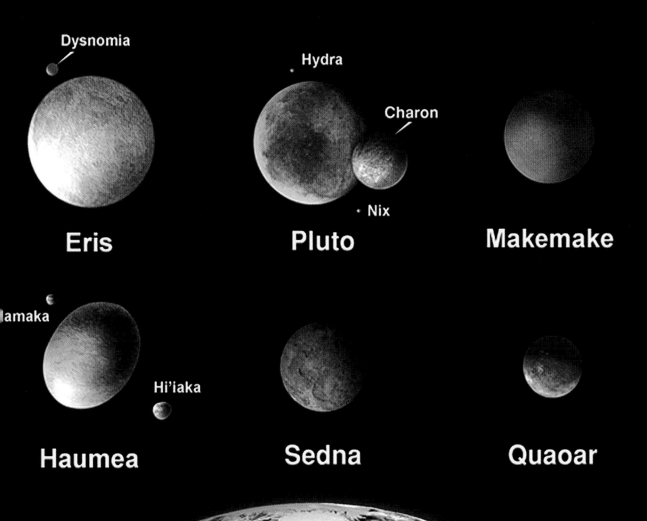

Habitable Super Earths?

The discovery of planets around other stars has naturally led to excitement and anticipation about potentially discovering other Earthlike worlds in our celestial neighborhood. The prospects have been daunting, however, as the **First Extrasolar Planets** were discovered around an exotic, high-energy pulsar supernova remnant, and most of the subsequent discoveries of planets around main sequence stars have been "hot Jupiters," massive gas-giant worlds orbiting extremely close in to their parent stars. More recently, some so-called super Earth planets (ranging from a few to 10–15 times the size of our world) have been discovered, but again, most are very close in to their parent stars.

In 2007, however, two members of a new class of planet were discovered— potentially habitable super Earths—among the system of at least six planets orbiting the star Gliese 581. The newly discovered planets, known as Gliese 581 c and d, are now inferred, based on radial velocity variations of the host star, to be around 5–10 Earth masses. Most important, these two planets orbit within Gliese 581's habitable zone—the distance from a star within which an Earthlike planet could maintain liquid water on its surface. In our solar system, the habitable zone extends from about Venus to Mars.

There is, of course, no guarantee that planets like Gliese 581 c and d, or others since discovered to orbit within their star's habitable zone, are actually habitable (or inhabited). The concept of the habitable zone merely acknowledges that life as we know it requires liquid water for its formation and sustainability. If these worlds are bathed in harmful radiation from intense solar flares, are heated to the melting point by tidal interactions with their star or with other planets, or have all their water tied up as frozen ice, then they might not truly be habitable. In addition, examples from our own solar system, such as **Europa**, **Titan**, and **Enceladus**, show that it can be possible for habitable worlds to exist outside the traditional habitable zone, if other energy sources are present that can help to keep liquid water stable on or under their surfaces. So far we know of only one perfect, "Goldilocks" planet for life as we know it—**Earth**. Many astronomers expect that soon, however, Earth 2.0, and others like it, will be found.

SEE ALSO Earth (c. 4.5 Billion BCE), Europa (1610), Titan (1655), Enceladus (1789), An Ocean on Europa? (1979), First Extrasolar Planets (1992), Planets around Other Suns (1995), Life on Mars? (1996), An Ocean on Ganymede? (2000), *Huygens* Lands on Titan (2005).

An artist's impression of the planetary system around the red dwarf star Gliese 581. The system includes at least three "super Earth" planets with masses from five to fifteen times that of Earth, two of which might orbit in the so-called habitable zone around the star.

Hanny's Voorwerp

The universe is teeming with galaxies. By some estimates, there may be 100 million galaxies in the observable universe, many of which are likely to be **Spiral Galaxies**, like our own **Milky Way**, but many of which are likely to fall into other categories, such as **Elliptical Galaxies** or irregular forms. Large-scale astronomical surveys using automated telescopes are now routinely acquiring digital images of many tens of millions of galaxies. There just aren't enough astronomers to analyze and classify all of them.

In response, and capitalizing on the global reach and power of the Internet, in 2007 a group of astronomy researchers from many institutions created the Galaxy Zoo, an online project that enlists volunteers worldwide to help classify galaxies from the burgeoning survey archives. Galaxy Zoo was inspired by the NASA Stardust mission's 2006 Stardust@home citizen science project to identify comet dust in photos of the sample return materials.

One of Galaxy Zoo's first discoveries came shortly after the project kicked off, when a Dutch schoolteacher and amateur astronomer named Hanny van Arkel noticed a strange, wispy structure in one of the images near the spiral galaxy IC 2497. Subsequent measurements by professional astronomers showed that the structure is enormous—larger than the Milky Way galaxy, and is about as far away as IC 2497 itself (about 650 million light-years distant). Not a molecular cloud or supernova remnant, and not like anything else previously known, the object was dubbed Hanny's Voorwerp (*voorwerp* is Dutch for "object").

Astronomers are still trying to understand what this object is. Some hypothesize that it is the remnants from a disrupted galaxy that are being ionized by radiation from a now-dead **Quasar**; others speculate that a black hole at the center of IC 2497 may be "beaming" radiation into gas surrounding IC 2497.

Whatever the explanation, Hanny's Voorwerp is a prime example of the power of citizen science. Tens of thousands of trained, enthusiastic volunteers are now being enlisted in Galaxy Zoo and other astronomy-related projects to study, for example, supernovae, extrasolar planets, solar weather, and planetary impact craters. These projects are now part of the Zooniverse, an online portal for citizen science. Join in!

SEE ALSO Milky Way (c. 13.3 Billion BCE), Elliptical Galaxies (1936), Spiral Galaxies (1959), Quasars (1963), *Stardust* Encounters Wild-2 (2004).

Hubble Space Telescope composite color image from April 2010 of Hanny's Voorwerp (Hanny's Object), the greenish-colored, wispy structure seen just below the spiral galaxy IC 2497 near the top of this view. This enigmatic object was discovered by an online volunteer "citizen scientist."

Kepler Mission

Around the turn of the twenty-first century, astronomers began perfecting and testing a variety of techniques for detecting **Extrasolar Planets**. The most common search method—radial velocity surveys—is most sensitive to giant (Jupiter-size) planets orbiting close in to their parent stars. Indeed, those kinds of (likely inhospitable) extrasolar planets have been overwhelmingly among the first to be discovered. Other methods can identify smaller, even Earth-size planets around other stars, but some of them, like **Pulsar** measurements and **Gravitational Lensing**, require either exotic environments likely inhospitable for life or one-time events that don't allow the detected planets to be further characterized.

Perhaps the most promising way to detect Earth-size planets in the neighborhood around their stars is to rely on luck: if the geometry is just right, such planets will occasionally pass in front of or transit their parent stars, dimming those stars' light by a tiny but detectable amount and in a predictable, periodic way. Finding Earth-size worlds that way is the goal of NASA's Kepler mission, named in honor of the Renaissance astronomer Johannes Kepler, discoverer of the fundamental **Laws of Planetary Motion**.

The *Kepler* satellite was launched in 2009 into an Earth-trailing solar orbit to conduct a simple mission: stare at 145,000 nearby main sequence stars for three and a

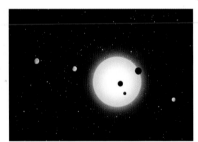

half years and monitor their light to search for periodic planetary transits. *Kepler*'s 42 charge-couple device (CCD) detectors form the largest camera (95 megapixels) ever launched into space, sensitive enough to detect changes in starlight as small as 0.004 percent, or 1 part in 25,000.

Initial results from the Kepler mission have been exciting. In the first six months, more than 1,200 candidate planets were detected around nearly 1,000 stars. Many of those are "hot Jupiters" in rapid, close-in orbits; however, more than 50 of them could be Earth-size planets in the habitable zone around their host stars. Subsequent measurements have increased the number of potential transiting planets to nearly 2,500 as of mid-2012. Required follow-up by ground-based telescopic observations are needed to confirm any discoveries of other nearby worlds like our own.

SEE ALSO Astronomy Goes Digital (1969), First Extrasolar Planets (1992), Planets around Other Suns (1995), Habitable Super Earths? (2007).

LEFT: *Artist's rendering of planets discovered around Kepler-11; six planets transit that Sun-like star.* RIGHT: *A sky map showing the field of view of the NASA Kepler mission, within the northern hemisphere constellation Cygnus. The boxes represent the different digital CCD detector fields.*

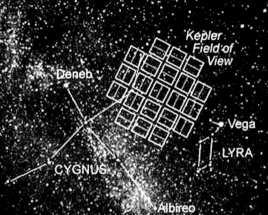

SOFIA

Astronomers like to build telescopes on high, isolated mountain peaks in order to get away from the city lights and to get above as much of Earth's atmosphere as possible. Smoke, smog, haze, water vapor, and other gases and aerosols that block light from reaching the surface, or that contaminate **Spectroscopic** measurements of the universe, limit the quality of scientific observations that can be done from the ground. Putting telescopes in space (such as the **Hubble Space Telescope**, or the Spitzer Infrared Observatory) is a great alternative, but those projects take decades to complete and cost hundreds of millions or several billion dollars to build.

As a way to gain some of the advantages of space-based observing platforms without as many costs or technical hurdles, NASA developed an airborne astronomy program around 1965. At first, they used Convair and Learjet aircraft to carry small telescopes to commercial-aircraft altitudes, then, starting in 1975, they used the Kuiper Airborne Observatory (KAO), a modified C-141A jet carrying a 36-inch (91-centimeter) telescope, to observe astronomical objects from altitudes up to 48,000 feet. Major scientific advances made possible by KAO include the discovery of the **Rings of Uranus** and a thin atmosphere on Pluto. The KAO was decommissioned in 1995 so that NASA could build the larger, more powerful Stratospheric Observatory for Airborne Astronomy, or SOFIA—a modified 747 aircraft carrying a powerful 100-inch (2.5-meter) reflecting telescope.

Initially plagued by some technical and cost-overrun problems (flying a 747 with a hole cut out of it is not simple), SOFIA finally began initial scientific flights in 2010. Flying at a nominal altitude of 41,000 feet (12,500 meters), SOFIA, like KAO, can get above almost all the water vapor in Earth's atmosphere, enabling a much wider range of infrared observations than is possible from ground-based telescopes. Initial science checkout observations in 2010 and 2011 acquired data on Jupiter's atmosphere, the **Orion Nebula**, and on the galaxy M82, where infrared observations can penetrate through much of the galaxy's dust to directly observe the formation of young stars. Now that routine science observations are beginning, future SOFIA observations will include studies of other star-forming regions, protoplanetary disks, exoplanets, and comets.

SEE ALSO First Astronomical Telescopes (1608), Orion Nebula "Discovered" (1610), Uranian Rings Discovered (1977), Hubble Space Telescope (1990), Giant Telescopes (1993), Spitzer Space Telescope (2003), Pluto Revealed! (2015).

NASA's Stratospheric Observatory for Infrared Astronomy (SOFIA) is a modified Boeing 747SP with a large retractable door built into the aft section over a 7- to 16-foot (2- to 5-meter) telescope built by the German Aerospace Center (DLR). SOFIA began initial science flight operations tests in 2010.

Rosetta Flies by 21 Lutetia

Spectroscopy and color measurements from telescopic observations during the second half of the twentieth century revealed that main belt and near-Earth asteroids can be grouped into a small alphabet soup of compositional classes. For example, asteroids that show colors and spectra indicating the presence of typical planet-forming volcanic minerals, like those found in stony meteorites, are known as S-type asteroids; darker objects with grayer colors and spectra that are more like those of carbonaceous (carbon-bearing) meteorites are known as C-type asteroids; objects with spectra similar to metallic meteorites are known as M-types, and so on. Dozens of different asteroid types have been proposed, depending on the classification scheme and research group.

Prior to 2010, spacecraft had encountered both S-type (for example, Eros, Gaspra, Ida, and Itokawa) and C-type (**253 Mathilde**) asteroids, but no others. Thus it was especially exciting when the European Space Agency's *Rosetta* spacecraft made a close flyby past the M-type asteroid 21 Lutetia on July 10, 2010. Rosetta is a comet rendezvous mission that was launched in 2004 and will rendezvous and deploy a lander onto the surface of periodic comet Churyumov-Gerasimenko in 2014. In addition, like many other space mission teams, the Rosetta team has been able to do some excellent "bonus" science by flying past other objects on the way.

The *Rosetta* images of Lutetia revealed it to be the largest asteroid then encountered by spacecraft (at 82 x 63 x 48 miles [132 x 101 x 76 kilometers] in size). It is also one of the densest (at 3.4 grams per cubic centimeter), suggesting a possible rocky, metallic composition consistent with its M-type classification. In terms of its visual appearance and geology, however, Lutetia shares many similarities with the other asteroids that have been photographed up close: a lumpy, irregular surface heavily covered by both relatively fresh and relatively degraded impact craters in a variety of sizes. Lutetia also shows evidence of a surface layer of fine-grained, mobile, impact-generated debris— what planetary scientists call a regolith. Why Lutetia's spectrum appears similar to metallic meteorites, and how such small objects with such low gravity (less than 0.3 percent of Earth's) can retain fine-grained regolith materials, are active areas of research and debate motivated by *Rosetta*'s flyby measurements.

SEE ALSO Ceres (1801), Vesta (1807), Asteroids Can Have Moons (1992), 253 Mathilde (1997), NEAR at Eros (2000), *Hayabusa* at Itokawa (2005).

A composite of images of all asteroids and comets that have been encountered by space missions as of late 2010. All 15 objects are shown at their correct relative sizes, providing a dramatic illustration of how much larger 21 Lutetia is than all previously encountered small bodies.

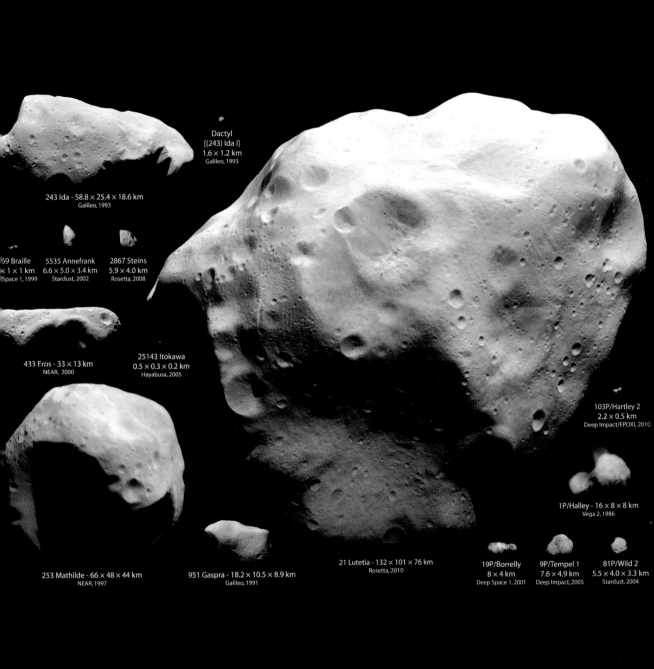

Dactyl
[(243) Ida I]
1.6 × 1.2 km
Galileo, 1993

243 Ida - 58.8 × 25.4 × 18.6 km
Galileo, 1993

59 Braille
× 1 × 1 km
Space 1, 1999

5535 Annefrank
6.6 × 5.0 × 3.4 km
Stardust, 2002

2867 Steins
5.9 × 4.0 km
Rosetta, 2008

433 Eros - 33 × 13 km
NEAR, 2000

25143 Itokawa
0.5 × 0.3 × 0.2 km
Hayabusa, 2005

103P/Hartley 2
2.2 × 0.5 km
Deep Impact/EPOXI, 2010

1P/Halley - 16 × 8 × 8 km
Vega 2, 1986

253 Mathilde - 66 × 48 × 44 km
NEAR, 1997

951 Gaspra - 18.2 × 10.5 × 8.9 km
Galileo, 1991

21 Lutetia - 132 × 101 × 76 km
Rosetta, 2010

19P/Borrelly
8 × 4 km
Deep Space 1, 2001

9P/Tempel 1
7.6 × 4.9 km
Deep Impact, 2005

81P/Wild 2
5.5 × 4.0 × 3.3 km
Stardust, 2004

Comet Hartley-2

After completing the successful 2005 mission to crash a projectile into comet **Tempel-1**, engineers responsible for NASA's *Deep Impact* spacecraft realized that there was enough fuel left onboard to operate the probe as a remote observatory for characterizing extrasolar planets using the transit method (which was also being used by the Kepler mission), and to potentially encounter a second comet nucleus as well. Deep Impact was thus recast as a new mission called EPOXI, for Extrasolar Planet Observation and Deep Impact Extended Investigation.

After three Earth flybys to provide gravity-assist trajectory tweaks, EPOXI was targeted to fly closely past the nucleus of the Earth-approaching comet Hartley-2 in November 2010. Discovered in 1986 by Australian astronomer Malcolm Hartley, Hartley-2 is a short-period comet that travels in an orbit between about 1.1 astronomical units (AU) and 5.9 AU every six and a half years. Short-period comets are further divided into Jupiter family comets, such as Hartley-2, with periods less than about 20 years, and Halley family comets (named after their most famous member), with periods from about 20 to 200 years. Many short-period comets are suspected to have once been long-period comets that had their orbits dramatically changed by a close encounter with one of the giant planets. Hartley-2, for example, may be a relatively primitive object from the **Öpik-Oort Cloud** that encountered Jupiter relatively recently.

The EPOXI data from the flyby support the idea of a primitive, outer-solar-system origin for Hartley-2. Powerful jets of ice, gas, and dust were seen spewing from the comet's 1.4-mile (2.3-kilometer) peanut-shaped nucleus in the dramatic EPOXI images, and **Spectroscopic** measurements showed that the ice is dominated by carbon dioxide (dry ice) rather than water. Initial studies also point to the possible presence of some organic molecules, such as methanol, in Hartley-2's jets and extended atmosphere.

At the rate that the comet is currently losing mass from the jets as well as from sublimation (evaporation of ice) of the surface in general, scientists predict that it might only survive for about another 100 orbits (700 years) or so before breaking up into smaller pieces. Thus it is likely that this little lump of primitive ices from the original **Solar Nebula** is indeed a recent interloper in the inner solar system.

SEE ALSO Solar Nebula (c. 5 Billion BCE), Halley's Comet (1682), Tunguska Explosion (1908), Öpik-Oort Cloud (1932), Comet SL-9 Slams into Jupiter (1994), "Great Comet" Hale-Bopp (1997), *Stardust* Encounters Wild-2 (2004), *Deep Impact*: Tempel-1 (2005), Kepler Mission (2009).

The nucleus of comet Hartley-2 photographed by the NASA EPOXI spacecraft during its November 4, 2010, flyby. Powerful jets of water vapor, other cometary gases, and dust are escaping from the comet's interior.

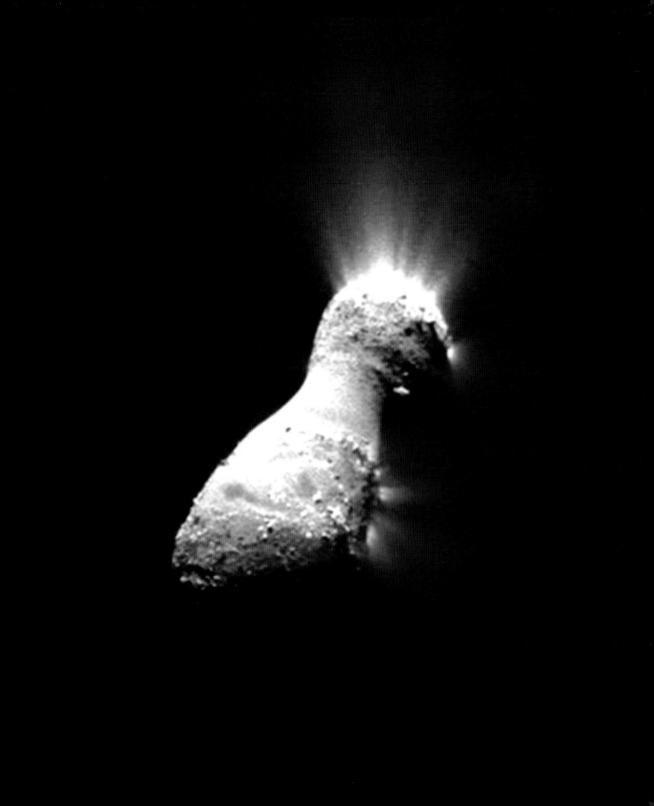

MESSENGER at Mercury

Mercury is the most difficult to observe of the classical planets because it is always close to the Sun. The planet has also been stubbornly resistant to spacecraft exploration, partly because of its challengingly hot thermal environment, where sunlight can be 5 to 10 times more intense than at Earth.

Only one space mission explored the innermost planet during the twentieth century. *Mariner 10*, launched in 1973, performed three close flybys past Mercury in 1974–1975. About half the planet was photographed, revealing a lunar-like, cratered surface and large-scale tectonic features that point to an early, molten surface that cooled and contracted over time. *Mariner 10* also discovered that Mercury has a strong magnetic field, likely related to the planet's large, partially molten core.

These intriguing discoveries compelled planetary scientists to propose a dedicated Mercury orbiter mission, which NASA selected for launch in 2004. After gravity-assist flybys of Earth, Venus, and three flybys of Mercury itself, the Mercury Surface, Space Environment, Geochemistry, and Ranging mission (*MESSENGER*, a contrived acronym in honor of Mercury's mythologic role as the messenger of the gods) finally

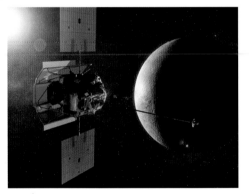

braked into a successful, elliptical orbit around Mercury in March 2011.

MESSENGER has completed mapping the entire planet, and has made exciting discoveries about ancient Mercurian volcanoes, craters, and tectonic landforms. A key goal as the mission continues is to determine if bright spots seen within permanently shadowed craters at Mercury's poles in **Arecibo** radar images are caused by preserved deposits of water ice (perhaps from cometary or asteroid impacts), or perhaps deposits of sulfur or other elements slowly being released from Mercury's interior.

MESSENGER's discoveries will also help influence the planning and operation of the next Mercury mission, the European Space Agency's BepiColombo orbiter, planned to launch in 2014 and settle into Mercury orbit in 2020.

SEE ALSO Mercury (c. 4.5 Billion BCE), Search for Vulcan (1859), Arecibo Radio Telescope (1963), Lunar Highlands (1972).

The first photograph taken from Mercury orbit, captured by the NASA MESSENGER spacecraft (seen in an artist's rendering in the inset) on March 29, 2011. The bright rayed crater near the top is called Debussy; the region from the middle to bottom of the photo is near Mercury's south pole.

Dawn at Vesta

The **Demotion of Pluto** from the domain of planethood in 2006 was accompanied by the promotion of one **Main Belt Asteroid**, 1 Ceres, and probably a second, 4 Vesta, to dwarf planet status. According to the current International Astronomical Union definition, a dwarf planet is a small body that has sufficient mass and self-gravity to have formed itself into a near-spherical shape. In the process, dwarf planets, like full-fledged planets, have probably differentiated into core, mantle, and crust, and thus may have had active interior and surface geologic processes during their histories.

Even the best Hubble Space Telescope (HST) views do not reveal much detail about **Ceres** and **Vesta**, and so more thorough exploration of these worlds requires close-up space missions. In 2007, NASA launched a mission that would do both. The novel mission, called Dawn (in honor of its primary targets, which formed at the dawn of the solar system), uses xenon ion thrusters to gently alter its trajectory to match that of Vesta (2011 encounter) and then Ceres (2015 encounter). If successful, Dawn will be the first mission to orbit one planetary object, then leave, and orbit a second.

Dawn went into orbit around Vesta (330 miles [530 kilometers] in diameter) in July 2011. High-resolution images reveal a heavily cratered surface, and confirmed the presence of an enormous south polar impact basin first seen in HST images, as well as a second older, overlapping basin. A series of enormous, deep grooves circle Vesta's equator, and appear to have been caused by the south polar impacts.

Dawn's **Spectroscopic** measurements confirm that Vesta is the source of a set of meteorites that came from a large, differentiated, volcanically active parent body. Mass and volume estimates reveal a density of about 3.4 grams per cubic centimeter—comparable to the densities of the Moon and Mars. Vesta thus appears to be a rare, preserved example of a protoplanet—an ancient, transitional solar-system body that is part asteroid and part planet, a frozen-in-time relic survivor of the formation of the terrestrial planets.

Dawn left Vesta in summer 2012 to head for an orbital rendezvous with the solar system's largest asteroid, Ceres, in 2015.

SEE ALSO Main Asteroid Belt (c. 4.5 Billion BCE), Ceres (1801), Vesta (1807), Asteroids Can Have Moons (1992), NEAR at Eros (2000), *Hayabusa* at Itokawa (2005), Demotion of Pluto (2006), *Rosetta* Flies by 21 Lutetia (2010).

The Dawn mission (the logo of which is shown at left) obtained this photo of Vesta after going into orbit around the large main belt asteroid in July 2011. The view is centered on the large and deep south polar impact basin called Rheasilvia and its enormous central peak.

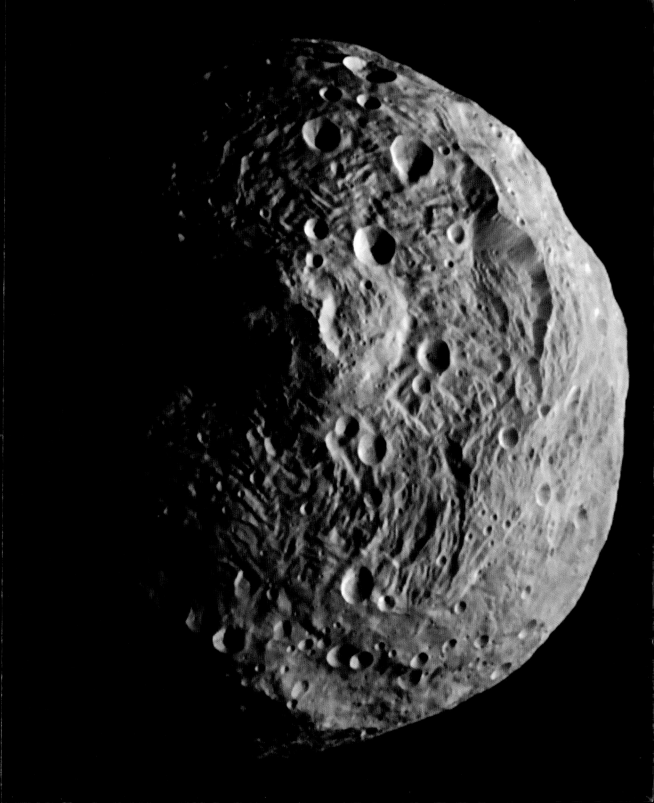

Mars Science Laboratory *Curiosity* Rover

More than 40 years of up-close exploration of **Mars** by orbiters, landers, and rovers has provided fascinating and compelling evidence that, by most accounts, the Red Planet was more Earthlike and more potentially habitable long ago than it is today. That evidence has come in many forms, including ancient river valley landscapes carved by liquid water flowing on the surface, and deposits of specific kinds of hydrated minerals that could only have formed in a more watery environment.

The Viking landers in the 1970s performed sensitive searches for organic molecules on Mars but came up empty, partly because their geologically boring landing sites had to be chosen before much was known about Mars's watery past. With the benefit of hindsight and an additional 35 years of study, NASA decided to search for traces of organic molecules, and potentially, past life, with a new Mars rover mission. Thus, in 2006, the Mars Science Laboratory rover mission was initiated, with a landing targeted at ancient sedimentary rocks in Gale crater, where liquid water and energy sources— two key requirements for life—are known to have existed based on the geology and mineralogy determined from orbit.

The rover, called *Curiosity*, launched in November 2011 and landed safely on Mars on August 6, 2012. It's a big space vehicle, about three times the size of the Mars rovers **Spirit** and **Opportunity**. And it carries an impressive payload: high-resolution color stereo cameras and a color microscope, a laser spectrometer to zap rocks and measure their composition, an X-ray instrument to identify minerals, and a very sensitive mass spectrometer to locate organic molecules in soil and rock samples.

An elaborate sky crane landing system delivered *Curiosity* safely to Mars, using retrorockets like those used by the *Viking* landers rather than the airbags used by the smaller **Mars Pathfinder**, *Spirit*, and *Opportunity* missions. If all goes well, NASA's newest robot astrobiologist will rove on Mars through at least 2014.

SEE ALSO Mars (4.5 Billion BCE), *Mars and Its Canals* (1906), First Mars Orbiters (1971), *Vikings* on Mars (1976), Life on Mars? (1996), First Rover on Mars (1997), Mars Global Surveyor (1997), *Spirit* and *Opportunity* on Mars (2004).

LEFT: *Technicians exercise the wheels of NASA's Mars Science Laboratory rover, named* Curiosity, *during initial driving tests at the Jet Propulsion Laboratory in July 2010.* RIGHT: Curiosity *delivers its first 360-degree color panorama of the Gale Crater landing site on August 8, 2012.*

Pluto Revealed!

Despite its 2006 demotion from full-fledged planet to dwarf planet, Pluto is still a fascinating and enigmatic place. During the 1980s a series of eclipses between Pluto and its large moon **Charon** enabled the basic details of the planet to be established from telescopic observations. Pluto is about 1,430 miles (2,300 kilometers) in diameter, or about 20 percent the size of Earth, but only 0.2 percent of Earth's mass because of its low density of 2 grams per cubic centimeter and thus its predominantly icy composition.

Further telescopic observations from the Kuiper Airborne Observatory revealed the presence of a thin nitrogen, methane, carbon monoxide atmosphere around Pluto, with a surface pressure at least 300,000 times less than Earth's atmosphere. **Spectroscopic** measurements of Pluto's surface show it to be composed of more than 98 percent nitrogen ice, with trace amounts of methane and carbon dioxide. Pluto seems in many ways to be very much like Neptune's large moon **Triton**, composed of nitrogen and other very low-temperature ices and surrounded by a thin, probably dynamically changing atmosphere.

Finding out what Pluto is really like is the job of the NASA *New Horizons* space probe. Launched in 2006, it is the fastest spacecraft ever sent out from Earth, with a velocity of nearly 37,000 miles per hour (16.5 kilometers per second). One can get a sense of the enormous scale of our solar system by realizing that at that velocity, *New Horizons* got to Jupiter in only about 13 months (a record), and then used the planet's gravity to accelerate itself to an even higher velocity. And yet, the solar system is so big that it will still take *New Horizons* more than eight years to get from Jupiter out to Pluto. Scientists think that it will be well worth the wait.

After that journey of more than nine years traveling from Earth to the outer reaches of the solar system, however, the probe will have only about 30 pressure-packed minutes to take its preprogrammed close-up pictures, spectra, and other measurements of Pluto and its atmosphere and moons before it zips on by. If all goes well, mission planners are then hoping to target the spacecraft to encounter one or more other **Kuiper Belt Objects** in the 2016–2020 time period, if a suitable candidate or candidates can be identified along the spacecraft's trajectory.

SEE ALSO Pluto and the Kuiper Belt (c. 4.5 Billion BCE), Triton (1846), Discovery of Pluto (1930), Charon (1978), Kuiper Belt Objects (1992), Demotion of Pluto (2006).

Artist's conception of the NASA New Horizons space probe approaching Pluto and its largest moon, Charon, in July 2015. The nuclear-powered spacecraft uses a radio antenna 7 feet (2.1 meters) in diameter to transmit and receive signals across 4.5 billion miles (7.5 billion kilometers) of interplanetary space.

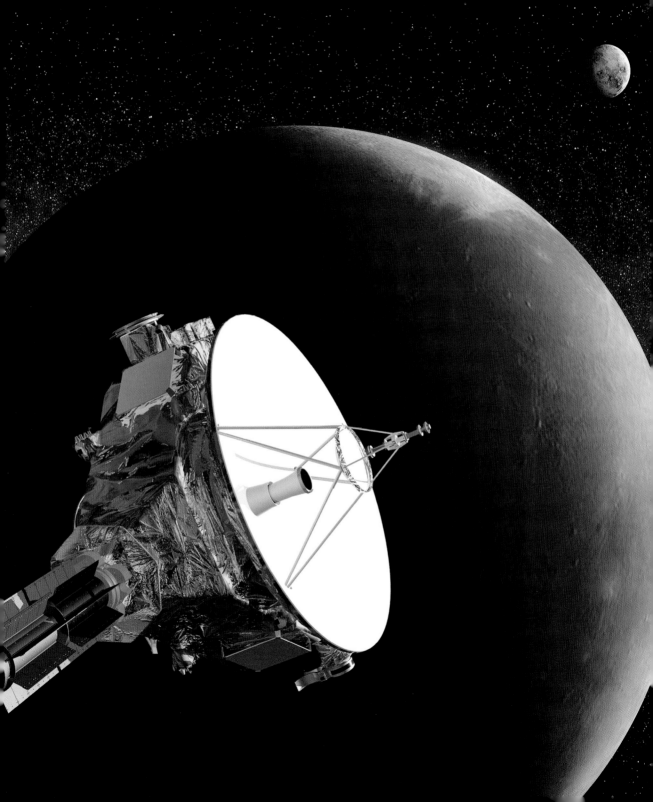

North American Solar Eclipse

Eclipses occur when one celestial body passes in front of another, as viewed from a particular location. Solar or lunar eclipses are the kinds of eclipses most familiar to people because they occur frequently enough, and are often dramatic enough, that they are memorable (or sometimes portentous) events.

A lunar eclipse occurs at full Moon, when the Sun, Earth, and Moon line up (in that order), and the Moon passes exactly behind Earth from the Sun and thus through Earth's shadow. The Moon's orbit is tilted relative to Earth's orbit around the Sun, however, and so the Moon only occasionally passes precisely through Earth's shadow. Most months at full Moon, the Moon passes a little above or a little below Earth's shadow, and, sadly, there is no eclipse.

A solar eclipse occurs at new Moon when the Sun, Moon, and Earth line up (in that order), and the Moon passes exactly between Earth and the Sun. Again, when the geometry is just right (rarely), the Moon's shadow can fall on the Earth. It is an incredible cosmic coincidence that the apparent angular size of the Moon in the sky is almost the same as the apparent angular size of the Sun (the Sun's diameter is about 400 times larger than the Moon's, but the Moon is about 400 times closer to Earth). The result is that very rarely can the Moon's disk completely cover the Sun's disk in the sky, resulting in a total solar eclipse.

Total solar eclipses are rare—any particular spot on Earth experiences one only every 370 years, on average. Some people, including many astronomers, are eclipse chasers, traveling to the predicted path of the Moon's shadow to view or take scientific data during these rare celestial events. The element **Helium**, for example, was discovered in 1868 in the Sun's extended atmosphere or corona, which is often much more visible during a solar eclipse.

The next major total solar eclipse visible from North America will occur on August 21, 2017, when the Moon's shadow will sweep across the United States, from Oregon to South Carolina. It will be a great chance to catch a Moon shadow, and the only chance in North America until April 2024!

SEE ALSO Astronomy in China (c. 2100 BCE), Earth Is Round! (c. 500 BCE), *De Sphaera* (c. 1230), Venus Transits the Sun (1639), Speed of Light (1676), Helium (1868), Kepler Mission (2009).

LEFT: *Path of the Moon's shadow across the United States during the August 21, 2017, total solar eclipse.* RIGHT: *A dramatic photograph, taken from the Russian Mir space station, of the Moon's shadow as it raced across the Earth at nearly 1,240 miles (2,000 kilometers) per hour during the August 11, 1999, total solar eclipse.*

James Webb Space Telescope

James E. Webb (1906–1992)

The power and beauty of space-based astronomy has been dramatically demonstrated by a variety of small, medium, and large space telescopes, culminating in the successful missions of NASA's four Great Observatories: the **Hubble Space Telescope** (HST), Compton Gamma-Ray Observatory, the **Spitzer Space Telescope**, and the **Chandra X-ray Observatory**. Like all complex spacecraft, however, these missions have finite lifetimes. Only HST has been capable of being repaired or upgraded by astronauts—but the retirement of the space shuttle in 2011 means the end of service calls to HST. NASA has thus been pondering its replacement for some time.

NASA's planned next-generation space telescope is known as the James Webb Space Telescope (JWST), after NASA's second administrator, James E. Webb, who oversaw the space agency during the Mercury, Gemini, and early Apollo astronaut programs. Planning for JWST actually began in 1989, the year before HST was launched. Over the course of more than 20 years the design has been revised numerous times. Currently the telescope is in final development, with a scheduled launch in 2018.

JWST will combine some capabilities from the Hubble Space Telescope (such as high-resolution imaging), the Keck Observatory (a precisely controlled segmented mirror design), and the Spitzer Space Telescope (sensitivity in the infrared) to become the scientific workhorse for space-based astronomy for at least a decade. Its 21-foot (6.5-meter) segmented primary mirror has six times the light-collecting area of HST, and the telescope is cooled to only 40 degrees above absolute zero to make it highly sensitive to faint, distant objects in the cosmos.

Astronomers have ambitious plans for JWST science across the entire range of visible through infrared astronomy and astrophysics research areas. Major science themes include studying the first stars and galaxies, formed after the early dark ages of the universe; studies of dark matter; new stars and their associated protoplanetary disks of gas and dust; the formation of planets; and the search for extrasolar planets and other cosmic environments conducive to life. JWST should be a spectacular astronomical discovery machine!

SEE ALSO First Astronomical Telescopes (1608), Hubble Space Telescope (1990), Gamma-Ray Astronomy (1991), Giant Telescopes (1993), Chandra X-ray Observatory (1999), Spitzer Space Telescope (2003).

Artist's conception of the James Webb Space Telescope, with its gold-coated primary mirror (21 feet [6.5 meters] in diameter) and deployed radiation shields, designed to protect the telescope from the contaminating light and heat of the Sun, Earth, and Moon.

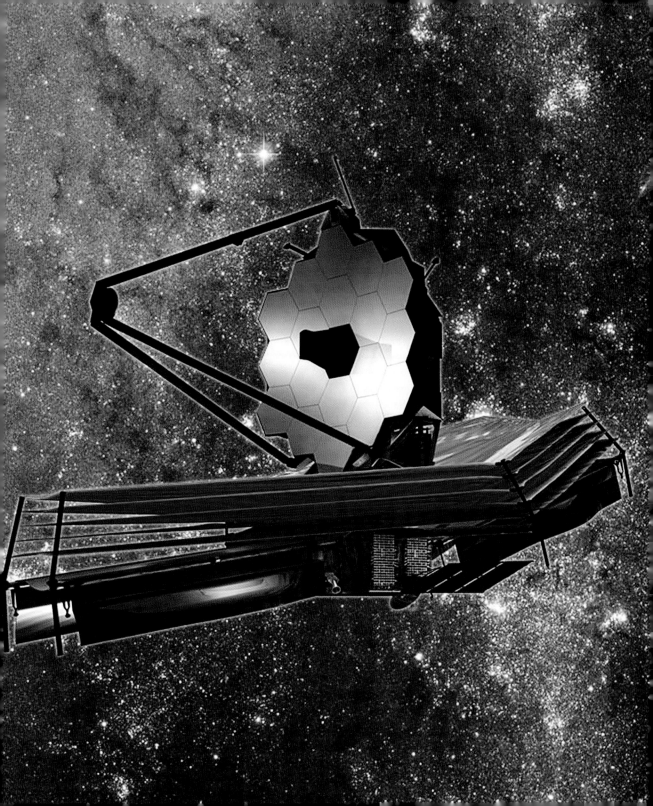

Apophis Near Miss

Dedicated telescopic surveys in the 1990s and 2000s have identified hundreds of thousands of new asteroids zipping around the solar system. Most are in the **Main Asteroid Belt** between Mars and Jupiter, but many are in other populations as well, such as **Jupiter's Trojan Asteroids** and three different populations of near-Earth asteroids (NEAs)—the Atens (orbiting closer to the Sun than Earth), Amors (orbiting farther out from Earth), and Apollos (with orbits that cross Earth's orbit). All three classes of NEAs pose potential impact hazards for Earth.

One of the most closely watched members of the NEA population is a small asteroid named 99942 Apophis. First discovered in 2004, its orbital parameters were calculated from follow-up telescope observations, including Arecibo radar measurements. Then, like hundreds of other NEAs, its parameters were entered into an automated computer program developed by astronomers to predict the future trajectories and Earth-impact probabilities of these asteroids. Apophis quickly set off some alarms because the calculations came back showing an approximately 1 in 37 chance that the asteroid would

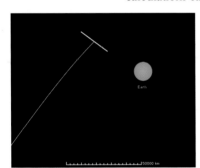

impact Earth on April 13, 2029. Apophis set the record for greatest impact risk yet identified, with a **Torino Scale** score of 4 out of 10.

Astronomers quickly organized more observing campaigns to refine the predictions of Apophis's orbit. The new data showed that the asteroid will get very close to Earth, only about two to three Earth diameters away—well within the orbits of **Geosynchronous Satellites**—but that it will not actually impact our planet. Apophis will pass close to Earth again in 2036, but its odds of hitting Earth then are down to 1 in 250,000, and its Torino Scale score has been downgraded to a 0.

Still, prudence is advised: an impact by a rocky asteroid 1,000 feet (300 meters) in diameter would not be devastating globally, but it would be bad news locally (as would, for example, a giant impact-generated tsunami). Apophis is named after the Egyptian god of destruction—let's hope that this hazardous little asteroid's name doesn't turn out to be accurate.

SEE ALSO Main Asteroid Belt (c. 4.5 Billion BCE), Ceres (1801), Vesta (1807), Jupiter's Trojan Asteroids (1906), Geosynchronous Satellites (1945), Arecibo Radio Telescope (1963), Torino Impact Hazard Scale (1999), NEAR at Eros (2000), *Hayabusa* at Itokawa (2005).

LEFT: *The detail shows a zoom of how close Apophis is predicted to get to our planet (the white bar represents the trajectory uncertainty).*RIGHT: *A trajectory plot, to scale, of the positions of the Earth and Moon and the predicted path of the near-Earth asteroid 99942 Apophis during its April 13, 2029, close approach.*

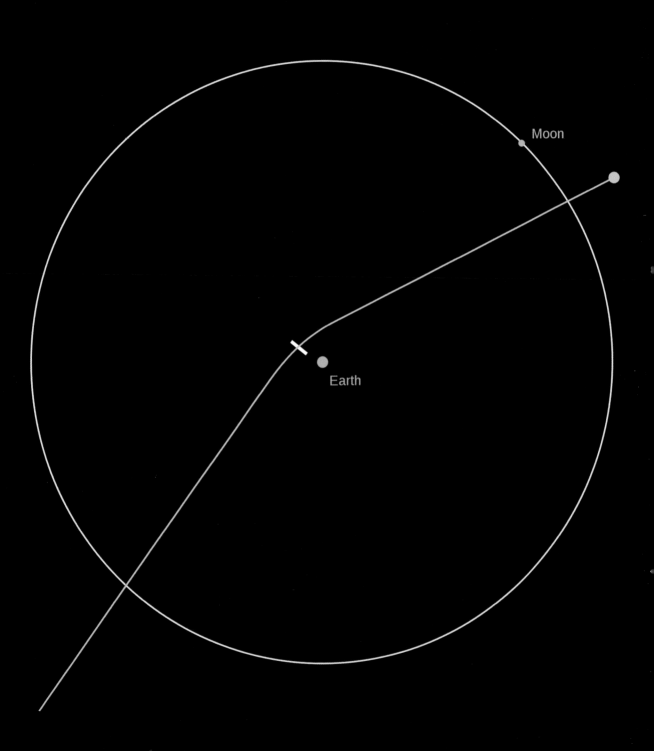

First Humans on Mars?

Decades of high-resolution observations from telescopes and robotic flybys, orbiters, landers, and rovers have continually increased the allure of **Mars** exploration. Evidence from the geology, mineralogy, and atmosphere of Mars points to a dramatic change in the planet's climate during its history. Today the Martian surface is cold, dry, and, as far as we can tell, inhospitable, but early Mars, for the first billion years after its formation, was a warmer and wetter planet. While it was probably never as warm and wet as our own planet, evidence suggests that it was more Earthlike and habitable.

But was or is Mars inhabited? Robotic missions have mapped the planet and documented the conditions at a variety of landing sites, but a serious search for past or present evidence of **Life on Mars** will be much harder. Like a forensic investigation, it will require piecing together clues from scattered fragments of evidence in order to see the big picture. The job will be comparable to what field geologists do here on the Earth, reconstructing the history of a region only after years of methodical field and laboratory work, combined with experience at other sites and, often, gut feelings and instincts. It will require more detailed geologic mapping, plus probable coring and deep drilling. In short, people, not just robots, will be needed to truly understand the place.

When will humans go to Mars? No one knows, but the imperative is growing, fueled by the discoveries of our robotic precursor emissaries. The mid-2030s is a guess, because first we need to build a new rocketry and life-support infrastructure to get humans beyond low Earth orbit and back out into deep space. That will take time.

More than 50 years ago president John F. Kennedy challenged Americans to dream big and take a bold risk by attempting to send astronauts to the Moon and back before 1970. The Apollo missions inspired a generation of scientific and technological innovation that has truly changed the world. Sending astronauts to Mars and back will be much riskier and much bolder. Can we once again rise to the challenge?

SEE ALSO Mars (c. 4.5 Billion BCE), *Mars and Its Canals* (1906), First on the Moon (1969), Second on the Moon (1969), Fra Mauro Formation (1971), Roving on the Moon (1971), Lunar Highlands (1972), Last on the Moon (1972), First Mars Orbiters (1971), *Vikings* on Mars (1976), Life on Mars? (1996), First Rover on Mars (1997), Mars Global Surveyor (1997), *Spirit* and *Opportunity* on Mars (2004).

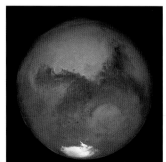

LEFT: *Hubble Space Telescope photo of Mars on August 26, 2003, at its closest approach to Earth in 60,000 years.* RIGHT: *NASA artist Pat Rawlings's conception of two astronaut explorers on Mars and their wheeled Mars roving vehicle, collecting samples and scouting for signs of past habitable environments.*

Sagittarius Dwarf Galaxy Collision

Planets can have satellites and asteroids can have satellites; it turns out that even galaxies can have satellites. Our own **Milky Way** galaxy probably has more than 20 satellite galaxies, including the Large and Small Magellanic Clouds. These dwarf galaxy companions are all smaller than our galaxy, and take hundreds of millions to billions of years to orbit the Milky Way. And it appears that at least one of them, known as the Sagittarius Dwarf **Elliptical Galaxy**, has collided with the Milky Way in the past and is on a course to slam into our galaxy again in about 100 million years.

The Sagittarius Dwarf Galaxy wasn't discovered until 1994 because much of it is hidden by the central bulge and disk of the Milky Way. It is made up of four main **Globular Clusters** and a bright arc of stars that makes a partial loop around the poles of the Milky Way. Astronomers believe that the loop outlines the path that the dwarf galaxy has taken through the plane of the Milky Way in the past, losing some mass and smearing out a bit more during each orbit. Eventually, after more passes through the plane of our galaxy, the stars in the Sagittarius Dwarf Galaxy will merge into and expand the size and mass of our already much larger Milky Way. These kinds of galactic collisions, mergers, and "cannibalistic" processes may in fact be the way that large spiral galaxies can grow so large—by consuming smaller, more ancient galaxies or galaxy clusters.

Some scientists believe that there is a connection between close galactic encounters among the Milky Way and its satellite galaxies and major extinction events or large-scale climate changes (for example, glaciations) on Earth. The idea is that these close encounters stir up distant comets and asteroids in the **Oort Clouds** around our Sun and nearby stars, causing more comets and asteroids to periodically hurtle inward and collide with the Earth, with potentially dramatic effects. The hypothesis is controversial, however, and difficult to test except with high-speed supercomputers. Still, the idea that not just passing stars but passing galaxies could have a profound effect on life on our planet is interesting because a detailed understanding of large-scale changes in climate and the fossil record could help reveal the past history of local galactic collisions. However, it is also humbling, because it makes us realize how lucky we are to be here at all.

SEE ALSO Milky Way (c. 13.3 Billion BCE), Globular Clusters (1665), Öpik-Oort Cloud (1932), Elliptical Galaxies (1936), Spiral Galaxies (1959).

2004 Hubble Space Telescope photograph of the colliding spiral galaxies NGC 2207 (top) and IC 2163 (bottom). The enormous amount of mass involved in galactic collisions generates enormous tidal forces; in this collision, the smaller galaxy is being stretched apart by the larger one.

Earth's Oceans Evaporate

The life cycle of a **Main Sequence** star like the Sun is quite predictable. Astronomers in the early twentieth century figured out the basic evolutionary track of stars like the Sun by observing countless numbers of similar stars at different stages of development. By the middle of the century, theories had also been worked out about the insides of the stars, and the **Nuclear Fusion** processes that make them shine. And thanks to primitive meteorite researchers and radioactive dating methods, we know the approximate age of the Sun (4.65 billion years) and thus can predict the next milestones of our star's life.

Hydrogen is converted into helium at the enormous temperatures and pressures in the Sun's core. Over time, then, the Sun's hydrogen supply is slowly decreasing. To keep its balance of gravitational (inward) versus radiational (outward) pressure—and thus to stay on the main sequence—the Sun's core is slowly getting hotter. This increases the rate of nuclear fusion in the core, offsetting the effect of the decreasing hydrogen supply and increasing the Sun's brightness over time. Astronomers estimate that the Sun's energy output is increasing by about 10 percent per billion years because of the decreasing supply of hydrogen.

Such a dramatic change in the Sun's energy output will have a correspondingly dramatic change on the Earth's climate. In tens to hundreds of millions of years, it will become warm enough that the oceans will start to permanently evaporate, turning our planet into a steamy world. Scientists further predict that within about a billion years, the slow breakdown of all that atmospheric water by sunlight and the subsequent escape of the liberated hydrogen will turn our planet into a bone-dry, inhospitable desert world. Unfortunately, the future is looking *too* bright.

But wait, it could be even worse: some long-term climate modelers think that our planet will become uninhabitable long before the oceans are completely dried up. As the climate gets hotter, more carbon dioxide will get trapped into carbonate rocks, leaving less for plants to use in photosynthesis. Within maybe a half billion years, then, much of the base of the food chain could collapse, making the biosphere overall unsustainable. It's not a cheery long-term prognosis, but maybe by then our species (or whatever it has become) will have found a new beautiful blue-water world to call home.

SEE ALSO Birth of the Sun (c. 4.6 billion BCE), Mira Variables (1596), Main Sequence (1910), Nuclear Fusion (1939), End of the Sun (~5.7 Billion).

An artist's conception of a "hot Jupiter," among the most common kind of extrasolar planets initially discovered in the solar neighborhood. A billion years from now, as the Sun continues to mature and grow hotter, the oceans will evaporate, and our planet become a "hot Earth."

Collision with Andromeda

Our **Milky Way** galaxy is indeed an island universe in the sense that it is an organized, isolated collection of perhaps 400 billion separate stars, plus gravitationally bound gas, dust, and **Dark Matter**. But our galaxy is also part of a larger collection of gravitationally bound neighbor galaxies dubbed the Local Group by astronomer Edwin Hubble. The Local Group consists of more than 30 galaxies, including the Milky Way and its satellite galaxies, such as the Large and Small Magellanic Clouds and the **Sagittarius Dwarf Elliptical Galaxy**, the **Andromeda** galaxy and its satellite galaxies, and others. Astronomers estimate that the Local Group spans about 10 million light-years across and has a combined mass exceeding 1 trillion solar masses.

The gravitational center of the Local Group is located somewhere between the Milky Way and Andromeda galaxies, which dominate the mass of the group. Astronomers have discovered that these two large **Spiral Galaxies** are moving toward each other, and that in the distant future, perhaps 3–5 billion years from now, they could actually collide, with the nature of the collision depending on the details of their velocities and their distributions of dark matter.

Collision is probably not the best word to describe the interactions of galaxies, however. Because they are actually made up of mostly empty space, it is unlikely that many stars would actually physically collide with each other. Rather, as the two galaxies essentially pass through each other, the gravitational and tidal forces between their associated stars and satellite galaxies will likely rip apart their beautiful spiral structures and could cause them to eventually merge into one larger irregular or elliptical supergalaxy.

The Local Group is in turn part of an even larger collection of galaxies in our part of the universe called the Virgo Supercluster, which consists of more than 100 interacting galaxy clusters like the Local Group, spanning more than 110 million light-years. The Virgo Supercluster is, in turn, just one of perhaps millions of similar superclusters that form the largest-scale structures in the observable universe—the **Walls of Galaxies,** often referred to as the cosmic web.

SEE ALSO Milky Way (c. 13.3 Billion BCE), Andromeda Sighted (964), Messier Catalog (1771), Cepheid Variables (1908), Dark Matter (1933), Spiral Galaxies (1959), Walls of Galaxies (1989), Sagittarius Dwarf Galaxy Collision (~100 Million).

The spectacular Andromeda galaxy, also known as Messier 31, is the closest spiral galaxy to the Milky Way. Even though it is more than 2 million light-years away, it is part of our "local group" of galaxies. Andromeda and the Milky Way appear to be heading toward a collision in the far future.

End of the Sun

The Sun's fate is sealed, and it is humbling and perhaps a little sad to realize that our glorious star will not shine forever. The Milky Way has billions of **Main Sequence** stars with the same stellar classification as our Sun, and we can study them all around us in different stages of their predictable life cycles. The destiny of a star is dictated by its initial mass; in the case of a star with the mass of our Sun, its destiny is a short, violent, energetic youth followed by a relatively long, stable, 10-billion-year middle age and then a relatively gentle, quiet death.

Radioactive dating of primitive meteorites as well as analysis of solar wind particles collected by missions like the *Genesis* spacecraft tell us that the **Sun** is about 4.65 billion years old, or about halfway through its life cycle on the main sequence. As it goes through middle age and uses up more of its hydrogen nuclear-fusion fuel supply, our star is slowly getting hotter—in about a billion years it will be hot enough to **Evaporate Earth's Oceans**. In 5 billion years or so all the Sun's hydrogen will be used up, and the core will contract and heat up further, expanding the Sun's outer atmospheric layers until it becomes a red giant star.

The red-giant Sun will swell to about 250 times its present size, engulfing and destroying the inner planets, including the Earth. As the Sun's helium and other heavier elements are also depleted, enormous pulsating death throes will jettison the Sun's outer layers (including all the atoms and former inhabitants of the now-vaporized Earth) into deep space as a **Planetary Nebula**, to be recycled into new stars. The remaining ember of the Sun's core will become a **White Dwarf** and slowly cool, eventually fading into the background oblivion of cold space.

Earth will be gone, but will life survive? If we can survive our current challenges and first become a multiplanet and then a multi-solar-system species, then perhaps our distant descendants—whatever species they become—will find a new habitable world around another, younger Sun-like star to call home.

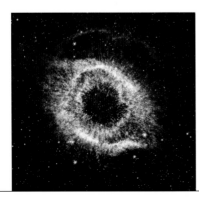

SEE ALSO Birth of the Sun (c. 4.6 Billion BCE), Planetary Nebulae (1764), Messier Catalog (1771), White Dwarfs (1862), Radioactivity (1896), Main Sequence (1910), Nuclear Fusion (1939), *Genesis* Catches Solar Wind (2001), Earth's Oceans Evaporate (~1 Billion).

LEFT: *Spitzer Space Telescope infrared image of the Helix Nebula, a planetary nebula formed during the death throes of a solar mass, such as a red giant star.* RIGHT: *Space artist Don Dixon's imagined view of the Moon transiting the disk of the swollen, red-giant Sun about 5 billion years in the future.*

Last of the Stars

Stellar evolution can be thought of as a giant cosmic recycling program. Clouds of gas and dust condense and contract through gravitational forces, eventually growing into spherical masses—stars—with deep internal pressures and temperatures high enough to initiate **Nuclear Fusion**. After their hydrogen or helium or other nuclear fuel is consumed, stars die in gentle to spectacular ways depending on their mass, expelling much of that mass out into deep space. Those stellar remains—clouds of gas and dust—can then condense and contract through gravitational forces to form new stars. It's a beautiful cycle of stellar life.

Each time a star dies, however, some significant fraction of its mass does not get recycled back into space, but remains behind as a slowly cooling **White Dwarf** star (in the case of low-mass stars) or another stellar remnant, such as a **Neutron Star** or a **Black Hole** (in the case of higher-mass stars). Thus, over time, all the material in the universe involved in star formation is eventually trapped in these final, nonrecycled stellar remnants. With the vast majority of stars living their lives as medium- to low-mass **Main Sequence** stars, the vast majority of the final trapped observable mass of the universe could end up in the form of white dwarf stars.

A typical main sequence star's lifetime is about 10 billion, or 10^{10} years, although lower-mass stars, like those near the theoretical lower limit of nuclear fusion (around 8 percent of the mass of the Sun), can have lifetimes as long as 10 trillion (10^{13}) years. Astronomers estimate (with considerable uncertainty) that by the time the universe is about 100 trillion (10^{14}) years old, or about 10,000 times as old as it is now, almost all the observable mass of the cosmos will have been trapped in white dwarfs, with a small amount trapped in red dwarfs and other remnants, such as neutron stars and black holes. Star formation will thus come to an end, and the universe will enter a very different, final era of development.

The cosmos will slowly begin to fade to black. After they cool, white dwarfs should theoretically become black dwarfs—stellar remnants with temperatures eventually approaching absolute zero. But no one really knows how long it will take the universe's last stars to shut off; in some theories, dark matter or weakly interacting nuclear forces may help keep these last embers of the once-glorious stars dimly shining for 10^{15} to 10^{25} years or more.

SEE ALSO White Dwarfs (1862), Main Sequence (1910), Neutron Stars (1933), Nuclear Fusion (1939), Black Holes (1965), End of the Sun (~5–7 Billion), Degenerate Era (~10^{17}–10^{37}), How Will the Universe End? (The End of Time).

Hubble Space Telescope photograph of ancient (12–13 billion years old) white dwarf stars in the Milky Way galaxy. The 2002 photo is from part of the globular cluster M4 in Scorpius.

Degenerate Era

In the far, far future, the universe will be a cold and dark place. Once star formation ends, the only sources of light and heat will be the compact, dense, remnants of stellar cores: **White Dwarfs**, **Neutron Stars**, and **Black Holes**, plus any slowly cooling remains of any red or brown dwarfs, or planets, moons, asteroids, comets, or cosmic dust that have survived the deaths of their stars.

Slowly, perhaps over 10^{15} to 10^{25} years, even the white dwarfs and remaining planetary objects will cool to absolute zero. But there is still likely to be some occasional action left in the universe nonetheless, because over even more enormous, almost incomprehensible expanses of cosmic time, cooling white dwarfs, black dwarfs, and other objects should encounter each other, as well as neutron stars and black holes, sometimes colliding and merging into larger-mass objects that could be massive enough to once again initiate **Nuclear Fusion** in their cores and shine again. The darkness of the distant future universe will likely be occasionally lit up by such lonely beacons.

Eventually, though, perhaps 10^{17} to 10^{37} years in the future, according to some theoretical models, all the mass of the universe will become concentrated into the densest, most massive compact objects: white dwarfs, neutron stars, and then black holes. Astronomers refer to this predicted future phase of the universe's development as the degenerate era, because the matter in these compact objects is predicted to be at such high densities that all their electrons will have been essentially stripped off their parent atoms (in physics language, the atoms will be "degenerate").

It's not clear how the universe will evolve during the degenerate era. The interactions between the extremely high-energy states of matter in compact objects and the still poorly understood properties and effects of **Dark Matter** and **Dark Energy**, which appear to dominate the matter-energy density of the universe, are basically unknown. Some cosmologists think that over these huge time periods, normal matter (such as protons) will decay, and that white dwarfs could assimilate dark matter and continue to shine for much longer than they would have otherwise, eventually merging with other compact objects into black holes.

SEE ALSO White Dwarfs (1862), Main Sequence (1910), Neutron Stars (1933), Dark Matter (1933), Nuclear Fusion (1939), Black Holes (1965), Dark Energy (1998), Last of the Stars (~10^{14}), How Will the Universe End? (The End of Time).

An artist's rendering of an isolated neutron star. Only a small number of such hugely dense, compact objects without an associated supernova remnant of gas and dust have been discovered in the current universe; however, isolated neutron stars might become common in the distant future.

Black Holes "Evaporate"

Stephen Hawking (b. 1942)

If the prevailing view of the far-distant future of the cosmos is correct, then all the universe's matter and energy (normal matter, **Dark Matter**, and **Dark Energy**) could ultimately end up trapped within **Black Holes**—compact stellar objects that are so massive that not even light can escape their gravitational pull. But what then? Would there be any way to know what happens next?

Possibly. Physicist Stephen Hawking and others have hypothesized that rapidly rotating black holes should create and emit particles (now dubbed Hawking radiation), and that these particles should decrease the mass and energy of such a black hole over time. Conceptually, the process has been described as black hole evaporation, and if it really occurs, it could have a profound influence on how the universe ends.

Black hole evaporation is a tricky concept. It relies on the fact that particles in the standard model of particle physics also have antiparticles, and on the fact that the very edge of the so-called event horizon of a black hole—the point past which no light or other information can escape from the black hole's gravitational field—is a strange and special place. Hawking postulated that if a particle–antiparticle pair, like an electron and a positron, for example, are created by some process right at the exact edge of the event horizon, it is possible for one of them to fall into the black hole and the other to escape. To an observer watching this happen, the black hole would appear to emit a particle and thus decrease ever so slightly in mass. The black hole itself gains energy (heats up) a tiny amount in the process, and if this occurs over a long enough period of time, Hawking proposed that black holes could explode in a violent **Burst of Gamma Rays**.

Modern gamma-ray satellites are searching for the predicted signals from such black hole evaporation events. If they are found, it could mean that during the extremely distant future, perhaps 10^{37} to 10^{100} years from now, even black holes could go away. The long-term fate of the universe could be that it all just fizzles out as a cold, dark, lonely soup of elementary particles such as photons, electrons, positrons, and neutrinos, hardly ever interacting with each other.

SEE ALSO White Dwarfs (1862), Main Sequence (1910), Neutron Stars (1933), Dark Matter (1933), Black Holes (1965), Hawking's "Extreme Physics" (1965), Gamma-Ray Bursts (1973), Dark Energy (1998), Last of the Stars (~10^{14}), Degenerate Era (~10^{17}–10^{37}), How Will the Universe End? (The End of Time).

An artist's conception of a spinning black hole, surrounded by a disk of ionized gas and dust falling into the massive star's enormous gravity well. Magnetic fields surrounding the black hole can further heat and ionize the surrounding gaseous disk.

How Will the Universe End?

The long history of astronomy and space exploration has been permeated with the drive to answer some of the biggest and deepest questions that we can ask: What is going on up there in the sky? Where did everything come from? How did life form? Are we alone? We are fortunate to live in a civilization and a time in human history when we have the luxury to actively seek the answers to such questions and the technology to try to do so.

A fitting place to end this journey through the major milestones of astronomy and space exploration is to come full circle from where we first began. In other words, the prevailing theory for the origin of our universe is that everything around us—all of space and time—was created 13.75 billion years ago in a gigantic, instantaneous explosion known as the **Big Bang**. The universe as we know it had a beginning. So an obvious big question to ask is, will the universe have an end? And, if so, when will it occur?

We know that the universe is expanding because we can measure the galaxies all receding from each other. Maybe this expansion will simply continue forever, perhaps fueled by the strangely repulsive and still almost completely mysterious force of **Dark Energy** accelerating the expansion of space, until the last of the stars fade away and even the **Black Holes** evaporate into the dark, quiet, cold of what astronomers call the heat death of the cosmos, perhaps 10^{100} years from now. Maybe.

Some cosmologists believe, though, that there might be a more violent fate in store for the universe. If the mass of everything in the universe is large enough that dark energy ultimately can't continue to accelerate the expansion of space, then the gravitational attraction between clusters of galaxies could slow and eventually reverse the expansion. Galaxies could begin moving toward each other, with all the mass of the cosmos eventually coming together again into one tiny, massive black hole singularity. What happens then? Another big bang? Or, perhaps, a big bounce?

The ultimate fate of the cosmos is of course unknown. Modern cosmologists are poking at the issue by actively trying to figure out if the universe is open (expanding forever), closed (destined to contract), or flat (in perfect balance). New observations and computer models may help figure out—to borrow a line from the modernist poet T. S. Eliot—whether the universe will end "not with a bang but a whimper."

SEE ALSO Big Bang (c. 13.7 Billion BCE), Hubble's Law (1929), Dark Matter (1933), Black Holes (1965), Dark Energy (1998) Age of the Universe (2001), Last of the Stars (~10^{14}), Degenerate Era (~10^{17}–10^{37}), Black Holes "Evaporate" (~10^{37}–10^{100}).

The galaxy cluster known as Abell S0740, photographed by the Hubble Space Telescope in 2005–2006. This cluster is about 450 million light-years away and contains a spectacular variety of galactic forms. What will happen to the universe's 100 billion galaxies in the far, far future?

Notes and Further Reading

I consulted a huge number of resources during the research for this book, including general historical and encyclopedic sources to verify much of the factual information and a variety of websites for additional details and follow-up (especially for many of the topics in which there are hot new results or substantial ongoing controversies). Below I have attempted to document as many of those resources as my notes and memory have allowed and to direct you to some additional resources. To save space, I use the *tinyurl.com* format for many Web sites; full URLs and other details can be found on my web site at *jimbell.sese.asu.edu/space-book*. The Internet is dynamic, and so some of the sites pointed to below may not be in use by the time this book appears in print.

Selecting just 250 milestones in the entire history of astronomy and space exploration is a daunting task, and my selections naturally reflect my own biases, knowledge, and experience. I would be delighted to consider suggestions for other topics to swap in for future editions of this book, and would also welcome any corrections or general feedback. Please feel free to contact me via Jim.Bell@asu.edu or *jimbell.sese.asu.edu/contact*.

General Reading

Beatty, J.K., C.C. Petersen, and A. Chaikin, eds. *The New Solar System*. Cambridge, UK: Cambridge Univ. Press, 1998.

Levy, D.H., ed. *The Scientific American Book of the Cosmos*. New York: St. Martin's Press, 2000.

Mitton, S., ed. *The Cambridge Encyclopaedia of Astronomy*. Cambridge, UK: Cambridge Univ. Press, 2001.

Moore, P., ed. *Astronomy Encyclopedia*. Oxford, UK: Oxford Univ. Press, 2002.

Weissman, P.R., L.A. McFadden, and T.V. Johnson, eds. *Encyclopedia of the Solar System*. San Diego, CA: Academic Press, 1999.

General-Interest Web sites

Curious about Astronomy? Ask an Astronomer: *curious.astro.cornell.edu*

Nine Planets: *nineplanets.org*

Views of the Solar System: *www.solarviews.com*

Bad Astronomy (blog): *blogs.discovermagazine. com/badastronomy*

Wikipedia: *www.wikipedia.org*

c. 13.7 Billion BCE, Big Bang

See "Misconceptions about the Big Bang" by C. Lineweaver and T. Davis (*Scientific American*, Feb. 2005) and "The First Few Microseconds" by M. Riordan and W. Zajc (*Scientific American*, Apr. 2006).

c. 13.7 Billion BCE, Recombination Era

Using data from the WMAP satellite, cosmologists determined the start of the recombination era to a stunningly accurate 380,000 years after the Big Bang. The results named *Science* magazine's "breakthrough of the year" for 2003 (see *Science*, Dec. 19, 2003).

c. 13.5 Billion BCE, First Stars

Cosmologist Volker Bromm and colleagues at the Univ. of Texas at Austin host tutorials, papers, and computer animations about the first stars and galaxies at *tinyurl.com/brdqoxx*.

c. 13.3 Billion BCE, Milky Way

A very nice series of maps and photographs of the Milky Way can be found at Atlas of the Universe: *tinyurl.com/2fooye*.

c. 5 Billion BCE, Solar Nebula

The generally accepted father of the modern solar nebular disk model is Soviet astronomer Victor Safronov (1917–1999); his book *Evolution of the Protoplanetary Cloud and the Formation of the Earth and the Planets* (NASA Tech. Trans. F-677, 1972) is a classic text.

c. 4.6 Billion BCE, Violent Proto-Sun

Australia Telescope Outreach and Education: *tinyurl.com/c4ey6en*.

c. 4.6 Billion BCE, Birth of the Sun

Spectacular photos, movies, and other information about the Sun can be found at the ESA/NASA Solar and Heliospheric Observatory (SOHO) web site: *tinyurl.com/thyo*.

c. 4.5 Billion BCE, Mercury

Strom, R.G., *Mercury: The Elusive Planet* (Cambridge, UK: Cambridge Univ. Press, 1987).

c. 4.5 Billion BCE, Venus

Find Venus at *nineplanets.org/venus.html*.

c. 4.5 Billion BCE, Earth

Dalrymple, G.B., *The Age of the Earth* (Stanford, CA: Stanford Univ. Press, 1994).

c. 4.5 Billion BCE, Mars

The Planetary Society, the world's largest public space-advocacy organization, hosts information devoted to the exploration of Mars: *tinyurl.com/cntykwg*.

c. 4.5 Billion BCE, Main Asteroid Belt

Details about more than a half million minor planets in the main asteroid belt and elsewhere in the solar system are compiled by the International Astronomical Union's Minor Planet Center at *tinyurl.com/d2scxfv*.

c. 4.5 Billion BCE, Jupiter

Bagenal, F., T.E. Dowling, and W.B. McKinnon, eds., *Jupiter: The Planet, Satellites, and Magnetosphere* (New York: Cambridge Univ. Press, 2007).

c. 4.5 Billion BCE, Saturn

Saturn's atmosphere is less dynamic than Jupiter's, but still reveals interesting and enigmatic features, such as a bright new storm system that became visible in 2010: *tinyurl.com/24prgxd*.

c. 4.5 Billion BCE, Uranus

A wonderful collection of photos of the Uranian atmosphere, rings, and moons can be found on the NASA Planetary Photojournal site for Uranus: *tinyurl.com/6pzdykv*.

c. 4.5 Billion BCE, Neptune

The idea that Uranus and Neptune have migrated outward since their formation is discussed in "The Chaotic Genesis of Planets" by D.N.C. Lin, *Scientific American*, May 2008.

c. 4.5 Billion BCE, Pluto and the Kuiper Belt

The fun 365 Days of Astronomy podcast site has an interesting entry on the Kuiper (rhymes with "viper") belt: *tinyurl.com/d6q9ckf*.

c. 4.5 Billion BCE, Birth of the Moon

Canup, R.M., and K. Righter, eds., *The Origin of the Earth and Moon* (Tucson: Univ. of Arizona Press, 2000).

c. 4.1 Billion BCE, Late Heavy Bombardment

The giant planets likely played a significant role in the late heavy bombardment: *tinyurl. com/csg6zh4*.

c. 3.8 Billion BCE, Life on Earth

Ricardo, A. and J.W. Szostak, "The Origin of Life on Earth," *Scientific American*, Sep. 2009.

550 Million BCE, Cambrian Explosion

Erwin, D.H., *Extinction: How Life on Earth Nearly Ended 250 Million Years Ago* (Princeton, NJ: Princeton Univ. Press, 2006).

65 Million BCE, Dinosaur-Killing Impact

For more details and references about the controversy surrounding the impact hypothesis, a great place to start is Wikipedia's K-T extinction event page: *tinyurl.com/mm2dz*.

200,000 BCE, *Homo Sapiens*

Seed magazine reporter Holly Capelo wrote an interesting summary of recent evidence that Paleolithic cave art may capture some aspects of ancient astronomical and celestial lore: *tinyurl.com/cvgtd6q*.

c. 50,000 BCE, Arizona Impact

Geologist David Rajmon has compiled an online database of the nearly two hundred known and suspected impact-crater sites on the Earth: *tinyurl.com/bqsgdsb*.

c. 5000 BCE, Birth of Cosmology

According to an official NASA definition, *cosmology* is the study of the structure and changes in the present universe, whereas the study of the origin and evolution of the early universe is technically called *cosmogony* (although no one I know uses the latter word).

c. 3000 BCE, Ancient Observatories

Newham, C.A., *The Astronomical Significance of Stonehenge* (Warminster, UK: Coates & Parker, 1993).

c. 2500 BCE, Egyptian Astronomy

I remember reading an early edition of E.C. Krupp's *Echoes of the Ancient Skies: The Astronomy of Lost Civilizations* (Mineola, NY: Dover, 2003) when I was young and being fascinated by how much the objects and motions of the sky meant to our distant ancestors.

c. 2100 BCE, Astronomy in China

Univ. of Maine professor Marilyn Shea highlights many ancient Chinese astronomers and astronomical instruments at *tinyurl.com/cxqtavp*.

c. 500 BCE, Earth Is Round!

If you're still not convinced that you're living on a rotating sphere, you can always stick your head in the sand and join other nonbelievers from the Flat Earth Society: *tinyurl.com/346e6c8*.

c. 400 BCE, Greek Geocentrism

According to Bakersfield College astronomy professor Nick Strobel (*tinyurl.com/blcrvgf*), Aristotle "had probably the most significant influence on many fields of studies (science, theology, philosophy, etc.) of any single person in history."

c. 400 BCE, Western Astrology

Andrew Fraknoi of Foothill College (Los Altos Hills, California) and the Astronomical Society of the Pacific have excellent pointers and resources for those who want to debunk astronomy-related pseudoscience: *tinyurl.com/yfbp4vy*.

c. 280 BCE, Sun-Centered Cosmos

Kragh, H.S., *Conceptions of Cosmos—From Myths to the Accelerating Universe: A History of Cosmology* (New York: Oxford Univ. Press, 2007).

c. 250 BCE, Eratosthenes Measures the Earth

Since 2000, Follow the Path of Eratosthenes has enabled students to reproduce his more than 2,200-year-old experiment on their own: *tinyurl.com/d7bd2k3*.

c. 150 BCE, Stellar Magnitude

Still confused about the backward stellar magnitude system used by astronomers? *Sky & Telescope* magazine contributor Alan MacRobert's article at *tinyurl.com/luxflk* might help.

c. 100 BCE, First Computer

See D.J. de Solla Price's "An Ancient Greek Computer" (*Scientific American*, June 1959) and Tony Freeth's "Decoding an Ancient Computer" (*Scientific American*, Dec. 2009).

45 BCE, Julian Calendar

Web Exhibits: *tinyurl.com/58ctv5*.

c. 150 Ptolemy's *Almagest*

See physics professor Dennis Duke's "Ancient Planetary Model Animations" (*tinyurl.com/blh7uql*).

185, Chinese Observe "Guest Star"

Astronomers using NASA's Spitzer Space Telescope and WISE satellite have pieced together the details of the supernova of 185 to explain its progression from the bright flash first observed to the roughly spherical remnants of gas and dust visible today (*tinyurl.com/88sosvy*).

c. 500, *Aryabhatiya*

Walter E. Clark's 1930 English translation of the *Aryabhatiya*: *tinyurl.com/chbvjet*.

c. 700, Finding Easter

You can check Venerable Bede's calculations of the date of Easter using modern computus methods at the Astronomical Society of South Australia's website: *tinyurl.com/9zsa*.

c. 825, Early Arabic Astronomy

A useful and educational introductory reference on early Arabic astronomy is Owen Gingerich's "Islamic Astronomy" (*Scientific American*, Apr. 1986).

c. 964, Andromeda Sighted

Persian astronomer 'Abd al-Rahmān al-Sūfī's *The Book of the Fixed Stars* is available via the World Digital Library: *tinyurl.com/cx7mkdr*.

c. 1000, Experimental Astrophysics

More details about the lives and work of al-Haytham and al-Bīrūnī can be found in recent articles by Prof. Jim Al-Khalili (*tinyurl.com/8q5k9c*), and author Richard Covington (*tinyurl.com/2wqe7t*).

c. 1000, Mayan Astronomy

A high-resolution version of the complete Dresden Codex can be downloaded from *tinyurl.com/d5f38vq*. See also Prof. Anthony Aveni's *Conversing with the Planets: How Science and Myth Invented the Cosmos* (New York: Kodansha International, 1994).

1054, "Daytime Star" Observed

Mitton, S., *The Crab Nebula* (New York: Charles Scribner's Sons, 1979).

c. 1230, *De Sphaera*

Learn a lot more about John of Sacrobosco at *tinyurl.com/cbbvrsd*.

c. 1260, Large Medieval Observatories

NASA's Ancient Observatories, Timeless Knowledge: *tinyurl.com/cl4busr*.

c. 1500, Early Calculus

The figure is adapted from Ramasubramanian, K., et al., "Modification of the Earlier Indian Planetary Theory by the Kerala Astronomers (c. 1500) and the Implied Heliocentric Picture of Planetary Motion" (*Current Science*, vol. 66, 784–790, 1994).

1543, Copernicus's *De Revolutionibus*

Helden, A.V.: *tinyurl.com/cebcm*.

1572, Brahe's "Nova Stella"

Thoren, V.E., *The Lord of Uraniborg: A Biography of Tycho Brahe* (New York: Cambridge Univ. Press, 1990).

1582, Gregorian Calendar

The US Naval Observatory's "Introduction to Calendars" (*tinyurl.com/d589vr8*) summarizes the six principal calendar systems currently in worldwide use.

1596, Mira Variables

Hoffleit, D.: *tinyurl.com/ct3mzgy*.

1600, Bruno's *On the Infinite Universe and Worlds*

Wikipedia's "Giordano Bruno" page (*tinyurl.com/ayqfd*) provides a starting point for more detailed study of the controversial friar, philosopher, and astronomer.

1608, First Astronomical Telescopes
The American Academy of Ophthalmology has an online history of spectacles at *tinyurl.com/bpbbqqn*; also see "The Telescope" from the Rice Univ. Galileo Project at *tinyurl.com/33gat4u*.

1610, Galileo's *Starry Messenger*
Museo Galileo provides details about Galileo's telescope at *tinyurl.com/d2n945d*. For more personal revelations, see Dava Sobel's *Galileo's Daughter: A Historical Memoir of Science, Faith, and Love* (New York: Walker & Co., 2011).

1610, Io
Lopes, R.M.C. and J.R. Spencer, eds., *Io After Galileo: A New View of Jupiter's Volcanic Moon*, edited by (Chichester, UK: Springer/Praxis, 2006).

1610, Europa
Spectacular views of Europa can be found on the NASA Planetary Photojournal search page: *tinyurl.com/cw7pz7w*.

1610, Ganymede
For more of the history of orbital resonances and celestial dynamics, see C.D. Murray and S.F. Dermott's *Solar System Dynamics* (New York: Cambridge Univ. Press, 2000).

1610, Callisto
Visit Dr. Paul Schenk's *3D House of Satellites* blog (*tinyurl.com/bssr43w*) to learn more about what Callisto and the other Galilean moons are like up close.

1610, Orion Nebula "Discovered"
To learn more about our distant ancestors' ideas about the Orion Nebula, check out E.C. Krupp's article "Igniting the Hearth," *Sky & Telescope* (Feb. 1999).

1619, Three Laws of Planetary Motion
More about Johannes Kepler can be found in C. Wilson's "How Did Kepler Discover His First Two Laws?" (*Scientific American*, March 1972) and O. Gingerich's *The Great Copernicus Chase and Other Adventures in Astronomical History* (Cambridge, MA: Sky Publishing, 1992).

1639, Venus Transits the Sun
A great popular-level account of the history of Venus transit observations is in W. Sheehan and J. Westfall's *The Transits of Venus* (Amherst, NY: Prometheus, 2004).

1650, Mizar-Alcor Sextuple System
Kaler, J.: *tinyurl.com/yezwdhv*; see also L. Ondra's "A New View of Mizar," originally published in *Sky & Telescope* (July 2004) and available at *tinyurl.com/bqjaeh4*.

1655, Titan
Huygens's *Systema Saturnium* is available from the Smithsonian Institution Libraries at *tinyurl.com/bpwdunv*.

1659, Saturn Has Rings
NASA Planetary Data System's "Saturn's Rings": *tinyurl.com/d28nu2n*; see also the previous note.

1665, Great Red Spot
A.P. Ingersoll reviews the history and science of the Great Red Spot in "Atmospheres of the Giant Planets," chapter 15 in *The New Solar System*, edited by J.K. Beatty, C.C. Petersen, and A. Chaikin (Cambridge, MA: Sky Publishing, 1999).

1665, Globular Clusters
National Optical Astronomy Observatories: *tinyurl.com/abjnve*.

1671, Iapetus
See the Cassini mission's "Iapetus" Web page: *tinyurl.com/7l6yghw*.

1672, Rhea
Details about the possible halo and ring around Rhea appear in G. H. Jones, et al., "The Dust Halo of Saturn's Largest Icy Moon, Rhea" (*Science, vol. 319*, 1380–1384, 2008).

1676, Speed of Light
See also S. Soter and N.D. Tyson (eds.), *Cosmic Horizons: Astronomy at the Cutting Edge* (New York: New Press, 2001).

1682, Halley's Comet
See Alan H. Cook's *Edmund Halley: Charting the Heavens and the Seas* (New York: Clarendon Press, 1998). Lists and orbital data for all known periodic comets are compiled by the IAU's Minor Planet Center at *tinyurl.com/28y8a5r*.

1684, Tethys
An animation of the surface geology of Tethys by C.J. Hamilton can be found at *tinyurl.com/bms9qbq*.

1684, Dione
NASA Planetary Photojournal: *tinyurl.com/c24cvnh*.

1684, Zodiacal Light
Details of the early history of zodiacal light observations can be found in C. E. Brame's "The Zodiacal Light" (*Popular Science Monthly*, July 1877), available at *tinyurl.com/bstncr3*.

1686, Origin of Tides
Excellent introductory discussions of tides can be found at "How Tides Work" on E. Siegel's blog *Starts with a Bang!* (*tinyurl.com/2axmfap*) and "Tidal Misconceptions" by D. Simanek

(*tinyurl.com/lhm5ac*), as well as pages 265–274 in V.D. Barger and M.G. Olsson's *Classical Mechanics: A Modern Perspective* (New York: McGraw-Hill, 1973).

1687, Newton's Laws of Gravity and Motion
Hawking, S., *On the Shoulders of Giants: The Great Works of Physics and Astronomy*, (Philadelphia: Running Press, 2002).

1718, Proper Motion of Stars
An accessible historical accounting of Halley's "Considerations on the Changes of the Latitudes of Some of the Principal Fixed Stars (1718)" can be found in R.G. Aitken's "Edmund Halley and Stellar Proper Motions" (*Astronomical Society of the Pacific Leaflets*, Oct. 1942), available at *tinyurl.com/c8mxavz*.

1757, Celestial Navigation
The bible of seagoing navigation and instrumentation is widely regarded to be Nathaniel Bowditch's *The American Practical Navigator*, first published in 1802 and available online at *tinyurl.com/c6pxcpl*.

1764, Planetary Nebulae
Details on Hubble Space Telescope images of the Cat's Eye and other planetary nebulae can be found at *tinyurl.com/cuoaxur*.

1771, Messier Catalog
Various compilations and links to "Messier marathon" sites can be found via the Paris Observatory (*tinyurl.com/bt5kq46*), and from the Students for the Exploration and Development of Space (*tinyurl.com/cqygeww*). Also, an English translation of Messier's original 1771 catalog of the first 45 objects is online at *tinyurl.com/c99ascl*.

1772, Lagrange Points
Astrophysicist Neil deGrasse Tyson discusses the history, physics, and space exploration potential of Lagrange points at *tinyurl.com/bmqhark*.

1781, Discovery of Uranus
Lemonick, M., *The Georgian Star: How William and Caroline Herschel Revolutionized Our Understanding of the Cosmos* (New York: W.W. Norton, 2009).

1787, Titania
A fascinating first-person account of the discovery of the first two moons of Uranus was published by William Herschel in 1787 as "An Account of the Discovery of Two Satellites Revolving Round the Georgian Planet" in *Philosophical Transactions of the Royal Society*, (January 1, 1787); it is online at *tinyurl.com/dyou62p*.

1787, Oberon

William Herschel's son, John, wrote about the orbits of Oberon and Titania and the obliquity of Uranus in "On the Satellites of Uranus," published in the *Monthly Notices of the Royal Astronomical Society* (March 14, 1834): *tinyurl.com/cy3nb8b*.

1789, Enceladus

"Cassini Images of Enceladus Suggest Geysers Erupt Liquid Water at the Moon's South Pole": *tinyurl.com/8k4d6g2*.

1789, Mimas

Details in the design and fabrication of the mirrors for Herschel's 40-foot (12-meter) telescope can be found in W. H. Steavenson's "Herschel's First 40-foot Speculum," published in *The Observatory* (vol. 50, 114–118, 1927): *tinyurl.com/8dyosha*.

1794, Meteorites Come from Space

Smith, C., S. Russell, and G. Benedix, *Meteorites* (Buffalo, NY: Firefly Books, 2011).

1795, Encke's Comet

J. Donald Fernie published an entertaining summary of Caroline Herschel's life and achievements, "The Inimitable Caroline," in the Nov./Dec. 2007 issue of *American Scientist*, online at *www.americanscientist.org/issues/pub/the-inimitable-caroline*.

1801, Ceres

Hubble Space Telescope photos of both 1 Ceres and 4 Vesta can be found on astronomy professor C. Seligman's website: *tinyurl.com/blemkol*.

1807, Vesta

An excellent recent scientific summary of asteroid research is *Asteroids III* (Tucson: Univ. of Arizona Press, 2002), edited by W.F. Bottke and colleagues. It includes a chapter by Prof. K. Keil, entitled "Geological History of Asteroid 4 Vesta: The Smallest Terrestrial Planet" (pp. 573–584), online at *tinyurl.com/blf4765*.

1814, Birth of Spectroscopy

An online biography of Joseph von Fraunhofer's life and achievements is at *tinyurl.com/c4bkbcv*.

1838, Stellar Parallax

The accompanying photo is a screen shot from a wonderful Web application by V. Bodurov that lets you view the stars in the Sun's neighborhood from any direction: *tinyurl.com/9htgzzq*.

1839, First Astrophotographs

The Hastings Historical society has detailed background about John Draper's and his son Henry's astronomical photography at *tinyurl.com/8fd5pmd* and *tinyurl.com/8hpad5n*.

1846, Discovery of Neptune

An engaging history of the discovery of Neptune is British astronomer Sir Patrick Moore's *The Planet Neptune: An Historical Survey Before Voyager* (New York: Wiley, 1996).

1846, Triton

For a modern summary of Triton's composition, geology, and possible origin, see D. Cruikshank, "Triton, Pluto, Centaurs, and Trans-Neptunian Bodies," in T. Encrenaz, et al., *The Outer Planets and Their Moons* (Norwell, MA: Springer, 2005, pp. 421–440).

1847, Miss Mitchell's Comet

More information about Maria Mitchell's life and legacy can be found at the Maria Mitchell Association (*tinyurl.com/9ke5z2y*), "founded in 1902 to preserve the legacy of Nantucket native astronomer, naturalist, librarian, and, above all, educator."

1848, Doppler Shift of Light

Wright, N.: *tinyurl.com/ygjz7t2*.

1848, Hyperion

Thomas, P.C., et al., "Hyperion's Sponge-like Appearance" (*Nature* vol. 448, pp. 50–56, 2007).

1851, Foucault's Pendulum

California Academy of Science: *tinyurl.com/yhn3g8*.

1851, Umbriel

Read a detailed account of the life of William Lassell in his 1880 obituary in *The Observatory* (vol. 3, pp. 586–590, 1880), online at *tinyurl.com/9qetb93*.

1857, Kirkwood Gaps

Fernie, J.D., "The American Kepler" (*American Scientist*, Sept./Oct. 1999: *tinyurl.com/8tmjpyt*.

1859, Solar Flares

"NASA Science News: A Super Solar Flare" (May 6, 2008): *tinyurl.com/32v6amx*.

1859, Search for Vulcan

Fontenrose, R., "In Search of Vulcan" (*J. History of Astronomy* vol. 4, p. 145, 1973): *tinyurl.com/95ua9fn*.

1862, White Dwarfs

An outstanding account of the history and legacy of telescope makers Alvan Clark and Sons is D.J. Warner and R.B. Ariail's *Alvan Clark & Sons: Artists in Optics* (Richmond, VA: Willmann-Bell, 1995).

1866, Source of the Leonid Meteors

Kronk, G.: *tinyurl.com/8zw8e8d*.

1868, Helium

The Wikipedia entry on helium at *tinyurl.com/n5of7* contains extensive details and references.

1877, Deimos and Phobos

A personal account of Asaph Hall's discovery of the moons of Mars is "The Discovery of the Satellites of Mars" (*Monthly Notices of the Royal Astronomical Society* vol. 38, pp. 205–209, 1878), online at *tinyurl.com/9cy46pc*. Details about the Mars rovers' observations of the solar transits of both moons can be found in an article that colleagues and I wrote: "Solar Eclipses of Phobos and Deimos Observed from the Surface of Mars" (*Nature* vol. 436, pp. 55–57, 2005).

1887, End of the Ether

Michelson and Morley's original paper was published as "On the Relative Motion of the Earth and the Luminiferous Ether" in the *American Journal of Science* (vol. 34, pp. 333–345, 1887): *tinyurl.com/92vz92u*.

1892, Amalthea

Barnard's description of his discovery was published in "Discovery and Observations of a Fifth Satellite to Jupiter" (*Astronomical Journal* vol. 12, pp. 81–85, 1892): *tinyurl.com/8bwe3fz*.

1893, Star Color = Star Temperature

Wilhelm Wien won the Nobel Prize in Physics in 1911 for his discoveries in light and energy; for a list of all past winners of the physics prize: *tinyurl.com/32r8ue*.

1895, Milky Way Dark Lanes

A summary of Max Wolf's astronomical career can be found in J.S. Tenn's "Max Wolf: The Twenty Fifth Bruce Medalist" (*Mercury*, July/Aug. 1994): *tinyurl.com/9sm5xt8*.

1896, Greenhouse Effect

United Nations' *Intergovernmental Panel on Climate Change Fourth Assessment Report*: *tinyurl.com/aprync*.

1896, Radioactivity

Hedman, M., *The Age of Everything: How Science Explores the Past* (Chicago: Univ. of Chicago Press, 2007).

1899, Phoebe

Cassini mission's "Phoebe" Web page: *tinyurl.com/9nh95kz*.

1900, Quantum Mechanics

New Scientist: *tinyurl.com/ca8lnx*.

1901, Pickering's "Harvard Computers"

Nelson, S., "The Harvard Computers" (*Nature* vol. 455, 36–37, Sept. 4, 2008).

1904, Himalia

Animated orbital views of the irregular satellites of all of the giant planets can be seen with the Univ. of Maryland's online Solar System Visualizer: *tinyurl.com/2acvd7*.

1905, Einstein's "Miracle Year"
Wikipedia's exhaustive entry on the life and career of Albert Einstein (*tinyurl.com/e9zvk*) is an outstanding place to learn more about the iconic physicist.

1906, Jupiter's Trojan Asteroids
Saturn, Neptune, and Mars (but, curiously, not Uranus) have also been found to have Trojan asteroids at their leading and trailing L4 and L5 points; even the moons Tethys and Dione have been found to have small Trojan satellites in their L4 and L5 points relative to Saturn. For details, see *tinyurl.com/yoklvg*.

1906, *Mars and Its Canals*
Lowell's *Mars and Its Canals* is available at *tinyurl.com/4lr2fql*.

1908, Tunguska Explosion
Artist and planetary scientist W.K. Hartmann has put together a fascinating account of eyewitness stories and artistic impressions about the Tunguska event at *tinyurl.com/95pjc2t*.

1908, Cepheid Variables and Standard Candles
Johnson, G., *Miss Leavitt's Stars: The Untold Story of the Woman Who Discovered How to Measure the Universe* (New York: W. W. Norton, 2005).

1910, Main Sequence
A fun online applet, "Stellar Evolution and the H-R Diagram," can be used to track the evolution of stars of different mass along and eventually off the main sequence: *tinyurl.com/b35942*.

1918, Size of the Milky Way
For details about the 1920 "Great Debate" between Harlow Shapley and his fellow American astronomer Heber Curtis (1872–1942) about the size of the universe, see *tinyurl.com/9afp4fn*.

1920, "Centaur" Asteroids
An up-to-date list of all known Centaurs and other "scattered-disk objects" is compiled by the IAU's Minor Planet Center at *tinyurl.com/99w9mrp*.

1924, Eddington's Mass-Luminosity Relation
Arthur Eddington's 1926 book, *The Internal Constitution of the Stars* (Cambridge, UK: Cambridge Univ. Press), became an instant astronomy classroom staple and inspiration for generations of astrophysicists.

1926, Liquid-Fueled Rocketry
Goddard's original 1919 book on rocketry, *A Method to Reach Extreme Altitudes* (Washington, D.C.: Smithsonian Institution Press), can be downloaded from *tinyurl.com/9tha5jc*.

1927, The Milky Way Rotates
Astronomer A. Ghez's Galactic Center Group hosts a wonderfully illustrated summary of views of our galaxy's center at different wavelengths: *tinyurl.com/9etp5wj*.

1929, Hubble's Law
Osterbrock, D.E., J.A. Gwinn, and R.S. Brashear, "Edwin Hubble and the Expanding Universe" (*Scientific American* vol 269, 84–89, July 1993).

1930, Discovery of Pluto
Tombaugh, C., "The Search for the Ninth Planet, Pluto" (*Astronomical Society of the Pacific Leaflets*, July 1946): *tinyurl.com/8redhe8*.

1931, Radio Astronomy
Karl Jansky's brother Cyril Jr.'s 1956 tale of the early history of Karl's discovery of "electrical disturbances apparently of extraterrestrial origin" is online as "My Brother Karl Jansky and His Discovery of Radio Waves from Beyond the Earth" at *tinyurl.com/rrst4*.

1932, The Öpik-Oort Cloud
Jan Oort's 1950 *Bulletin of the Astronomical Institutes of the Netherlands* article, from which the Oort Cloud gets its name, expands on Ernst Öpik's original 1932 hypothesis: *tinyurl.com/99tcy9w*.

1933, Neutron Stars
Details about the Hubble Space Telescope's 1997 identification of a lone neutron star, "Hubble Sees a Neutron Star Alone in Space," are at *tinyurl.com/cstllk2*.

1933, Dark Matter
N.D. Tyson and S. Soter provide some more details about the "irascible character" Fritz Zwicky in their profile at *tinyurl.com/c45z6l3*.

1936, Elliptical Galaxies
Edwin Hubble's classic 1936 book, *The Realm of the Nebulae* (New Haven, CT: Yale Univ. Press), is based on a series of lectures that he gave at Yale in 1935 on his observations and interpretations of "island universes."

1939, Nuclear Fusion
In his essay on the history of stellar nuclear fusion, "How the Sun Shines" (*tinyurl.com/bocbkj4*), astronomer J. Bahcall wrote of Hans Bethe's 1939 paper *Energy Production in Stars*, "If you are a physicist and only have time to read one paper in the subject, this is the paper to read."

1945, Geosynchronous Satellites
Arthur C. Clarke's 1945 prophetic *Wireless World* magazine article about the future of communications satellites, and other articles and documents about the early space program, appear in a volume edited by space historian J. Logsdon: *Exploring the Unknown: Selected Documents in the History of the U.S. Civil Space Program* (*tinyurl.com/bruoxsd*).

1948, Miranda
Planetary scientist P. Schenk has created some spectacular movies and views of the dramatic and weird topography on tiny Miranda: *tinyurl.com/cr9cm3g*.

1955, Jupiter's Magnetic Field
See L. Garcia's article on the Radio Jove website: *tinyurl.com/csy4rch*.

1956, Neutrino Astronomy
Gelmini, G.B., A. Kusenko, and T.J. Weile, "Through Neutrino Eyes: Ghostly Particles Become Astronomical Tools" (*Scientific American*, May 2010).

1957, *Sputnik 1*
For an entertaining glimpse of the America shocked by *Sputnik* and then spurred on to reach the Moon, check out H. Hickam's *Rocket Boys* (New York: Delacorte Press, 1998) and the related 1999 film *October Sky* (Universal Pictures).

1958, Earth's Radiation Belts
More details about the phenomenally successful Explorer small satellite program (with 93 launches between 1958 and 2012) can be found at *tinyurl.com/qp34s*.

1958, NASA and the Deep Space Network
For details on NASA space science missions being tracked by the DSN, see *tinyurl.com/5ucc4c* and *tinyurl.com/7ebsjx3*.

1959, Far Side of the Moon
The *far side* of the Moon is not (usually) the same as the *dark side*. The Moon goes through a cycle of day and night, so, as on Earth, the lit side and the dark side are constantly changing. At full Moon is the far side also the dark side; at new Moon, the near side becomes the dark side. Check out P. Plait's explanation at *tinyurl.com/ya4vf3w*.

1959, Spiral Galaxies
Spiral galaxy and dark matter researcher V. Rubin is profiled in the Astronomical Society of the Pacific's "Women in Astronomy" website: *tinyurl.com/6e8r54*.

1960, SETI
Kaplan, F., "An Alien Concept" (*Nature* vol. 461, 345–346, Sep. 17, 2009).

1961, First Humans in Space
In honor of Yuri Gagarin's status as the first person to travel into space, every April 12 since

Photo Credits

© Don Dixon/Cosmographica: 27, 55, 319, 509

© William K. Hartmann (Planetary Science Inst.): 263, 271, 443

Courtesy of Prints and Photographs Division, Library of Congress: p. 246 LC-USZ62-102506; p. 251 LC-B2-1250-11; p. 252 LC-USZ62-115881; p. 260 LC-USZ62-128068; p. 272 LC-B2-6358-11

NASA/ESA Images: p. 5 NASA/JSC; p. 17 NASA/JPL-Caltech/MSSS; p. 21 NASA/WMAP Science Team; p. 29 J. Morse/STScI, and NASA; p. 31 NASA/GSFC/SDO; p. 33 NASA/JHUAPL/Carnegie Inst. Washington; p. 35 ESA/VIRTIS & VMC teams; p. 37 NASA/GSFC/NOAA/USGS; p. 39 S. Lee (U. Colorado), J. Bell (Cornell U.), M. Wolff (SSI), NASA; pp. 43, 45 NASA/JPL/Space Science Institute; p. 49 NASA/JPL; p. 61 NASA/Don Davis; pp. 84-85 NASA/STScI; p. 92 NASA/JPL-Caltech/UCLA/CXC/SAO; p. 93 NASA/JPL; p. 101 NASA/JPL-Caltech; p. 107 NASA, ESA, J. Hester (Arizona State U.); p. 116 X-ray: NASA/CXC/SAO, Infrared: NASA/JPL-Caltech; Optical: MPIA, Calar Alto, O.Krause et al.; p. 129 NASA/JPL; p. 131 NASA/JPL/Ted Stryk; pp. 133, 135 NASA/JPL; p. 137 NASA, ESA, M. Robberto (STScI/ESA) and the Hubble Space Telescope Orion Treasury Project Team; p. 141 NASA/Goddard Space Flight Center/SDO; p. 142 ESO Online Digital Sky Survey; p. 145 NASA/JPL/SSI; p. 147 NASA and The Hubble Heritage Team (STScI/AURA); p. 149 NASA/JPL/Björn Jónsson; p. 151 NASA and The Hubble Heritage Team (STScI/AURA); pp. 152-155 NASA/JPL/SSI; p. 157 NASA, ESA, and E. Karkoschka (U. Arizona); p. 158 Halley Multicolor Camera Team, Giotto Project, ESA; p. 159 NASA/JPL; p. 160 NASA/JPL/SSI/Sean Walker; pp. 161-163 NASA/JPL/SSI; p. 165 ESO/Y. Beletsky; p. 171 NASA/F.M. Walter (SUNY Stoneybrook); p. 174 ESO; p. 175 NASA/ESA/HEIC/Hubble Heritage Team; p. 179 NASA/WMAP Science Team; pp. 183-184 NASA/JPL; pp. 186-187, 189 NASA/JPL; p. 193 NASA/JPL-Caltech/M. Kelley(Univ. Minnesota); p. 195 NASA, ESA, J. Parker (Southwest Research Inst), P. Thomas (Cornell U.), L. McFadden (U. Maryland-College Park), and M. Mutchler and Z. Levay (STScI); p. 197 NASA, ESA, and L. McFadden (U. Maryland-College Park); p. 199 N.A.Sharp, NOAO/NSO/Kitt Peak FTS/AURA/NSF; p. 205 NASA/JPL; p. 207 NASA/JPL/USGS; p. 213 NASA/JPL/SSI; pp. 217-218 NASA/JPL; p. 223 NASA/GSFC/SDO; p. 227 NASA/SAO/CXC; p. 233 NASA/JPL-Caltech/U. Arizona; p. 234 NASA/JPL/Cornell U.; p. 235 NASA/JPL-Caltech/U. Arizona; p. 239 Michael Carroll and NASA; p. 241 NASA, ESA, and J. Anderson and R. van der Marel (STScI); p. 245 NASA/JPL, Scripps/NOAA/ESRL; p. 249 NASA/JPL/SSI; p. 265 Dr. Wendy L. Freedman, Observatories of the Carnegie Institution, Washington, NASA; p. 267 ESO; p. 269 NASA, ESA, The Hubble Heritage Team, (STScI/AURA) and A. Riess (STScI); p. 270 NASA; p. 273 NASA/SDO; p. 275 NASA; p. 277 2MASS, a joint project of the U Mass. and the Infrared Processing and Analysis Center CALTECH, funded by NASA/NSF. Atlas image mosaics by E. Kopan, R. Cutri, S. Van Dyk (IPAC); p. 279 NASA, ESA, S. Beckwith (STScI), and the HUDF Team; p. 283 National Radio Astronomy Observatory; p. 285 NASA/JPL-Caltech; p. 287 Fred Walter (SUNY Stony Brook) and NASA; p. 289 X-ray: NASA/CXC/M. Markevitch et al. Optical: NASA/STScI, Magellan/U. Arizona/D. Clowe et al. Lensing Map: NASA/STScI, ESO WFI, Magellan/U. Arizona/D. Clowe et al.; p. 291 NASA, ESA, Hubble Heritage Team (STScI/AURA); p. 292 SOHO/ESA/NASA; p. 293 NASA; pp. 294-295 NASA/JPL; pp. 296-297, NASA/JPL; p. 303 NASA; p. 305 USAF/J. Strang; p. 307 NASA/JPL; p. 309 NASA; p. 311 NASA, ESA, S. Beckwith (STScI), and The Hubble Heritage Team (STScI/AURA); p. 313 T.A. Rector (U. Alaska-Anchorage, NRAO/AUI/NSF), NOAO/AURA/NSF) and B.A. Wolpa (NOAO/AURA/NSF); p. 323 NASA/ESA/Felix Mirabel; p. 327 ESA/HFI/LFI; p. 331 NASA/CXC/HST/ASU/J. Hester et al.; pp. 334-337 NASA; p. 341 Alexander Krot;

p. 342 NSSDC Photo Gallery; p. 347 NASA/JSC; pp. 348-349 NSSDC Photo Gallery; pp. 351-355 NASA; p. 357 NASA/Swift/Mary Pat Hrybyk-Keith and John Jones; pp. 358-359 NASA; pp. 360-365 NASA/JPL; p. 366 NASA, ESA, H. Weaver (JHUAPL), A. Stern (SwRI), and the HST Pluto Companion Search Team; p. 367 NASA, ESA and G. Bacon (STScI); p. 368 ESA; p. 369 NASA/JPL-Caltech/J. Huchra (Harvard-Smithsonian CfA); p. 370 NASA/JPL; p. 371 NASA/JPL/USGS; p. 372 NASA/JPL/Cornell; p. 373 NASA/JPL/LASP; p. 375 NASA/JPL; p. 377 NASA, ESA, Richard Ellis (Caltech) and Jean Paul Kneib (Observatoire Midi-Pyrénées, France); p. 379 NASA/ARC/JPL; pp. 381-383 NASA/JPL; p. 385 NASA; p. 387 NASA/JPL; p. 388 Don Davis/NASA; p. 389 NASA; p. 391 ESO/A-M. Lagrange et al.; p. 393 NASA/JPL; p. 395 NASA, ESA, P. Challis and R. Kirshner (Harvard-Smithsonian CfA); p. 397 P. Cinzano et al., DMSP Satellites, RAS; p. 398 NASA/JPL/USGS; p. 399 NASA/JPL; p. 401 M. Blanton and the Sloan Digital Sky Survey; p. 403 NASA/STSci; p. 405 NASA/JPL/USGS; 407 NASA/MSFC; p. 409 NASA/GSFC; p. 411 NASA/JPL-Caltech/R. Hurt (SSC); p. 415 NASA/JPL; p. 418 Mount Stromlo and Siding Springs Observatory; p. 419 Hubble Space Telescope Comet Team and NASA; p. 421 S. Kulkarni (Caltech), D. Golimowski (JHU) and NASA; p. 423 NASA, ESA, and G. Bacon (STScI); p. 424 NASA/JPL; p. 425 NASA/JPL/DLR; p. 427 NASA/JSC; p. 431 NASA/JPL/JHUAPL; p. 433 NASA/JPL/IMP Team; p. 434 NASA/Corby Waste; p. 435 NASA/JPL/MSSS; p. 437 NASA/STS-119 Shuttle Crew; p. 439 NASA, ESA, E. Jullo (JPL), P. Natarajan (Yale U.), and J.-P. Kneib (Laboratoire d'Astrophysique de Marseille, CNRS, France); p. 444 NASA/MSFC; p. 445 X-ray: NASA/CXC/SAO/J. Hughes et al, Optical: NASA/ESA/Hubble Heritage Team (STScI/AURA); p. 447 NASA/JPL; p. 448 NASA/JHUAPL; p. 449 NEAR Project/JHUAPL/NASA; p. 453 NASA/WMAP Science Team; p. 454 NASA/JPL; p. 455 NASA/ESA/SOHO; p. 456 NASA/JPL-Caltech; p. 457 NASA/JPL-Caltech/J. Stauffer (SSC/Caltech); p. 458 NASA/JPL/Cornell/USGS; p. 459 NASA/JPL-Caltech/USGS/JPL; p. 460 NASA/JPL; p. 461 Fernando Garcia Navarro and the Cassini Imaging Team, ISS, JPL, ESA, and NASA; pp. 462-463 NASA/JPL-Caltech; p. 465 NASA/JPL-Caltech/U. Maryland; 466 ESA; p. 467 NASA/JPL; pp. 470-471 NASA/JPL/SSI; p. 473 NASA; p. 475 ESO10; p. 477 NASA, ESA, W. Keel (U. Alabama), and the Galaxy Zoo Team); p. 478 NASA/Tim Pyle; p. 479 Carter Roberts/Eastbay Astronomical Society; p. 481 NASA/Jim Ross; p. 483 Montage by Emily Lakdawalla (The Planetary Society). Ida, Dactyl, Braille, Annefrank, Gaspra, Borrelly: NASA/JPL/Ted Stryk. Steins: ESA/OSIRIS team. Eros: NASA/JHUAPL. Itokawa: ISAS/JAXA/Emily Lakdawalla. Mathilde: NASA/JHUAPL/Ted Stryk. Lutetia: ESA/OSIRIS team/Emily Lakdawalla. Halley: Russian Academy of Sciences/Ted Stryk. Tempel 1, Hartley 2: NASA/JPL/UMD. Wild 2: NASA/JPL; p. 485 NASA/JPL-Caltech/U. Maryland; p. 486 NASA; p. 487 NASA/JHUAPL/Carnegie Inst.; p. 488 NASA/JPL; p. 489 NASA; p. 490 NASA/JPL-Caltech; p. 491 NASA/JPL-Caltech/MSSS; p. 493 JHUAPL/SwRI; p. 497 NASA/STScI; p. 500 NASA, J. Bell (Cornell U.) and M. Wolff (SSI); p. 501 NASA/Pat Rawlings, SAIC; p. 503 NASA, ESA, and The Hubble Heritage Team (STScI); p. 505 NASA/JPL-Caltech; p. 508 NASA/JPL-Caltech/J. Hora (Harvard-Smithsonian CfA); p. 511 NASA and H. Richer (U. British Columbia); p. 515 XMM-Newton, ESA, NASA/Dana Berry; p. 517 NASA, ESA, Hubble Heritage Team (STScI/AURA)

Other Images: p. 19 Moonrunner Designs Ltd; p. 23 © J. L. Johnson, T. H. Greif, P. Navratil, V. Bromm, and Texas Advanced Computing Center; p. 25 © Mila Zinkova; p. 41 MDF@Wikimedia/NASA/JPL/IAU Minor Planet Center; p. 47 © Lawrence Sromovsky, U. Wisconsin-Madison/W. M. Keck Observatory; p. 53 © Joe Tucciarone; p. 55 © Julian Baum/Take 27 Ltd; p. 57 © Barry Sutton; p. 59 © M. Lellouch/Grand Canyon Nat'l Park; p. 63 HTO/Wikimedia; p. 65 © AirPhoto/

Jim Wark; p. 67 © Inst. for Biblical & Scientific Studies; p. 69 © Kristian Resset; p. 71 © Ricardo Liberato; p. 73 Utagawa Kuniyoshi; p. 74 © Pedro Ré; p. 75 © Orbis World Globes: www.earthball.com; p. 77 Peter Apian, *Cosmographia*, Antwerp, 1524; p. 79 Erhardt Schön (1515); p. 81 ibiblio.org/Aristarchus; p. 83 Brian Brondel/Wikimedia; p. 86 Marsyas/Wikimedia; p. 87 © Mogi Vicentini; p. 89 Kruosio/Wikimedia; p. 91 John Day (1559); p. 95 Mukerjee/Wikimedia; p. 97 © Graham Rigg of South Shields Daily Photo; p. 99 Seyyed Hossein Nasr (1976) *Islamic Science: An Illustrated Study*, World of Islam Festival Publishing Company; p. 100 al-Sufi/*Book of Fixed Stars* (c. 964); p. 103 © INTERFOTO PRESSEBILDAGENTUR / ALAMY; p. 105 Adamt/Wikimedia; p. 106 Alex Marentes/flickr; p. 109 Sacrobosco, *Tractatus de Sphaera* (1550)/Mario Taddei ancient books collection; p. 110 takwing. kwong/Wikimedia; p. 111 © Alex Ostrovski; p. 113 Ramasubramanian et al. (1994); p. 114 Wikimedia; p. 115 Andreas Cellarius, *Harmonia Macrocosmica* (1661); p. 117 Det Nationalhistoriske Museum på Frederiksborg Slot; p. 119 Collections of the Prince of Liechtenstein, Vaduz - Vienna; p. 121 M. Weiss (CXC); p. 123 Ettore Ferrari (1845–1929); p. 125 H. J. Detouche (1854–1913); pp. 126-128, 130, 132, 134 Wikimedia; p. 138 Wikimedia; p. 139 *Mysterium Cosmographicum* (1596); p. 143 © Lucasfilm, Ltd.; p. 146 *Systema Saturnium* (1659); p. 156 Kirstine Meyer, *Om Ole Rømers Opdagelse af Lysets Tøven* (1915); p. 167 © Robert Wilson; p. 168 Andrew Dunn; p. 169 © Istockphoto.com/Duncan Walker; p. 172 NOAA; p. 173 Wikimedia; p. 177 © Thierry Lombry; p. 180 Mike Young; p. 181 W.M. Keck Observatory; p. 185 Courtesy of Institute of Astronomy Library, University of Cambridge; p. 188 *Leisure Hour* (1894); p. 191 Meteorite Recon/Wikimedia; p. 192 *The Scientific Papers of Sir William Herschel* (1829); p. 200 Courtesy of Institute of Astronomy Library, University of Cambridge; p. 201 © Vladimir Bodurov: http://blog.bodurov.com/Nearest-Stars-3D-Map; p. 203 NYU Archives; p. 209 © Nantucket Historical Assoc./flickr; p. 211 Pfalstad/Wikimedia; p. 215 Daniel Sancho/Wikimedia; p. 219 © Royal Astronomical Society/Science Source; p. 221 © Minor Planet Center; p. 225 Reyk/Wikimedia; p. 229 *Bilderatlas der Sternenwelt* (E. Weiss, 1888); p. 231 © Miloslav Druckmüller, Institute of Mathematics, Brno University of Technology, Czech Republic; p. 237 Holgar Motzkau, Prolineserver/Wikimedia, Wall & Wall, *Introductory Physics, A Problem Solving Approach*, p. 559 (P. Hewitt); p. 243 © Tyler Nordgren; p. 247 © Bettmann/CORBIS; p. 250 orci/Wikimedia, Chris Heilman, skatebiker/Wikimedia; p. 253 © M. Lemke and C. S. Jeffery; p. 255 Univ. Maryland; p. 257 Ferdinand Schmutzer (1870–1928); p. 259 © Guillermo Abramson/Celestia freeware (abramson@cab.cnea.gov.ar); p. 261 Courtesy of Lowell Observatory Archives; p. 262 Wikimedia; p. 264 AAVSO; p. 278 Hale Observatories, courtesy AIP Emilio Segre Visual Archives (Huntington Library); p. 281 Courtesy of Lowell Observatory Archives; p. 290 Hubble, *The Realm of the Nebulae*, p. 45, Yale University Press; p. 299 © John Spencer; p. 300 R. Svoboda (UC Davis) and K. Gordan (LSU); p. 301 STFC; p. 308 Silvercat/Wikimedia; p. 315 Courtesy of A. Siddiqi (Fordham U.); p. 317 Courtesy of the NAIC - Arecibo Observatory, a facility of the NSF; p. 321 © J. Chennamangalum; p. 324 Steward Cohen (2001); p. 329 EugeneZelenko/Wikimedia; p. 333 Jon Sullivan; p. 339 © Istockphoto.com/Sergii Shcherbakov; p. 343 © Don P. Mitchell; p. 345 Goodvint/Wikimedia; p. 400 © Massimo Ramella; p. 413 © Michael Carroll; p. 417 © LaurieHatch.com; p. 429 E. Kolmhofer, H. Raab; Johannes-Kepler-Observatory, Linz, Austria; p. 441 © Ian Britton/freefoto.com; p. 451 © Royal Swedish Academy of Sciences; p. 469 © ISAS/JAXA; p. 494 © Michael Zeiler, eclipse-maps.com; p. 495 © Mir 27 Crew/CNES; pp. 498-499 Marco Polo/Wikimedia; p. 507 Adam Evans/flickr; p. 513 © Casey Reed, Courtesy of Penn State U.

2001 has been celebrated as "Yuri's Night" at space-related parties and events around the world. Find out more about the next Yuri's Night at *yurisnight.net*.

1963, Arecibo Radio Telescope
Arecibo Observatory: *tinyurl.com/9roxj3j*.

1963, Quasars
An introduction to Hubble Space Telescope imaging and spectroscopy of quasars and their host galaxies can be found at *tinyurl. com/8qtve6j*.

1964, Cosmic Microwave Background
Since its founding in 1925, Bell Laboratories has been a great example of private industry promoting scientific and technological advancement. In addition to the discovery of the cosmic microwave background radiation and the invention of radio astronomy, Bell Labs also pioneered the transistor, the laser, solar cells, and the first telecommunications satellite.

1965, Black Holes
An entertaining and educational account of the science and mystery of black holes can be found in astrophysicist N.D. Tyson's *Death by Black Hole: And Other Cosmic Quandaries* (New York: W. W. Norton, 2007).

1965, Hawking's "Extreme Physics"
Hawking's best-selling *A Brief History of Time: From the Big Bang to Black Holes* (1988) and *The Universe in a Nutshell* (2001), both published by Bantam Books, are excellent general-audience introductions to modern cosmology and the exotic world of black holes, wormholes, and other extreme physics concepts.

1965, Microwave Astronomy
Gravity Probe B: *tinyurl.com/97fq5pe*.

1966, *Venera* 3 Reaches Venus
Mitchell, D.P.: *tinyurl.com/3nud9*.

1967, Pulsars
Max Planck Institute for Gravitational Physics Einstein Online: *tinyurl.com/8rqoc6u*.

1967, Study of Extremophiles
T. Brock's call for expanding the search for habitable environments on Earth was published in "Life at High Temperatures" (*Science* vol. 158, Nov. 1967, pp. 1012–1019).

1969, First on the Moon
Apollo Lunar Surface Journal: *tinyurl. com/2bmqcq*.

1969, Second on the Moon
Not much has been published about the Soviet Union's failed human lunar exploration program. A summary of the Soviet efforts is available from space-history researcher M. Lindroos at *tinyurl.com/8j2nj4q*.

1969, Astronomy Goes Digital
Willard Boyle and George Smith shared a part of the 2009 Nobel Prize in Physics for their invention of the CCD. They describe their pioneering work at *tinyurl.com/ydlehwe*.

1970, Organic Molecules in Murchison Meteorite
Rosenthal, A.M., "Murchison's Amino Acids: Tainted Evidence?" (*Astrobiology*, February 12, 2003): *tinyurl.com/9ha432o*.

1970, *Venera* 7 Lands on Venus
The National Space Science Data Center maintains a chronological list of Venus space exploration missions at *tinyurl.com/8taqi9x*.

1970, Lunar Robotic Sample Return
The NASA Lunar Reconnaissance Orbiter Camera team has made a priority of taking photos of "anthropogenic targets" such as the *Luna*, *Surveyor*, and *Apollo* landers; see images and details at *tinyurl.com/8gotnwy*.

1971, Fra Mauro Formation
Chaiken, A., *A Man on the Moon: The Voyages of the Apollo Astronauts* (New York: Penguin, 1998).

1971, First Mars Orbiters
Hartmann, W.K. and O. Raper, *The New Mars: The Discoveries of* Mariner 9 (NASA Special Publication 337, 1974).

1971, Roving on the Moon
Historical documents and technical schematics about the Apollo lunar rovers can be found in *A Brief History of the Lunar Roving Vehicle* (*tinyurl. com/8nxezlh*) and "The Lunar Roving Vehicle — Historical Perspective" (*tinyurl.com/997dad8*).

1972, Lunar Highlands
A spectacular series of virtual-reality animated panoramas of all the Apollo landing sites can be explored online at *moonpans.com/vr/*.

1972, Last on the Moon
See Wikipedia's "Apollo program" web page at *tinyurl.com/ynrjsz*.

1973, Gamma-Ray Bursts
The original 1973 *Astrophysical Journal* paper announcing the discovery of GRBs, "Observations of Gamma-Ray Bursts of Cosmic Origin" by R.W. Klebesadel, I.B. Strong, and R.A. Olson, is online at *tinyurl.com/9dhw9ot*.

1973, *Pioneer 10* at Jupiter
Details about the plaque carried by *Pioneer 10* and *11*, as well as the Golden Record carried by *Voyager 1* and *2*, can be found in C. Sagan's *Murmurs of Earth: The Voyager Interstellar Record* (New York: Ballantine, 1978).

1976, *Vikings* on Mars
M. Carr's beautifully illustrated *The Surface of*

Mars (New Haven, CT: Yale Univ. Press, 1981) was a definitive summary of Martian geology up until the mid-1990s.

1977, *Voyager* "Grand Tour" Begins
Swift, D., *Voyager Tales: Personal Views of the Grand Tour* (Reston, VA: American Institute of Aeronautics and Astronautics, 1997).

1977, Uranian Rings Discovered
Elliot, J. and R. Kerr, *Rings: Discoveries from Galileo to Voyager* (Cambridge, MA: MIT Press, 1987).

1978, Charon
Charon was the last major (nonirregular) satellite of a classical planet to be discovered by telescope. For a chronological list of all of the satellites that have been discovered in our solar system (by telescope or spacecraft), see *tinyurl. com/3uuj6t*.

1978, Ultraviolet Astronomy
NASA's "Imagine the Universe!" website (*tinyurl.com/ux7i*) has an illustrated primer on types of electromagnetic radiation.

1979, Active Volcanoes on Io
R. Lopes and M. Carroll, *Alien Volcanoes* (Baltimore. Johns Hopkins Univ. Press, 2008).

1979, Jovian Rings
NASA's Planetary Rings Node: *pds-rings.seti.org*.

1979, An Ocean on Europa?
Greenberg, R.J., *Europa: The Ocean Moon* (New York: Springer, 2005).

1979, Gravitational Lensing
Wikipedia's "Gravitational lens" page at *tinyurl.com/ola3h* contains visualizations and animations that further explain the concept.

1979, *Pioneer 11* at Saturn
NASA Special Publication 349 (1977), called *Pioneer Odyssey*, is a richly illustrated history of the *Pioneer 10* and *11* projects: *tinyurl. com/9lnp9ex*.

1980, *Cosmos: A Personal Voyage*
Joining the Planetary Society (*www.planetary. org*), a nonprofit, public space-advocacy and education organization founded by C. Sagan, B. Murray, and L. Friedman in 1980, is a great way to directly participate in planetary and space exploration.

1980, 1981, *Voyager* Saturn Encounters
Pyne, S., *Voyager: Seeking New Worlds in the Third Great Age of Discovery* (New York: Viking, 2010).

1981, Space Shuttle
With the retirement of the space shuttle fleet in 2011, what comes next for US human spaceflight? A presidential commission in 2009 recommended that NASA take a "flexible path"

to future destinations, enabling missions to the Moon, Mars, or asteroids: *tinyurl.com/ygcz243.*

1982, Rings around Neptune
The current scholarly bible for the latest information on ring science is L. Esposito's *Planetary Rings* (New York: Cambridge Univ. Press, 2006).

1983, *Pioneer 10* beyond Neptune
You can monitor information on the five NASA spacecraft on their way out of our solar system at *tinyurl.com/8jlw3sm.*

1984, Circumstellar Disks
Circumstellar Disk Learning Site: *tinyurl. com/94879ma.*

1989, *Voyager 2* at Uranus
Voyager 2 is as yet the only spacecraft to visit Uranus. However, in 2011, NASA's Planetary Decadal Survey for 2013–2022 called for a possible Uranus orbiter mission to follow up *Voyager 2*'s discoveries. Download the survey's report from *tinyurl.com/3j8qcjb.*

1987, Supernova 1987A
A spectacular time-lapse movie of so-called "light echoes" from Supernova 1987A between 1996 and 2002 is posted at *tinyurl. com/9x8zuu6.*

1988, Light Pollution
Learn more about the International Dark-Sky Association (and join!) at *www.darksky.org.*

1989, *Voyager 2* at Neptune
The 1995 Univ. of Arizona Space Science Series book *Neptune and Triton* (D.P. Cruikshank, ed.) will likely remain an authoritative source on the Neptune system for a long time, as no new missions to the eighth planet are planned in the near future.

1989, Walls of Galaxies
Astronomer S.D. Landy introduces the concept of large-scale cosmic structures (including walls of galaxies) in "Mapping the Universe" (*Scientific American*, June 1999).

1990, Hubble Space Telescope
The Hubble Site (*hubblesite.org*) is a one-stop Internet shop for an amazing collection of information, stories, and pictures of the cosmos.

1990, Venus Mapped by *Magellan*
D. Grinspoon, who worked closely with *Magellan* Venus mission data, presents a fun account of Earth's "twin" planet in *Venus Revealed: A New Look Below the Clouds of Our Mysterious Twin Planet* (New York: Basic Books, 1998).

1991, Gamma-Ray Astronomy
NASA's "Imagine the Universe!" website offers a great "History of Gamma-Ray Astronomy" at *tinyurl.com/ceoaa83.*

1992, Mapping the Cosmic Microwave Background
Two leading COBE scientists, J. Mather and G. Smoot, received the 2006 Nobel Prize in Physics for helping to create a new era of precision observational cosmology using space-based missions.

1992, First Extrasolar Planets
Twelve more candidate planets have now been detected around eleven other pulsars besides those around PSR B1257+12. See the Extrasolar Planet Encyclopaedia at *tinyurl. com/39qusq* for updates and details.

1992, Kuiper Belt Objects
The IAU's Minor Planet Center is now tracking more than 1,250 trans-Neptunian objects in the Kuiper belt: *tinyurl.com/9zxhsbz.*

1992, Asteroids Can Have Moons
Finding a moon around an asteroid allows astronomers to determine the mass and density of the asteroid (using Kepler's laws), which gives clues about composition (ice, rock, metal) and interior structure (coherent rock or rubble pile).

1993, Giant Telescopes
Wikipedia hosts a list of the world's largest optical telescopes, both historical and modern, at *tinyurl.com/cnfuo4p.*

1994, Comet SL-9 Slams into Jupiter
Spencer, J. and J. Mitton, *The Great Comet Crash: The Collision of Comet Shoemaker-Levy 9 and Jupiter* (New York: Cambridge Univ. Press, 1995).

1994, Brown Dwarfs
Weather patterns on brown dwarfs could be quite wild—rain made of liquid iron falling through an atmosphere made of vaporized rock, for example. For details, see J. Bryner's "Wild Weather: Iron Rain on Failed Stars" at *tinyurl.com/bn2q4jg.*

1995, Planets around Other Suns
Extrasolar Planets Encyclopaedia's "Interactive Extra-solar Planets Catalog," *tinyurl. com/32bozw.*

1995, *Galileo* Orbits Jupiter
Meltzer, M., *Mission to Jupiter: A History of the Galileo Project* (NASA Special Publication 4231, 2007): *tinyurl.com/3gfnqge.*

1996, Life on Mars?
National Space Science Data Center: *tinyurl. com/6gjhsug.*

1997, "Great Comet" Hale-Bopp
G.W. Kronk's "Cometography": *tinyurl. com/8q6scg5.*

1997, 253 Mathilde
Since Mathilde is jet black and likely contains

a significant amount of carbon, the naming theme chosen for craters and other features on its surface was coal fields and coal basins on Earth. See *tinyurl.com/3rnenrp* for a list of the naming themes used on all solar system bodies studied thus far.

1997, First Rover on Mars
To get a feel for the *Sojourner* rover in action, check out the time-lapse rover "movies" created by planetary scientist J. Maki and the Mars Pathfinder team at *tinyurl.com/976hnvs.*

1997, Mars Global Surveyor
The team that built and operated MGS Mars Orbiter Camera (MOC) created a spectacular collection of greatest-hits photos at *tinyurl. com/8ezruqa.*

1998, International Space Station
An animation showing the assembly sequence for the ISS between 1998 and 2011 can be found at *tinyurl.com/d4plha.*

1998, Dark Energy
See *tinyurl.com/yv7q7d* at the Hubble Site and the April 2009 *Scientific American* article "Does Dark Energy Really Exist?" by T. Clifton and P.G. Ferreira.

1999, Earth's Rotation Speeds Up
Wikipedia's "Leap second" page at *tinyurl. com/b4oar* provides a fascinating and detailed account of the history and controversy surrounding this curious feature of modern timekeeping.

1999, Torino Impact Hazard Scale
More details about the Torino Impact Hazard Scale and the more recent Palermo Technical Impact Hazard Scale can be found at *tinyurl. com/kwt3tg* and *tinyurl.com/94lg6dx*, respectively.

1999, Chandra X-ray Observatory
Chandra X-ray Observatory Center: *tinyurl. com/j84ul.*

2000, An Ocean on Ganymede?
To me, Ganymede is a planet—larger than Mercury; differentiated into core, mantle, and crust; with a deep ocean and magnetic field—that just happens to be in orbit around Jupiter. It's no wonder that the European Space Agency has decided to launch a dedicated Ganymede orbiter mission in 2022 called the Jupiter Icy Moons Explorer. See *tinyurl.com/7nbred7* for details.

2000, NEAR at Eros
National Space Science Data Center: *tinyurl. com/cpvjrkv.*

2001, Solar Neutrino Problem
An interesting history of neutrinos and the 1970–2002 astronomical detective story known as the "solar neutrino problem" is

A.B. McDonald, J.R. Klein, and D.L. Wark's "Solving the Solar Neutrino Problem" (*Scientific American*, April 2003).

2001, Age of the Universe
Besides WMAP and HST, other methods of estimating the age of the universe come from estimating ages of the oldest stars in globular clusters and the oldest white dwarfs, as well as radioactive dating of meteorites combined with modeling of the time for heavy elements to form in supernova explosions. All the methods give results in the range of 10 to 20 billion years. See "How Old Is the Universe" by J.C. Villanueva at *tinyurl.com/97qw7mu*.

2001, *Genesis* Catches Solar Wind
Burnett, D. and the Genesis Science Team, "Solar Composition from the Genesis Discovery mission" (*Proceedings of the National Academy of Sciences*, May 9, 2011): *tinyurl.com/8lck3ra*.

2003, Spitzer Space Telescope
Spitzer Web page at NASA's Jet Propulsion Laboratory: *tinyurl.com/44ys3*.

2004, *Spirit* and *Opportunity* on Mars
My coffee table book *Postcards from Mars* (New York: Dutton, 2006) and my stereo-viewer book *Mars 3-D* (New York: Sterling, 2008) showcase the stories and photographic highlights from the *Spirit* and *Opportunity* rover missions.

2004, *Cassini* Explores Saturn
Dougherty, M.K., L.W. Esposito, and S.M. Krimigis, eds., *Saturn from Cassini-Huygens* (New York: Springer, 2009).

2004, *Stardust* Encounters Wild-2
The December 15, 2006, issue of *Science* magazine presents the first detailed analysis of the chemistry and mineralogy of the *Stardust* samples.

2005, *Deep Impact*: Tempel-1
A collection of animations of the *Deep Impact* projectile's crash into Tempel-1 is posted at *tinyurl.com/8fsh7qx*.

2005, *Huygens* Lands on Titan
Lorenz, R. and C. Sotin, "The Moon That Would Be a Planet," (*Scientific American vol. 302, 36–43, Mar. 2010).

2005, *Hayabusa* at Itokawa
Planetary Science Research Discoveries: *tinyurl.com/8wxxq3x*.

2005, Shepherd Moons
The Cassini website at www.ciclops.org contains spectacular images of rings and shepherd moons, as well as great descriptions from team members about the work being done with *Cassini* images.

2006, Demotion of Pluto
The new IAU definition of "planet" is described in detail at *tinyurl.com/qfrdxc*, along with a discussion of the ensuing debate and controversy.

2007, Habitable Super Earths?
Excitement continues to mount about the potential habitability of Gliese 581 d in particular, based on new computer models of its possible climate. For details, see "First Habitable Exoplanet? Climate Simulation Reveals New Candidate That Could Support Earth-Like Life" at *tinyurl.com/424vjmk*.

2007, Hanny's Voorwerp
Check out—or join—Galaxy Zoo at *tinyurl.com/2v3fjk* and the Stardust@Home project at *tinyurl.com/8636s*.

2009, Kepler Mission
Kepler mission home page at NASA's Ames Research Center: *kepler.nasa.gov*.

2010, SOFIA
SOFIA Science Center: *sofia.usra.edu*.

2010, *Rosetta* Flies by Lutetia
Planetary Society blogger Emily Lakdawalla describes the details of her mosaic comparing Lutetia with the other asteroids and comets visited by spacecraft at *tinyurl.com/csjulym*.

2010, Comet Hartley-2
Details regarding analysis of *Deep Impact* and observations of Hartley-2 are continually updated at *tinyurl.com/2ebtgxm*.

2011, *MESSENGER* at Mercury
MESSENGER mission home page: *messenger.jhuapl.edu*.

2011, *Dawn* at Vesta
See my articles "Dawn's Early Light: A Vesta Fiesta!" and "Protoplanet Closeup" in the November 2011 and September 2012 issues of *Sky & Telescope* for details, and see *dawn.jpl.nasa.gov* for the latest images.

2012, Mars Science Laboratory *Curiosity* Rover
Photos and movies of the *Curiosity* rover being built and tested, a time-lapse movie of the rover's successful "sky crane" landing on Mars in August 2012, and the latest scientific results can be viewed from the mission's main website: *tinyurl.com/8h94w65*.

2015, Pluto Revealed!
New Horizons team website: *pluto.jhuapl.edu*.

2017, North American Solar Eclipse
Eclipse scientist F. Espenak keeps a detailed, updated set of web pages on upcoming solar and lunar eclipses and planetary transits at *tinyurl.com/6cqw2c*.

2018, James Webb Space Telescope
Gardner, J.P. and colleagues, *Space Science Reviews* (vol. 123, pp. 485–606, 2006): *tinyurl.com/d7elwth*.

2029, Apophis Near Miss
Apophis's 2036 close approach to Earth depends on where it passes the Earth and Moon in 2029 and how its trajectory responds to subtle variations in the Earth's and Moon's gravity fields, which cannot be perfectly modeled in the computer.

~2035–2050, First Humans on Mars?
There are no insurmountable technical or engineering challenges to starting a human mission to Mars. The major obstacle appears to be the lack of sufficient government funding.

~100 Million, Sagittarius Dwarf Galaxy Collision
Reich, E.S., "How Does Your Galaxy Grow?" (*New Scientist* vol 2717, July 17, 2009).

~1 Billion, Earth's Oceans Evaporate
Kasting, J. and colleagues, "Earth's Oceans Destined to Leave in Billion Years": *tinyurl.com/8t28g6x*.

3–5 Billion, Collision with Andromeda
A spectacular computer-animated simulation of the collision and merger of the Milky Way and Andromeda-like galaxies is posted on the Hubble Space Telescope's website at *tinyurl.com/2mfudk*.

5–7 Billion, End of the Sun
Kaler, J.B., *Stars* (New York: Scientific American Library, 1992).

~10^{14}, Last of the Stars
Ideas about the end of the so-called stelliferous era of star formation vary; see T. Darnell's "The Decay of Heaven" (*tinyurl.com/8r4na6n*).

~10^{17}–10^{37}, Degenerate Era
Adams, F.C., and G. Laughlin, *The Five Ages of the Universe: Inside the Physics of Eternity* (New York: Free Press, 1999).

~10^{37}–10^{100}, Black Holes "Evaporate"
For some mind-bending ideas about creating evaporating black holes in the laboratory, check out "Quantum Black Holes" by astrophysicists B.J. Carr and S.B. Giddings in the May 2005 issue of *Scientific American*.

How Will the Universe End?
If you've never contemplated the end of time as a potential tourist destination, see D. Adams's *The Restaurant at the End of the Universe*, originally published in 1980—after reviewing *The Hitchhiker's Guide to the Galaxy*, first published in 1979.

Index

Note: Page numbers in **bold** indicate primary discussions.